AF389500

Feature Articles on Nutrition and Obesity Management

Feature Articles on Nutrition and Obesity Management

Editor

Javier Gómez-Ambrosi

MDPI • Basel • Beijing • Wuhan • Barcelona • Belgrade • Manchester • Tokyo • Cluj • Tianjin

Editor
Javier Gómez-Ambrosi
Clínica Universidad de
Navarra
Pamplona
Spain

Editorial Office
MDPI
St. Alban-Anlage 66
4052 Basel, Switzerland

This is a reprint of articles from the Special Issue published online in the open access journal *Nutrients* (ISSN 2072-6643) (available at: http://www.mdpi.com).

For citation purposes, cite each article independently as indicated on the article page online and as indicated below:

LastName, A.A.; LastName, B.B.; LastName, C.C. Article Title. *Journal Name* **Year**, *Volume Number*, Page Range.

ISBN 978-3-0365-8102-6 (Hbk)
ISBN 978-3-0365-8103-3 (PDF)

Cover image courtesy of Beatriz Ramírez

Contents

About the Editor

Javier Gómez-Ambrosi

Javier Gómez-Ambrosi is a Researcher at the Metabolic Research Laboratory of the Clínica Universidad de Navarra and Associate Professor at the School of Medicine in the University of Navarra, Pamplona, Spain. His main area of research is obesity and related morbidities, from a clinical and molecular point of view. His research combines basic research in experimental animals and cells with the clinical setting, trying to disentangle the pathophysiological mechanisms responsible of the impact of excess adiposity on the development of comorbidities. He has published more than 200 articles (h-index 58) and has been the PI in more than 20 research projects.

Editorial

Recent Progress in the Management of Obesity

Javier Gómez-Ambrosi [1,2,3]

1   Metabolic Research Laboratory, Clínica Universidad de Navarra, 31008 Pamplona, Spain; jagomez@unav.es
2   Centro de Investigación Biomédica en Red-Fisiopatología de la Obesidad y Nutrición (CIBEROBN),
    Instituto de Salud Carlos III, 31008 Pamplona, Spain
3   Obesity and Adipobiology Group, Instituto de Investigación Sanitaria de Navarra (IdiSNA),
    31008 Pamplona, Spain

Citation: Gómez-Ambrosi, J. Recent Progress in the Management of Obesity. *Nutrients* **2023**, *15*, 2651. https://doi.org/10.3390/nu15122651

Received: 23 May 2023
Accepted: 25 May 2023
Published: 6 June 2023

Obesity represents the most prevalent metabolic disease nowadays, posing a significant public health risk. This situation has led to a better understanding of the systems that regulate body weight and energy homeostasis. Obesity shortens life expectancy by increasing the risk of developing comorbidities such as type 2 diabetes (T2D), cardiovascular disease, fatty liver disease, and several types of cancer, among other conditions [1]. Reduced calorie intake and increased energy expenditure have traditionally been the cornerstones of the therapeutic strategy for patients living with obesity. Obesity-related comorbidities can significantly improve even with a small amount of weight loss [2]. This Special Issue includes some of the most notable progress achieved in recent years in the treatment of patients with obesity.

A better understanding of the ethiopathology of obesity should represent the pillar on which to base a good management for this condition. In this sense, in recent years, we have expanded our knowledge about the wide array of drivers that can facilitate or contribute to the development of obesity. Compiling most of these factors, the review by Catalán et al. summarizes many of the obesogens that may explain the increasing prevalence of obesity worldwide [3]. Besides "classical" direct causes, such as genetic and behavioral determinants of energy intake and expenditure, the review includes some less appreciated drivers of the excess adiposity epidemic, such as the microbiota, infectobesity, the influence of chronobiology, and the roles of endocrine disrupters, urban planning and climate change. Their review evidences the relevance of the "exposome" in the development and perpetuation of the obesity epidemic [3]. Archer and Lavie bring an interesting perspective according to which effective management strategies need a personalized approach that takes into account the subtyping of obesity phenotypes. They distinguish between acquired and inherited obesity. The former refers to the development of excessive adiposity after puberty; because acquired obesity is behavioral in origin, it can be responsive to dietary and exercise-based therapies. On the other hand, inherited obesity includes all types of obesities that occur before pubescence (infancy and childhood) and are present at birth, which would be less susceptible to treatment [4]. Having accessible tools that allow us to properly phenotype patients with obesity considering their cardiometabolic risk is essential to establish the most appropriate treatment [1]. In this sense, Sanchez et al. [5] describe the use of the measurement of skin autofluorescence (SAF), a non-invasive estimator of advanced glycation endproducts (AGEs), in patients with obesity. Although SAF correlates with body fat percentage estimated with the CUN-BAE [6], it is not increased in individuals with obesity, being more related to the presence of cardiometabolic risk factors. The authors suggest that SAF may represent a useful tool for the identification of individuals with unhealthy obesity, opening the door to new approaches to managing obesity in clinical practice [5].

A change in dietary habits is still the first step in the treatment of obesity. With a focus on components of the Mediterranean diet (MD) that may help to maintain proper mitochondrial function, Portincasa's group extensively reviews the benefits of this diet, providing cellular and animal models as well as clinical trials in individuals with metabolic syndrome assessing the efficacy of MD components on mitochondrial structure and activity [7]. On the other hand, a high intake of ultra-processed food (UPF) has been related to an increased risk of obesity and obesity-associated comorbidities. It has been debated for a long time whether ultra-processing itself is harmful or if UPFs just have a reduced nutritional content. Dicken and Batterham, in an exhaustive review, demonstrate that, consistently across different studies, adjustment for fat, sugar and sodium intake, or for adherence to a variety of healthy or unhealthy dietary patterns, has a very limited impact on the detrimental relationship between UPF intake and a diverse range of health-related outcomes [8]. These findings cast doubt on the claim that the negative effects of UPFs can be entirely attributable to their nutritional content and clearly suggest that features of ultra-processing are significant determinants that have an impact on obesity and on health in general [8].

Phase angle (PA) could be used as marker of health status in relation to nutrition, including in patients with obesity, to monitor the efficacy of weight loss and skeletal muscle mass preservation [9]. Basiri and colleagues show in this Special Issue that a treatment with nutritional supplements and diet education in addition to the standard care in patients with overweight or obesity and diabetic foot ulcers has positive ponderal and metabolic effects, including a tendency towards a lower decrease in PA. Given that an increase in PA is associated with a reduction in the risk of mortality in patients with diabetes, their findings may be considered clinically relevant [10].

Having tools capable of reliably predicting weight loss throughout a nutritional intervention has been shown to be very useful during dietary treatment in patients with overweight or obesity. Markovikj et al. report that a modification of the Wishnofsky equation, described several decades ago to determine the body mass loss in a dietary intervention based on the timeframe of energy intake reduction, accurately predicts weight loss in 100 adults with overweight or obesity under a ketogenic diet [11].

When lifestyle modification fails, and before considering bariatric surgery, pharmacological interventions should be considered as an alternative therapy for weight loss. In this sense, achieving a normal weight via long-term drug therapy with appropriate tolerability and safety has remained a difficult challenge until recently. However, in recent years, new drugs or combinations of thereof, for example semaglutide and tirzepatide, providing mean weight loss well above 10% and improving cardiovascular outcomes in patients with T2D give hope for the future [12]. The scoping review by le Roux's group reports that the results of the Semaglutide Treatment Effect in People with Obesity (STEP) trials confirm the efficacy of once weekly 2.4 mg semaglutide on weight loss in patients with obesity [13]. Although semaglutide produced some gastrointestinal-related side effects, it was in general safe and well tolerated. Given the effectiveness of the drug, the authors wonder if nutritional therapy may have to be redefined and indicated to achieve better health instead for weight loss [13]. An original study included in the Special Issue carried out by the same group tried to delve into the mechanisms by which the duodenal-jejunal bypass liner (endobarrier) induces more pronounced weight loss than a conventional dietary treatment in patients with obesity and T2D. They conclude that the greater weight loss was due to mechanisms other than a reduction in energy intake or a change in food preferences [14].

The outbreak of the COVID-19 pandemic and the lockdown that accompanied it had a very notable impact on our lives, as well as on our health [15]. Due to the lockdown, health providers were forced to increase the use of telehealth and telemedicine. Gilardini and colleagues investigated the interest of patients with obesity in taking part in a remotely delivered multidisciplinary program for weight loss [16]. According to their findings, males and elder people were more reluctant than females and younger people to be involved in an online nutritional intervention. They also conclude that the use of telemedicine in the management of obesity could reduce lost workdays and patient travel time, increasing the number of subjects who could receive treatment and improving treatment adherence [16].

Finally, Abeltino et al. describe the usefulness of Personalized Metabolic Avatar (PMA) to predict the response to a diet [17]. By means of deep learning, they develop a data-driven metabolic model, derived from the information provided by smart bands and impedance balances, which allows simulations of diet programs, allowing the setting of customized targets for obtaining optimal weight [17,18].

We have witnessed progress in the treatment of obesity in recent years with, for example, the advances mentioned above. However, much remains to be done and further research must be carried out to improve and optimize the management of patients with obesity and to increase their quality of life.

Funding: Supported by ISCIII–FEDER (PI20/00080) and CIBEROBN, ISCIII, Spain; and PC098-099 MEPERTROBE and Department of Health 58/2021, Gobierno de Navarra-FEDER, Spain.

Conflicts of Interest: The author declares no conflict of interest.

References

1. Salmón-Gómez, L.; Catalán, V.; Frühbeck, G.; Gómez-Ambrosi, J. Relevance of body composition in phenotyping the obesities. *Rev. Endocr. Metab. Disord.* 2023, in press. [CrossRef]
2. Perdomo, C.M.; Cohen, R.V.; Sumithran, P.; Clement, K.; Frühbeck, G. Contemporary medical, device, and surgical therapies for obesity in adults. *Lancet* **2023**, *401*, 1116–1130. [CrossRef] [PubMed]
3. Catalán, V.; Avilés-Olmos, I.; Rodríguez, A.; Becerril, S.; Fernández-Formoso, J.A.; Kiortsis, D.; Portincasa, P.; Gómez-Ambrosi, J.; Frühbeck, G. Time to consider the "Exposome Hypothesis" in the development of the obesity pandemic. *Nutrients* **2022**, *14*, 1597. [CrossRef] [PubMed]
4. Archer, E.; Lavie, C.J. Obesity subtyping: The etiology, prevention, and management of acquired versus inherited obese phenotypes. *Nutrients* **2022**, *14*, 2286. [CrossRef] [PubMed]
5. Sanchez, E.; Sanchez, M.; Lopez-Cano, C.; Bermudez-Lopez, M.; Valdivielso, J.M.; Farras-Salles, C.; Pamplona, R.; Torres, G.; Mauricio, D.; Castro, E.; et al. Is there a link between obesity indices and skin autofluorescence? A response from the ILERVAS project. *Nutrients* **2023**, *15*, 203. [CrossRef] [PubMed]
6. Gómez-Ambrosi, J.; Silva, C.; Catalán, V.; Rodríguez, A.; Galofré, J.C.; Escalada, J.; Valentí, V.; Rotellar, F.; Romero, S.; Ramírez, B.; et al. Clinical usefulness of a new equation for estimating body fat. *Diabetes Care* **2012**, *35*, 383–388. [CrossRef] [PubMed]
7. Khalil, M.; Shanmugam, H.; Abdallah, H.; John Britto, J.S.; Galerati, I.; Gómez-Ambrosi, J.; Frühbeck, G.; Portincasa, P. The potential of the mediterranean diet to improve mitochondrial function in experimental models of obesity and metabolic syndrome. *Nutrients* **2022**, *14*, 3112. [CrossRef] [PubMed]
8. Dicken, S.J.; Batterham, R.L. The role of diet quality in mediating the association between ultra-processed food intake, obesity and health-related outcomes: A review of prospective cohort studies. *Nutrients* **2022**, *14*, 23. [CrossRef] [PubMed]
9. Garcia-Almeida, J.M.; Garcia-Garcia, C.; Ballesteros-Pomar, M.D.; Olveira, G.; Lopez-Gomez, J.J.; Bellido, V.; Breton Lesmes, I.; Burgos, R.; Sanz-Paris, A.; Matia-Martin, P.; et al. Expert consensus on morphofunctional assessment in disease-related malnutrition. Grade Review and Delphi Study. *Nutrients* **2023**, *15*, 612. [CrossRef] [PubMed]
10. Basiri, R.; Spicer, M.T.; Ledermann, T.; Arjmandi, B.H. Effects of nutrition intervention on blood glucose, body composition, and phase angle in obese and overweight patients with diabetic foot ulcers. *Nutrients* **2022**, *14*, 3564. [CrossRef] [PubMed]
11. Markovikj, G.; Knights, V.; Kljusuric, J.G. Ketogenic Diet applied in weight reduction of overweight and obese individuals with progress prediction by use of the Modified Wishnofsky Equation. *Nutrients* **2023**, *15*, 927. [CrossRef] [PubMed]
12. Müller, T.D.; Blüher, M.; Tschöp, M.H.; DiMarchi, R.D. Anti-obesity drug discovery: Advances and challenges. *Nat. Rev. Drug Discov.* **2022**, *21*, 201–223. [CrossRef] [PubMed]
13. Alabduljabbar, K.; Al-Najim, W.; le Roux, C.W. The impact once-weekly semaglutide 2.4 mg will have on clinical practice: A focus on the STEP trials. *Nutrients* **2022**, *14*, 2217. [CrossRef] [PubMed]
14. Aldhwayan, M.M.; Al-Najim, W.; Ruban, A.; Glaysher, M.A.; Johnson, B.; Chhina, N.; Dimitriadis, G.K.; Prechtl, C.G.; Johnson, N.A.; Byrne, J.P.; et al. Does bypass of the proximal small intestine impact food intake, preference, and taste function in humans? An experimental medicine study using the duodenal-jejunal bypass liner. *Nutrients* **2022**, *14*, 2141. [CrossRef]
15. Yárnoz-Esquíroz, P.; Chopitea, A.; Olazarán, L.; Aguas-Ayesa, M.; Silva, C.; Vilalta-Lacarra, A.; Escalada, J.; Gil-Bazo, I.; Frühbeck, G.; Gómez-Ambrosi, J. Impact on the nutritional status and inflammation of patients with cancer hospitalized after the SARS-CoV-2 lockdown. *Nutrients* **2022**, *14*, 2754. [CrossRef] [PubMed]
16. Gilardini, L.; Cancello, R.; Cavaggioni, L.; Bruno, A.; Novelli, M.; Mambrini, S.P.; Castelnuovo, G.; Bertoli, S. Are people with obesity attracted to multidisciplinary telemedicine approach for weight management? *Nutrients* **2022**, *14*, 1579. [CrossRef]

17. Abeltino, A.; Bianchetti, G.; Serantoni, C.; Ardito, C.F.; Malta, D.; De Spirito, M.; Maulucci, G. Personalized metabolic avatar: A data driven model of metabolism for weight variation forecasting and diet plan evaluation. *Nutrients* **2022**, *14*, 3520. [CrossRef] [PubMed]
18. Abeltino, A.; Bianchetti, G.; Serantoni, C.; Riente, A.; De Spirito, M.; Maulucci, G. Putting the personalized metabolic avatar into production: A comparison between deep-learning and statistical models for weight prediction. *Nutrients* **2023**, *15*, 1199. [CrossRef] [PubMed]

Review

Time to Consider the "Exposome Hypothesis" in the Development of the Obesity Pandemic

Victoria Catalán [1,2,3], Iciar Avilés-Olmos [4], Amaia Rodríguez [1,2,3], Sara Becerril [1,2,3], José Antonio Fernández-Formoso [2], Dimitrios Kiortsis [5], Piero Portincasa [6], Javier Gómez-Ambrosi [1,2,3,*] and Gema Frühbeck [1,2,3,7,*]

[1] Metabolic Research Laboratory, Clínica Universidad de Navarra, 31008 Pamplona, Spain; vcatalan@unav.es (V.C.); arodmur@unav.es (A.R.); sbecman@unav.es (S.B.)

[2] CIBER Fisiopatología de la Obesidad y Nutrición (CIBEROBN), ISCIII, 31008 Pamplona, Spain; tono.fernandez@ciberisciii.es

[3] Obesity and Adipobiology Group, Instituto de Investigación Sanitaria de Navarra (IdiSNA), 31008 Pamplona, Spain

[4] Department of Neurology, Clínica Universidad de Navarra, 31008 Pamplona, Spain; iaviles@unav.es

[5] Department of Nuclear Medicine, Medical School, University of Ioannina, 45110 Ioannina, Greece; dkiorts@uoi.gr

[6] Clinica Medica "A. Murri", Department of Biomedical Sciences and Human Oncology, University of Bari Medical School, 70124 Bari, Italy; piero.portincasa@uniba.it

[7] Department of Endocrinology & Nutrition, Clínica Universidad de Navarra, 31008 Pamplona, Spain

* Correspondence: jagomez@unav.es (J.G.-A.); gfruhbeck@unav.es (G.F.); Tel.: +34-948-425600 (ext. 806567) (J.G.-A.); +34-948-255400 (ext. 4484) (G.F.)

Citation: Catalán, V.; Avilés-Olmos, I.; Rodríguez, A.; Becerril, S.; Fernández-Formoso, J.A.; Kiortsis, D.; Portincasa, P.; Gómez-Ambrosi, J.; Frühbeck, G. Time to Consider the "Exposome Hypothesis" in the Development of the Obesity Pandemic. *Nutrients* 2022, 14, 1597. https://doi.org/10.3390/nu14081597

Academic Editor: Jean-François Landrier

Received: 2 March 2022
Accepted: 8 April 2022
Published: 12 April 2022

Publisher's Note: MDPI stays neutral with regard to jurisdictional claims in published maps and institutional affiliations.

Abstract: The obesity epidemic shows no signs of abatement. Genetics and overnutrition together with a dramatic decline in physical activity are the alleged main causes for this pandemic. While they undoubtedly represent the main contributors to the obesity problem, they are not able to fully explain all cases and current trends. In this context, a body of knowledge related to exposure to as yet underappreciated obesogenic factors, which can be referred to as the "exposome", merits detailed analysis. Contrarily to the genome, the "exposome" is subject to a great dynamism and variability, which unfolds throughout the individual's lifetime. The development of precise ways of capturing the full exposure spectrum of a person is extraordinarily demanding. Data derived from epidemiological studies linking excess weight with elevated ambient temperatures, in utero, and intergenerational effects as well as epigenetics, microorganisms, microbiota, sleep curtailment, and endocrine disruptors, among others, suggests the possibility that they may work alone or synergistically as several alternative putative contributors to this global epidemic. This narrative review reports the available evidence on as yet underappreciated drivers of the obesity epidemic. Broadly based interventions are needed to better identify these drivers at the same time as stimulating reflection on the potential relevance of the "exposome" in the development and perpetuation of the obesity epidemic.

Keywords: obesogens; "exposome"; environment; epigenetics; microbiota; antibiotics; viral infection; sleep; endocrine disruptors; brown adipose tissue; thermogenesis

1. Introduction

If practitioners are asked about the current key public health challenges, in addition to the COVID-19 pandemic, many will mention obesity among the top priorities. The prevalence of obesity has tripled during the last decades, imposing an enormous burden not only on people's health, but also on society at large with obesity increasing world-wide [1–3]. Risk factor exposure, relative risk, and imputable disease burden have been addressed in a comprehensive and standardized way by the Global Burden of Diseases, Injuries, and Risk Factors Study [4]. A rigorous analysis of the trends and specific levels of risk factor exposure together with a quantitative assessment of the plausible human health effects is of utmost importance. In this context, deep knowledge is required about when current efforts are being inadequate as opposed to when public health initiatives are showing fruitful effects. Identifying the ecological factors and external drivers of change that are currently tipping the balance may prove extraordinarily useful. This approach represents a biomedical challenge and public health need. Thus, it is worthwhile considering the conceptual basis to better understand alterations at the population level as well as their potential interaction with the surrounding with an innovative perspective on, as yet, underappreciated but conceivable factors. A search for original articles and reviews published between January 1990 and February 2022 focusing on causes and contributors was performed in PubMed and MEDLINE using the following search terms (or combination of terms): "obesity", "epidemic or pandemic", "comorbidity or comorbidities", "outcomes", "mortality", "drivers", "sedentarism", "physical inactivity", "environment or environmental", "antibiotics", "microbiota", "genetics", "epigenetics", "viral infection", "infectobesity", "sleep", "chronobiology", "obesogens", "endocrine disrupters", "thermogenesis", "urban planning", "climate change" and "exposome". Only English-language, full-text articles were included. Additional articles that were identified from the bibliographies of the retrieved articles were also used, as well as selected very recent references from March 2022. Articles in journals with explicit policies governing conflicts-of-interest, and stringent peer-review processes were favored. Data from larger replicated studies with longer periods of observation, when possible, were systematically chosen to be presented. More weight was given to randomized controlled trials, prospective case–control studies, meta-analyses and systematic reviews.

Up-to-date our thinking on the obesity epidemic has focused mainly on direct causes, such as genetic and behavioral determinants of energy intake and expenditure [5,6]. The combination of increased sedentarism and life expectancy have contributed to the obesity epidemic and its comorbidities with people exhibiting a poorer physical function [7–9]. Exercise produces extraordinarily complex physiological responses at the same time as inducing changes in cellular energy balance, leading to intensity-dependent activation of AMP-activated protein kinase (AMPK) in skeletal muscle [7,10], via effects on diverse intramuscular and hormonal factors adaptations to increased physical activity include amelioration of the cardiorespiratory fitness, as shown by an augmented maximal oxygen uptake together with an elevated muscle oxidative capacity promoted by an increased mitochondrial biogenesis and angiogenesis. Elicited signals include enhanced catecholamine signaling, sarcoplasmic calcium release, changes in mechanical stretch and force, metabolic alterations, disruptions to the redox state and acid–base balance, increased muscle temperature, and increased circulating adrenaline concentrations. These signals operate on transmembrane receptors, thereby activating downstream signaling pathways, or directly stimulate the release of exercise-responsive signaling molecules. Interestingly, exercise stimulates the secretion of metabolites, extracellular vesicles, and myokines that enable crosstalk with other organs, like adipose tissue, pancreas, liver, heart, gut, and brain as well as the vascular and immune systems.

When focusing on the time scale, two quite diverse influences can be distinguished that exert their effects on ingestive behavior, as well as on other aspects of energy homeostasis [11]. The evolutionary time frame, on the one hand, determines the selection of metabolic and behavioral traits embedded within a concrete genome. Famine, as a continuous peril to survival, has led to the selection of the so-called "thrifty genes". Within a given environmental context, this thriftiness can be manifested at different levels, such as (i) the 'energy-sparing' metabolism to increase efficiency (metabolic), (ii) the proclivity to quick adipose tissue accretion (adipogenic), (iii) the capability to slow down or even switch off non-essential processes (physiologic), (iv) the propensity to hastily swallow available food (gluttony), (v) the proneness towards sedentarism to spare or conserve energy (sloth), and, finally, (vi) behavioural adaptations that can even result in selfish hoarding to warrant survival (Figure 1).

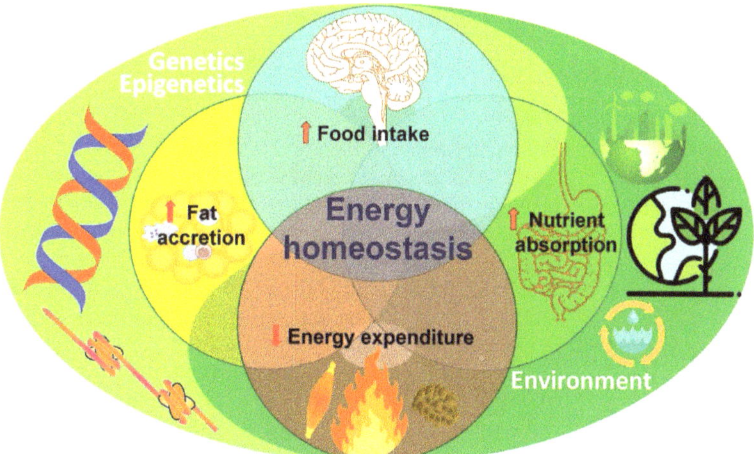

Figure 1. Schematic diagram of the factors involved in energy homeostasis. The classical Venn diagram shows how in obesity the intersection between increased food intake, nutrient absorption, and fat accumulation, together with decreased energy expenditure, the main factors determining energy homeostasis, are simultaneously under the broader influence of the environment as well as genetics and epigenetics.

The life-course time frame, on the other hand, is responsible for determining the phenotype. The early embryo's nutritional environment can exert major influences on its survival as well as on its short- and long-term physiological milieu. Thereafter, fetuses are still susceptible to nutritional intake, determined via the utero-placental unit and the maternal energetic supply. Through childhood, the adaptive plasticity is maintained and continues into adolescence and adulthood. Thereby, satiety and appetite, which encompass ingestive behavior, underlie a huge array of adaptations aimed first at survival. Thus, our "thrifty genes", the "nutrition transition", and the "technology-driven sedentariness" have been the main causes blamed, with regard to the obesity epidemic. However, recent mounting evidence obtained in diverse scientific settings is challenging this view. This narrative review reports the available evidence on the potential relevance of the "exposome" and the impact of yet underappreciated drivers of the obesity pandemic.

2. Emerging Evidence Working as Warning Signs

The past half-century has witnessed a particularly rapid increase in obesity, localized initially in high-income countries and urban settings, but also spreading, subsequently, to both low- and middle- income countries, as well as rural areas [3,12]. In this context, a conceptual framework may need to be put forward, focusing on more profound drivers embedded within society together with their interaction with biological, psychological, and socioeconomic processes.

2.1. Genetics

Rare, severe, early-onset monogenic obesity is often opposed to common or polygenic obesity as polarized and quite distinct entities. Studies for both forms of obesity, however, report shared genetic and biological underpinnings, thereby highlighting the pivotal role of the brain in body weight control [6]. New insights come from genome-wide association studies (GWAS) which are characterized by advanced sequencing technology in huge sample sizes. Moreover, cross-disciplinary post-GWAS approaches, combining novel analytical techniques and omics technologies, are opening new ways of understanding, and fostering the translation of genetic loci into meaningful biological pathways.

Genome-wide association scans for obesity-related traits have shown small size effects of the implicated genes that can be even reversed by physical activity [13–17]. Additionally, obesity appears to spread more through social than family ties [18], thereby further decreasing the relative relevance of genetics. On the contrary, the human genome is regulated via epigenetics whereby concrete ecological exposures bear risk for excess weight and associated comorbidities [19–24]. Given that survival of organisms is determined by the adequacy of nutrient intake to parallel energy expenditure, excess adiposity originally emerged as an advantageous developmental plasticity adaptation encompassing both intrauterine and intergenerational effects that bear maladaptive consequences in the current inappropriate scenario. While maternal nutrition and metabolism were well-established critical determinants of adult offspring health, adverse offspring outcomes are also reportedly associated with the father's diet [25,26], thereby indicating non-genetic inheritance of paternal influence. In this sense, men with moderate obesity display distinct DNA methylation profiles as well as small non-coding RNA expression in sperm [27]. However, it is unknown to what extent epigenetic influences on gametes impact on the metabolic profile of the progeny. Moreover, lately, reproductive performance changes have taken place, including higher fertility among people with elevated fatness and increasing maternal age [28]. A noteworthy point is that the mother's age influences excess weight risk via its impact on birth weight, whereby older women are at risk of delivering either larger or smaller babies as would be expected according to their gestational age, a circumstance that, in turn, augments the chances of originating adults with excess weight. In fact, the pregnant mother's age and body mass index (BMI), as well as the father's, together with the natal weight, the post-natal weight, and fat depot gain profiles reportedly exert an impact on the offspring's life [29].

Assortative mating, i.e., the non-random mating of people as regards their phenotype and cultural factors, may have further contributed to the obesity epidemic [30]. The shift in the development of obesity earlier in time allows the univocal identification of partners with a specific phenotype concerning weight already in the late teens and early twenties [31]. Thus, the increase in excess weight evidenced recently in descendants may also relate to the impact of both simple and complex interactions on the non-random coupling of people based on BMI. People with high adiposity may go out with people with a similar phenotype and may be more comfortable as well as be attracted by persons with the same physical characteristics rather than by those with a normal weight. In addition, sharing the same sociocultural interests among people with similar BMI may also take place. Whilst matching of couples with excess body fat may accentuate the genetic susceptibility in the progeny, the underlying mechanism is still unclear [32]. Interestingly, married couples formed by people with elevated BMI already at school age have been shown to tend to increase alongside the excess weight pandemic, that, in turn, can elevate the progeny's susceptibility to obesity [32].

2.2. Microbiome

The gut microbiome has also proven to be a key player in energy homeostasis [33,34], whereby specific gut microbial communities may be contemplated as another plausible factor for obesity development. Broad modifications in the gut microbiome have been evidenced in people with excess weight, which are reactive to changes in body weight [35–38]. Although a huge interindividual variability has been observed, in obesity, an overall reduction in microbial diversity, together with a particular decreased amount of *Bacteroidetes* at the same time as a consequent elevation of *Firmicutes*, have been reported. More precisely, observational obesity studies indicate less gut bacterial diversification with augmented levels of *Bacteroides fragilis*, *Fusobacterium*, *Lactobacillus reuteri*, and *Staphylococcus aureus*, at the same time as a lower representation of *Lactobacillus plantarum*, *Methanobrevibacter*, *Akkermansia muciniphila*, *Dysosmobacter welbionis*, and *Bifidobacterium animalis* in people living with obesity as compared to non-obese persons [38,39]. Mechanistically, the microbiome of people living with obesity has been associated with increasing energy-harvesting efficiency from the diet and alterations in gut permeability leading to metabolic endotoxemia, as well as changes in host gene expression that regulate inflammation, insulin resistance, fat storage, and fatty liver [40–42]. Latest findings indicate that microbiomes obtained from people with normal weight and obesity are different in how they interact with the host and its metabolism [43].

2.3. Infectobesity

Infection is getting more attention as a possible cause or inducing factor of obesity. The supporting findings come from both epidemiological data and the biological plausibility derived from the direct roles of some viral agents on reprogramming of the host's metabolism towards adipogenesis. Over the past decades, evidence has been growing with regard to an increased incidence in children and adults living with obesity of both nosocomial and community-acquired infections, suggesting that specific infections may be involved in the development of obesity [44]. More recently, the COVID-19 syndemic has further shown how people living with obesity are more likely to become infected with the coronavirus SARS-CoV-2 and exhibit an elevated risk of hospitalization, complications, and mortality, in probable relation to an altered immune response to infection, a chronic low-grade inflammation, together with an increased cardiometabolic risk [45–48].

Viral infections, as well as by other microorganisms, have been put forward as a plausible explanation for the excess weight epidemic with the concept of "infectobesity" harbouring the possibility that some viruses and microbes may wield an etiological role in the development of obesity [49–52]. The specific impact of excess weight on the risk of infections and the immune response triggered by infections has been addressed in a small number of studies in the population with obesity [44,53,54]. It is noteworthy that obesity augments the susceptibility to infections via an impaired immune response [55]. In addition, excess weight can also affect the pharmacokinetics of antimicrobial drugs as well as the response to vaccines [56,57]. A direct role on the host's metabolism reprogramming towards adipogenesis has been put forward as a causative or inducing factor of obesity. The existence of circulating antibodies against certain infectious agents (e.g., *Chlamydia pneumoniae* and adenovirus-36) has been associated with the suffering of excess weight [58,59]. Viral agents involved in the genesis of obesity can be classified into five main categories expanding from *Adenoviruses* and *Herpes* viruses to phages, slow viruses of transmissible spongiform encephalopathies, and other encephalitides, as well as hepatitides. Of all the viruses analyzed, adenovirus-36 (Ad-36) emerged as an appropriate candidate, according to clinical and modelling data [60]. Although mechanisms by which this adenovirus may prompt excess weight development need to be fully unraveled, it has been postulated that weight gain occurs via a direct adipogenic effect, whereby Ad-36 enters adipocytes modifying enzymatic and transcriptional factors leading to triacylglycerol accretion, increased oxidative stress, inflammation, and differentiation of preadipocytes into mature adipocytes [61,62]. A potential link between Ad-36 and obesity-related nonalcoholic fatty liver disease (NAFLD) development relies on leptin gene expression and insulin sensitivity reduction, glucose uptake increase, lipogenic and pro-inflammatory pathway activation in adipose tissue, and macrophage chemoattractant protein-1 elevation [63]. The possibility of the exchange of components of the microbiota, including the virome and virobiota, should not be discarded. In this context, the gut microbiota reportedly sustains intrinsic interferon signaling [63].

Of note, under persistent viral infections, the adaptation of the host's metabolism and immunity may be jeopardized. In addition to fructose-rich diets, decreased insulin sensitivity, chronic systemic low-grade inflammation and mitochondrial alterations, and gastrointestinal microbiota are reportedly involved in the development and worsening of NAFLD [64–66]. Due to the affected hepatic metabolism, the secretion of organokines (adipokines, myokines, hepatokines, and osteokines, among others) can be altered [67]. Changes in the secretion pattern of hepatokines can indirectly or directly contribute to aggravating NAFLD. In particular, reciprocal alterations with a decrease in fibroblast growth factor (FGF) 19 and an increase in FGF21 concentrations have been reported in obesity [68,69]. Plausible organ-specific changes in the reactiveness to the FGFs are characteristic in excess weight with adipose and hepatic changes taking differing directions in β-Klotho expression.

2.4. Chronobiology

Energy balance conservation constitutes a dynamic process with circadian rhythmicity acting as a "timekeeper" playing a decisive role in systemic homeostasis [70]. Under physiological circumstances, clock-primed biological functions synchronize to anticipate daily demands to warrant survival. Light exposure, physical activity, and sleep patterns, as well as meal timing and composition are common factors involved in energy homeostasis. It is noteworthy that the disruption or desynchronization of these factors can favor the genesis of a wide number of non-communicable diseases (NCDs), among them obesity and its comorbidities [71]. Chronological features delineate the integration in time of prediction by clock genes and metabolic and bioenergetics reactions to nutrients, whereby molecular chronotypes might be further participating in the genesis of obesity.

The internal clock makes the organism ready for regular physiological functions, such as eating and sleeping, with alterations in clock priming causing disturbances in biological rhythms and metabolism [72]. The worldwide obesity prevalence increases and metabolism alterations concur with sleep debt together with an increase in shift work as well as night exposure to light [73–76]. Sleep curtailment, as well as alterations in the chronobiology, foster elevations in BMI and sabotage dietary efforts to diminish adiposity [77]. Lack of sleep was reportedly followed by augmented hunger, elevated circulating ghrelin concentrations, and decreased circulating leptin levels, when their energy intake was restricted, as opposed to when people were in positive energy balance. Moreover, reduced sleep reportedly impacts on numerous neuroendocrine signals coordinating substrate use such as the concentrations of catecholamines, thyroid, cortisol, and growth hormone. Sleep privation and sleep alterations relate to maladaptation of the hypothalamic–pituitary–adrenal axis, translating into increased production of glucocorticoids [78,79], which can compromise the immune system [80] and increase abdominal obesity in the long term [81]. It has been recently shown that people with excess weight curtailing their sleep regularly experienced a negative energy balance by extending their sleep duration in a real-life scenario [82]. A better knowledge of the interaction between circadian rhythm disturbance and energy homeostasis may help to explain the pathophysiological processes fundamental to weight gain, thereby paving the path towards identifying novel therapeutic approaches.

2.5. Endocrine Disrupters–Obesogens

The hypothesis relating to the evolutionary origination of well-being and sickness stems from decades ago [83,84]. Subsequently, diverse epidemiological studies evidenced the relation between maternal obesity while pregnant and the possibility of the progeny to develop certain chronic adult diseases or NCDs. Among the plausible underlying mechanisms, early developmental insulin resistance stands out. Additional factors include an increased placental nutrient transfer and fetal exposure to endocrine-disrupting chemicals (EDCs), which cross the placenta, exhibit diverse tissular bioaccumulation levels, and show gender-specific vulnerability, with male fetuses being more vulnerable than female ones [85,86]. In utero environment modifications may also underlie the transmittable epigenetic changes that can endure over various generations, thereby supporting the rationale for disease development later in life. Accumulating evidence shows that EDCs interfere with endocrine regulation and metabolism, leading to lifestyle-related cardiometabolic risk factors [87,88].

Interestingly, EDCs are widespread in the environment and our daily life, with exposure encompassing the air and foods, as well as habitual items as close as personal care products [89]. Whilst the effects of individual compounds have been extensively studied, the combination of chemicals needs to be analyzed in more detail to better understand the realistic landscape of exposure to EDCs. While a dose-response relationship has not been clearly established and may not always be predictable, accumulating evidence shows that already low exposures taking place in daily life may exert a notable impact on the individual's susceptibility [90]. Moreover, in utero EDCs exposure exerts transgenerational effects reaching even the F4 generation [23]. EDCs impinge on pre- and postnatal growth, metabolism, body weight control, thyroid function, sexual development, puberty, and reproduction, among others. Though the exact mechanisms of how phenotypic features are transferred from an exposed organism to the progeny remain largely unknown strong evidence is mounting regarding a variety of epigenetic mechanisms including differential methylation of both DNA and histones, together with histone retention, non-coding RNAs expression and deposition, as well as chromatin organization and structure changes [23].

Obesity is positively associated with the exposure to EDCs [87,91]. The hypothesis of obesogens in the environment purports that pollutants of a chemical nature have the capacity to induce excess weight modifying metabolism and homeostatic set-points, affecting appetite regulation, altering lipid metabolism to stimulate adipocyte hypertrophy, and promoting adipogenic pathways aimed at fat cell hyperplasia, thereby predisposing, initiating or exacerbating weight gain [92,93]. Phthalates, per- and polyfluoroalkyl substances, polycyclic aromatic hydrocarbons, bisphenol A (BPA), heavy metals (cadmium, arsenic and mercury), and pesticides are well-known EDCs [94]. Important concepts regarding the potential impact of EDC include window and duration of exposure, role of combinations or mixtures, transgenerational effects, and epigenetic mechanisms. EDCs interrupt hormonal signaling, alter adipocyte differentiation, and interfere with metabolism, in particular during early developmental stages for several generations [94]. Various EDCs like BPA, diethylstilbestrol, phthalates and organotins, to mention a few, can interfere with signaling by targeting pathways of nuclear hormone receptors (glucocorticoid receptors, sex steroid, retinoid X receptor, and peroxisome proliferator-activated receptor γ) relevant to adipocyte proliferation and differentiation. At the adipocyte level, this is achieved by disrupting body weight homeostasis promoting long-term obesogenic changes with the epidemiological impact that can be multiplied when the interference takes place in moments of particular sensitivity like the fetal period and childhood. Thus, individuals exposed to obesogens may be preprogrammed towards an adipogenic fate worsened by socioeconomic circumstances favoring unhealthy diets as well as insufficient physical activity that promote poor diet and inadequate exercise and struggle lifelong to maintain a healthy weight. It is of note that BPA, polybrominated diphenyl ethers, phthalates, together with perfluoro products have been steadily increasing their levels in humans establishing a specific connection among adipogenic phenotypes with exposure and transcriptional network control [95].

The metabolism of xenobiotics is commonly viewed as a process of detoxification, but occasionally the metabolites of some compounds, which are usually inert or harmless, can become biologically active [96]. EDCs, in addition to stimulating adipogenesis and lipogenesis, can also repress lipolytic signaling, thereby inducing altered phenotypes [97]. Neurohormonal regulation of lipolytic rate classically underlies catecholamine-induced activation and insulin-stimulated suppression [98]. However, a large number of lipolytic mediators include mitogen-activated protein kinase, AMP-activated protein kinase, atrial natriuretic peptides, adipokines, and structural membrane proteins [99–105]. Among the latter ones, aquaglyceroporins (AQP3, AQP7, AQP9, and AQP10) represent a subfamily of aquaporins participating in glycerol movement across cell membranes. Due to their glycerol permeability, aquaglyceroporins are involved in energy balance. Glycerol influx and efflux control in metabolically relevant organs by aquaglyceroporins plays a pivotal role with the dysregulation of these glycerol channels being associated with metabolic diseases, such as obesity, insulin resistance, non-alcoholic fatty liver disease, and cardiac hypertrophy [106]. In fact, glycerol embodies a key metabolite as a substrate for de novo synthesis of triacylglycerols and glucose as well as an energy substrate for ATP production via mitochondrial oxidative phosphorylation. Noteworthy, the control of glycerol release by aquaglyceroporins in adipocytes plays a pivotal role in energy homeostasis reportedly associated with NCDs, such as insulin resistance, and obesity [107]. The potential interference of EDCs with a number of lipolytic factors deserves further analysis. Furthermore, EDCs also disrupt activity of brown and beige fat, the thermogenic adipose tissues [108].

Given the habitual exposure to multiple EDCs, the assessment of public health effects is complicated. In this respect, special care during pregnancy and childhood would be desirable. Sound knowledge about plausible mechanistic explanations on how specific exposures in a given environment translate into making individuals more susceptible to suffer some diseases like obesity [23]. Adequate determination of the surrounding toxicology together with its derived health risks might be achieved via advanced computational and prediction tools, and investigations of both systematic and integrative approaches further validating novel reliable metabolic biomarkers. Additionally, integration efforts aimed at mimicking the surrounding's specific circumstances are needed in new studies pursuing the evaluation of the effects of EDCs.

2.6. Urban Planning

Interestingly, urban environment characteristics may also contain upstream drivers of obesity [109,110]. Nonetheless, consideration of the simultaneous combination of environmental factors is not normally addressed. When looking at the same time at 86 elements characterizing the urban "exposome" relating to BMI via geocoded exposures including individual home addresses, traffic noise, air pollution, built environment, and green-space, as well as neighborhood socio-demographic factors, relevant insight can be obtained. Exposure-obesity associations were identified after adjustment for individual socio-demographic characteristics. Associations of BMI with the mean neighborhood house cost, food facilities within a close reach, oxidation capacity of particulate elements, air pollution, low-income neighborhoods, and one-person households exhibited the strongest consistency [109]. BMIs were more elevated in low-income neighborhoods, in people with lower mean house cost, lower proportion of single-people households, and areas with lower numbers of healthy food facilities. The holistic analysis of the obesogens of the environment emphasizes the mounting information as regards the relevance of socioeconomics, urban planning, and air pollution as regards the neighborhood.

2.7. Climate Change

Global warming is a well-known public health challenge and bidirectional influences regarding adiposity and global warming have been established [111]. Since 1950, carbon emissions worldwide have increased at an exponential rate. Transport, construction, manufacturing, housing, forestry, and agriculture modifications, together with the world population increase in important obesity rates, can be considered as principal contributors to carbon emissions. With increasing atmospheric temperature, less adaptive thermogenesis can be expected in people with obesity who may simultaneously be less physically active, at the same time as increasing their carbon footprint. Thus, over the last centuries environmental influences like an increase in ambient temperature in relation to climate change and global warming together with transportation, temperature insulation of both edifices, and individuals have decreased the necessity of people to generate energy by inducing thermogenesis. Therefore, it is important to consider the environmental impact of the rising obesity rates to learn more about how to tackle the excess weight pandemic, at the same time as how to minimize energy consumption, food waste, greenhouse gas emissions, in general, and carbon footprint, in particular. Of note, the Mediterranean diet, which is characterized by low in meat intake, reportedly reduces by 72% greenhouse gas emissions, by 58% land use, and by 52% energy consumption [111].

In this context, it is important to consider that human fat consists mainly of white adipose tissue (WAT) and brown adipose tissue (BAT) [112]. Whereas WAT stores energy surplus and releases it according to the needs of the organism, BAT converts it to heat playing a role in body temperature control [113–117]. Patches of brown-like adipocytes that appear in WAT constitute beige fat [118]. Like BAT, beige fat also represents a further thermogenic adipose depot with increased levels of thermogenic genes and respiration rates. Interestingly, beige adipocytes, also termed brite (derived from the contraction of "brown-in-white") cells, resemble white adipocytes in the basal state, but are rich in mitochondria and release heat when activated in response to thermogenic stimuli [119]. Thus, beige or brite adipocytes exhibit a distinct gene expression pattern to that of brown or white fat cells. The worldwide temperature increase might be also playing a role in the obesity epidemic via a concomitant reduction in BAT activity [120]. BAT as well as beige fat have been greatly underestimated in adults. For many years the BAT contribution in adults to energy expenditure both in terms of amount and effectiveness was presumed to be trivial due to the presence of only marginal brown fat depots [114]. BAT and beige fat express uncoupling protein 1 (UCP1), which rapidly generates heat when activated. UCP1 is stimulated by cold-exposure and diet leading to increased activity of the sympathetic nervous system as well as oxidation of huge quantities of glucose and lipids. The identification of functional BAT in adult humans that can be stimulated by cold exposure has changed our understanding of cellular bioenergetics, especially with regard to adaptive thermogenesis in humans [113–117].

Whilst BAT research has mainly addressed its participation in non-shivering thermogenesis, the identification of its highly dynamic secretory capacity has revealed its endocrine and paracrine function via the release of "batokines" [121]. These plentiful BAT-derived molecules impinge on the physiology of diverse cell types and multiple organ systems like adipose tissue, skeletal muscle, liver, and cardiovascular system, among others [122]. Interestingly, the variety of signaling molecules encompassed by batokines extends from peptides and lipids to metabolites and microRNAs [123]. Further research in humans aimed at delineating the role of batokines beyond the BAT-mediated energy expenditure is required. Among the endocrine batokines peptide factors like adiponectin, FGF21, interleukin-6, neuregulin-4, myostatin, and phospholipid transfer protein, as well as some microRNAs like miR-92a and miR-99b, stand out, whereas the lipids include lipokines, bioactive compounds, derived from adipose tissue, that regulate diverse molecular signaling pathways. Recently, an oxylipin, 12,13-dihydroxy-9Z-octadecenoic acid (12,13-diHOME), has attracted interest. The elevation in serum 12,13-diHOME has been associated with improved metabolic health with the action of this molecule appears to be mediated by brown adipose tissue (BAT). Its circulating concentrations are negatively correlated with BMI and insulin sensitivity. Exposure to cold and physical exercise result in an increase in circulating levels of 12,13-diHOME, which promotes browning of WAT and stimulates fatty acid absorption by BAT via stimulating the translocation of the fatty acid transporters CD36 and FATP1 to the cell membrane [124,125]. Moreover, the existence of other as yet unidentified factors involved in energy balance regulation should not be discarded [126,127].

2.8. Plurality of Obesity Epidemics

A noteworthy, elegant cross-species analysis has clearly shown a plurality of epidemics of excess weight among domestic mammals, even without the presence of the elements characteristically conceived as the main predetermining factors of the obesity epidemic via their impact on lifestyle habits like diet and physical activity [128]. These findings indicate that excess weight genesis over the last decades depends on the confluence of additional yet underappreciated environmental influences.

3. The "Exposome" as a Plausible Underlying Mechanism of Action

The word "exposome" stands for the assessment over the whole life of a person of all the exposures and its relationship to disease. This concept has been fostered by the success in mapping the human genome [129,130]. Of note, the exposure of a person starts at conception and in utero, continuing over childhood and adolescence (Figure 2). Job-related insults as well as influences from leisure time and the environment further accumulate during adulthood progressing up to senescence. Many single nucleotide polymorphisms

(SNPs) are genetic variants of low penetrance involved in the control of food intake, body weight, and lipid metabolism, among others.

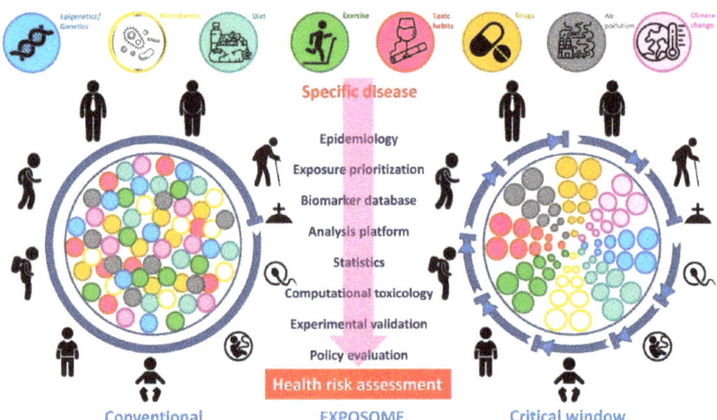

Figure 2. Conventional versus critical window "exposome" views in assessment of health risk in humans. The comprehensive portrait of an individual's "exposome" evolves throughout the lifetime with the possibility of prioritized exposure factors, during specific time-points or critical windows as opposed to a standard, random or linear exposure. The diagram emphasizes the relevance of measurements at different time-points (modified from Fang et al. [131]).

Despite their low penetrance, the SNPs' high prevalence implies a potential substantial contribution to the disease burden at the population level. This means that in a concrete exposure scenario the majority of SNPs, although being of low penetrance, will emerge because of strong environmental influences. While exposures of the surrounding exhibit an exceedingly relevant protagonism in the development of NCDs, a clear association is not easy to unravel. The "exposome" will be best deciphered by obtaining deeper knowledge on how dietary and lifestyle exposures interplay with the individual's unique genetic, epigenetic, and physiologic characteristics translate into disease. In this scenario, the "exposome" can be contemplated from a conventional point of view, in which insults are randomly distributed along the whole lifecycle, or with the lens of the critical window exposure, in which insults are non-randomly allocated to specific time-periods during life [131]. Improvement in disease etiology identification at the population level will come from complementing the emphasis on genotyping by a detailed analysis of the plentiful environmental exposures [132,133], with its accurate assessment remaining a formidable and pending demand in obesity assessment. Moreover, the development of methods that accurately capture both the external environment as well as the internal chemical background of the individual are urgently needed (Figure 3). In order to complement the "genome" with its matching "exposome" the same precision for an individual's environmental exposure as we have for the subject's genome should be pursued.

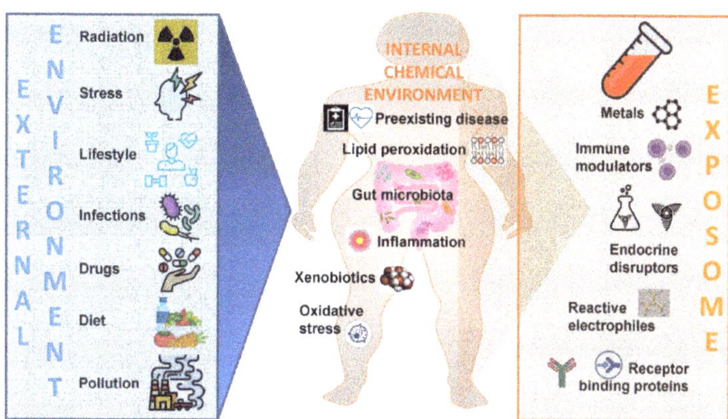

Figure 3. Characterization of the "exposome". The "exposome" of a given person represents the combined exposures from all external sources that reach the internal chemical environment. Specific biomarkers or potential signatures of the "exposome" might be detected in the bloodstream.

Need for an Integral Consideration of the Collective Impact of Simultaneously Acting Drivers

Contrarily to the genome, the "exposome" is subject to a great dynamism and variability, which unfolds throughout the individual's lifetime. The development of precise ways of determination that capture the full exposure spectrum of a person is extraordinarily demanding. These considerations are particularly relevant for children and adolescents with obesity, given that the increased exposure is expected to translate into larger adverse effects than weight gain only during adulthood [134,135]. Furthermore, the concept of epigenetics comprises the study of changes in the organism caused by alterations in gene expression rather than modifications of the genetic code itself [136,137]. Interestingly, epigenetic marks can be affected by air pollution, organic pollutants, exposure to benzene, metals, and electromagnetic radiation. Other potential environmental stressors capable of changing the epigenetic landscape include chemical and xenobiotic compounds present in the atmosphere or water.

Moreover, while responses to certain specific exposures are invariable, to other external insults responses may change ("resposome"), with disparity depending on genome and epigenome changes (Figure 4). While some alterations reveal chronicity in exposure, certain cases reveal a latent response, based on "priming" for a late pathogenesis via epigenetic changes.

Analysis of the current human "exposome" emphasizes the challenges represented by the concepts of lifelong exposure and the need to compute all environmental factors in order to obtain the whole real life exposomic scenario [131]. To overcome these limitations and establish the relation between human health and the "exposome" focusing on critical-window periods can be combined with data- and hypothesis-driven exposomics. Moreover, analysis of high-throughput and multidimensional data of both internal and external exposure factors are welcome [131]. Useful tools to analyze the "exposome" and foster exposomics should comprise different steps, i.e., (i) the development of biomarkers capturing exposure effect, susceptibility to exposure, and disease progression; (ii) the application of advances that integrate systems biology with environmental big data; and (iii) exploratory data mining to analyze the relationships between exposure effects, and other factors that ultimately lead to obesity development and thereby provide potential mechanistic information (Figure 5). Artificial intelligence will broadly reshape medicine, thereby improving the experiences of both patients and clinicians. In fact, artificial intelligence is already being applied in an ever-increasing number of medical fields moving from what might have been considered speculation years ago to reality right now. Progress in data analysis, including image deconvolutions, non-image data sources, unconventional problem formulations, sophisticated algorithms, and human–artificial intelligence collaborations, will reduce the gap between research and clinical practice. While these challenges are being addressed, artificial intelligence will develop exponentially, making healthcare more accessible, efficient, and accurate for patients worldwide [138].

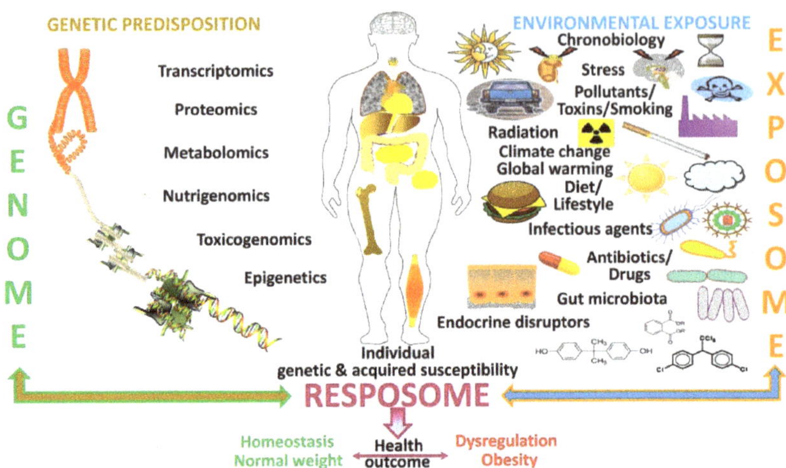

Figure 4. Factors influencing the resposome. Schematic diagram on how the genetic predisposition (genome) interacts with the environmental exposure ("exposome") to influence an individual's genetic and acquired susceptibility shaping its responses (resposome) that yield the ultimate health outcome as regards body weight control.

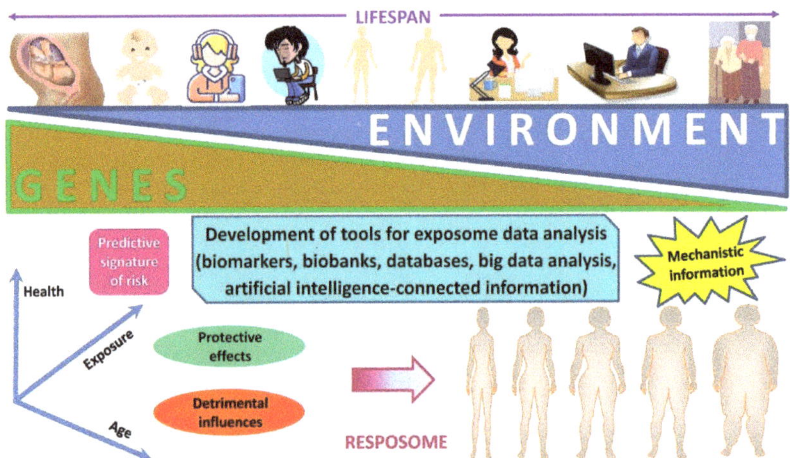

Figure 5. Evolution of the individual's genetic and environmental framework across the lifespan. Over a lifetime, genetic and environmental influences may change reciprocally with acute and chronic exposures translating into a specific information with predictive interest as well as effective biomarkers that may provide mechanistic insight of pragmatic application.

In order to be particularly helpful, "exposome" assessment should combine GWAS together with epigenome-wide association trials and detailed metabolic-endocrinological phenotyping of the individuals. Moreover, these combined analyses should be applied at multiple time-points to establish the potential interaction effect. The large amount of data on exposures provided by these projects hinders the interpretation of their relationship with health outcomes and omics. In this regard, similar or parallel databases to genetics

(OMIN, dbSNP, or TCGA) may be developed for exposomics. Together with handling and archiving large data volumes, the lack of standard nomenclature, the quality of output from each analytical platform, or the heterogeneity of data constitute important issues to be resolved. Given the important public health problem posed by the rise in NCDs like obesity, the presented proposal of integration of elements that constitute the "exposome" will strengthen the better comprehension of the intricate underlying mechanisms, thereby opening pathways to innovative preventive and therapeutic strategies.

4. Conclusions

The more simplistic energy balance model of obesity has been surpassed by epidemiological, biological, psychological, and socioeconomic evidence. Far-reaching holistic modelling of obesity is required in order to establish effective interventions aimed at its efficient treatment and better prevention. The origins of excess weight are rooted in an extremely complicated biological network, set within a similarly intricate societal and environmental organization (Figure 6), which needs to be carefully considered.

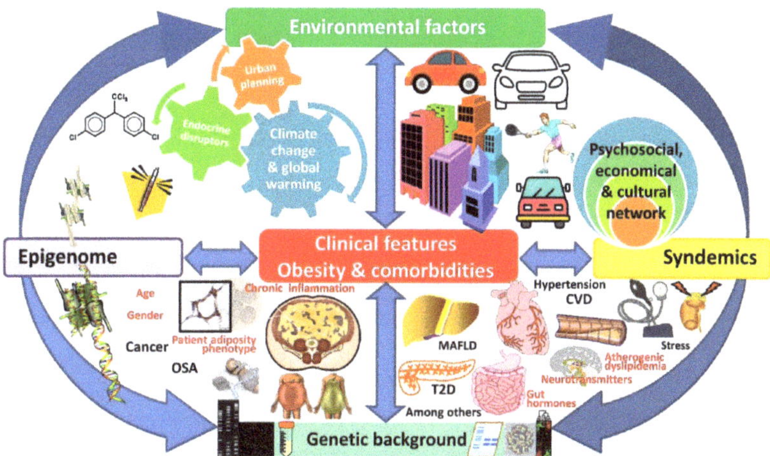

Figure 6. Multidimensional view of the complex interaction of the main drivers involved in excess weight development and obesity-associated comorbidities. OSA, obstructive sleep apnea; MAFLD, metabolic-associated fatty liver disease; T2D, type 2 diabetes; CVD, cardiovascular diseases.

Analysis of alternative and less researched etiologies is needed. The gut microbiome, circadian rhythms, and infectobesity, to mention only a few, constitute other candidate alternate etiologies. More multidisciplinary, translational research must analyze the intricacies of such alternate etiologies, as well as develop unprecedented stratagems for fending off a multifactorial and plurietiological pathology via, for example, prioritization of root cause interrogation and group risk assessment. Knowledge gaps persist in this relevant area whereby a comprehensive, leveraged patient-centered research would be welcome. Due to the struggle in the coming years to override the key factors steering the present excess weight epidemic, an inclusive, detailed, pro-active, durable program and fresh perspectives to unravel the whole panoply of causative factors is needed to outline a feasible counter reply to manage the defiance imposed by the pandemic. A comprehensive understanding of the causative factors of obesity might provide more effective management approaches.

Author Contributions: Conceptualization and first draft, G.F. Writing and editing, V.C., I.A.-O., A.R., S.B., J.A.F.-F., D.K., P.P., J.G.-A., G.F. All authors have read and agreed to the published version of the manuscript.

Funding: This work was funded by the Spanish Institute of Health ISCIII (Subdirección General de Evaluación and Fondos FEDER project PI19/00785, PI19/00990, PI20/00080 and PI20/00927), and CIBEROBN. Funding sources had no role in manuscript writing or the decision to submit it for publication.

Institutional Review Board Statement: Not applicable.

Informed Consent Statement: Not applicable.

Data Availability Statement: Not applicable.

Conflicts of Interest: The authors declare no conflict of interest.

References

1. Finucane, M.M.; Stevens, G.A.; Cowan, M.J.; Danaei, G.; Lin, J.K.; Paciorek, C.J.; Singh, G.M.; Gutierrez, H.R.; Lu, Y.; Bahalim, A.N.; et al. National, regional, and global trends in body-mass index since 1980: Systematic analysis of health examination surveys and epidemiological studies with 960 country-years and 9.1 million participants. *Lancet* **2011**, *377*, 557–567. [CrossRef]
2. NCD Risk Factor Collaboration. Worldwide trends in body-mass index, underweight, overweight, and obesity from 1975 to 2016: A pooled analysis of 2416 population-based measurement studies in 128.9 million children, adolescents, and adults. *Lancet* **2017**, *390*, 2627–2642. [CrossRef]
3. NCD Risk Factor Collaboration. Rising rural body-mass index is the main driver of the global obesity epidemic in adults. *Nature* **2019**, *569*, 260–264. [CrossRef] [PubMed]
4. Collaborators, G.R.F. Global burden of 87 risk factors in 204 countries and territories, 1990–2019: A systematic analysis for the Global Burden of Disease Study 2019. *Lancet* **2020**, *396*, 1223–1249.
5. Hill, J.O. Understanding and addressing the epidemic of obesity: An energy balance perspective. *Endocr. Rev.* **2006**, *27*, 750–761. [CrossRef]
6. Loos, R.J.F.; Yeo, G.S.H. The genetics of obesity: From discovery to biology. *Nat. Rev. Genet.* **2022**, *23*, 120–133. [CrossRef]
7. McGee, S.L.; Hargreaves, M. Exercise adaptations: Molecular mechanisms and potential targets for therapeutic benefit. *Nat. Rev. Endocrinol.* **2020**, *16*, 495–505. [CrossRef]
8. Dixon, B.N.; Ugwoaba, U.A.; Brockmann, A.N.; Ross, K.M. Associations between the built environment and dietary intake, physical activity, and obesity: A scoping review of reviews. *Obes. Rev.* **2021**, *22*, e13171. [CrossRef]
9. Woessner, M.N.; Tacey, A.; Levinger-Limor, A.; Parker, A.G.; Levinger, P.; Levinger, I. The evolution of technology and physical inactivity: The good, the bad, and the way forward. *Front. Public Health* **2021**, *9*, 655491. [CrossRef]
10. Rodríguez, A.; Becerril, S.; Ezquerro, S.; Méndez-Giménez, L.; Frühbeck, G. Crosstalk between adipokines and myokines in fat browning. *Acta. Physiol.* **2017**, *219*, 362–381. [CrossRef]
11. Prentice, A.M. Early influences on human energy regulation: Thrifty genotypes and thrifty phenotypes. *Physiol. Behav.* **2005**, *86*, 640–645. [CrossRef] [PubMed]
12. Zukiewicz-Sobczak, W.; Wroblewska, P.; Zwolinski, J.; Chmielewska-Badora, J.; Adamczuk, P.; Krasowska, E.; Zagorski, J.; Oniszczuk, A.; Piatek, J.; Silny, W. Obesity and poverty paradox in developed countries. *Ann. Agric. Environ. Med.* **2014**, *21*, 590–594. [CrossRef] [PubMed]
13. Speliotes, E.K.; Willer, C.J.; Berndt, S.I.; Monda, K.L.; Thorleifsson, G.; Jackson, A.U.; Allen, H.L.; Lindgren, C.M.; Luan, J.; Magi, R.; et al. Association analyses of 249,796 individuals reveal 18 new loci associated with body mass index. *Nat. Genet.* **2010**, *42*, 937–948. [CrossRef]
14. Li, S.; Zhao, J.H.; Luan, J.; Luben, R.N.; Rodwell, S.A.; Khaw, K.T.; Ong, K.K.; Wareham, N.J.; Loos, R.J. Cumulative effects and predictive value of common obesity-susceptibility variants identified by genome-wide association studies. *Am. J. Clin. Nutr.* **2010**, *91*, 184–190. [CrossRef] [PubMed]
15. Bray, M.S.; Loos, R.J.; McCaffery, J.M.; Ling, C.; Franks, P.W.; Weinstock, G.M.; Snyder, M.P.; Vassy, J.L.; Agurs-Collins, T. NIH working group report-using genomic information to guide weight management: From universal to precision treatment. *Obesity* **2016**, *24*, 14–22. [CrossRef]
16. Wahl, S.; Drong, A.; Lehne, B.; Loh, M.; Scott, W.R.; Kunze, S.; Tsai, P.C.; Ried, J.S.; Zhang, W.; Yang, Y.; et al. Epigenome-wide association study of body mass index, and the adverse outcomes of adiposity. *Nature* **2017**, *541*, 81–86. [CrossRef]
17. Graham, S.E.; Clarke, S.L.; Wu, K.H.; Kanoni, S.; Zajac, G.J.M.; Ramdas, S.; Surakka, I.; Ntalla, I.; Vedantam, S.; Winkler, T.W.; et al. The power of genetic diversity in genome-wide association studies of lipids. *Nature* **2021**, *600*, 675–679. [CrossRef]
18. Christakis, N.A.; Fowler, J.H. The spread of obesity in a large social network over 32 years. *N. Engl. J. Med.* **2007**, *357*, 370–379. [CrossRef]
19. Cole, T.J.; Power, C.; Moore, G.E. Intergenerational obesity involves both the father and the mother. *Am. J. Clin. Nutr.* **2008**, *87*, 1535–1536. [CrossRef]
20. Pollard, T.M.; Rousham, E.K.; Colls, R. Intergenerational and familial approaches to obesity and related conditions. *Ann. Hum. Biol.* **2011**, *38*, 385–389. [CrossRef]
21. Adamo, K.B.; Ferraro, Z.M.; Brett, K.E. Can we modify the intrauterine environment to halt the intergenerational cycle of obesity? *Int. J. Environ. Res. Public Health* **2012**, *9*, 1263–1307. [CrossRef] [PubMed]

22. Johnson, P.C.; Logue, J.; McConnachie, A.; Abu-Rmeileh, N.M.; Hart, C.; Upton, M.N.; Lean, M.; Sattar, N.; Watt, G. Intergenerational change and familial aggregation of body mass index. *Eur. J. Epidemiol.* **2012**, *27*, 53–61. [CrossRef] [PubMed]

23. Mohajer, N.; Joloya, E.M.; Seo, J.; Shioda, T.; Blumberg, B. Epigenetic transgenerational inheritance of the effects of obesogen exposure. *Front. Endocrinol.* **2021**, *12*, 787580. [CrossRef]

24. Ling, C.; Rönn, T. Epigenetics in human obesity and type 2 diabetes. *Cell Metab.* **2019**, *29*, 1028–1044. [CrossRef] [PubMed]

25. Ng, S.-F.; Lin, R.C.Y.; Laybutt, D.R.; Barres, R.; Owens, J.A.; Morris, M.J. Chronic high-fat diet in fathers programs b-cell dysfunction in female rat offspring. *Nature* **2010**, *467*, 963–966. [CrossRef] [PubMed]

26. Carone, B.R.; Fauquier, L.; Habib, N.; Shea, J.M.; Hart, C.E.; Li, R.; Bock, C.; Li, C.; Gu, H.; Zamore, P.D.; et al. Paternally induced transgenerational environmental reprogramming of metabolic gene expression in mammals. *Cell* **2010**, *143*, 1084–1096. [CrossRef] [PubMed]

27. Donkin, I.; Versteyhe, S.; Ingerslev, L.R.; Qian, K.; Mechta, M.; Nordkap, L.; Mortensen, B.; Appel, E.V.; Jorgensen, N.; Kristiansen, V.B.; et al. Obesity and bariatric surgery drive epigenetic variation of spermatozoa in humans. *Cell Metab.* **2016**, *23*, 369–378. [CrossRef]

28. McAllister, E.J.; Dhurandhar, N.V.; Keith, S.W.; Aronne, L.J.; Barger, J.; Baskin, M.; Benca, R.M.; Biggio, J.; Boggiano, M.M.; Eisenmann, J.C.; et al. Ten putative contributors to the obesity epidemic. *Crit. Rev. Food Sci. Nutr.* **2009**, *49*, 868–913. [CrossRef]

29. Faienza, M.F.; Wang, D.Q.; Frühbeck, G.; Garruti, G.; Portincasa, P. The dangerous link between childhood and adulthood predictors of obesity and metabolic syndrome. *Intern. Emerg. Med.* **2016**, *11*, 175–182. [CrossRef]

30. Heitmann, B.L.; Westerterp, K.R.; Loos, R.J.; Sørensen, T.I.; O'Dea, K.; McLean, P.; Jensen, T.K.; Eisenmann, J.; Speakman, J.R.; Simpson, S.J.; et al. Obesity: Lessons from evolution and the environment. *Obes. Rev.* **2012**, *13*, 910–922. [CrossRef]

31. Speakman, J.R.; Djafarian, K.; Stewart, J.; Jackson, D.M. Assortative mating for obesity. *Am. J. Clin. Nutr.* **2007**, *86*, 316–323. [CrossRef] [PubMed]

32. Ajslev, T.A.; Angquist, L.; Silventoinen, K.; Gamborg, M.; Allison, D.B.; Baker, J.L.; Sorensen, T.I. Assortative marriages by body mass index have increased simultaneously with the obesity epidemic. *Front. Genet.* **2012**, *3*, 125. [CrossRef] [PubMed]

33. Turnbaugh, P.J.; Ley, R.E.; Mahowald, M.A.; Magrini, V.; Mardis, E.R.; Gordon, J.I. An obesity-associated gut microbiome with increased capacity for energy harvest. *Nature* **2006**, *444*, 1027–1131. [CrossRef] [PubMed]

34. Pinart, M.; Dötsch, A.; Schlicht, K.; Laudes, M.; Bouwman, J.; Forslund, S.K.; Pischon, T.; Nimptsch, K. Gut microbiome composition in obese and non-obese persons: A systematic review and meta-analysis. *Nutrients* **2021**, *14*, 12. [CrossRef]

35. Furet, J.P.; Kong, L.C.; Tap, J.; Poitou, C.; Basdevant, A.; Bouillot, J.L.; Mariat, D.; Corthier, G.; Dore, J.; Henegar, C.; et al. Differential adaptation of human gut microbiota to bariatric surgery-induced weight loss. Links with metabolic and low-grade inflammation markers. *Diabetes* **2010**, *59*, 3049–3057. [CrossRef]

36. Liou, A.P.; Paziuk, M.; Luevano, J.M., Jr.; Machineni, S.; Turnbaugh, P.J.; Kaplan, L.M. Conserved shifts in the gut microbiota due to gastric bypass reduce host weight and adiposity. *Sci. Transl. Med.* **2013**, *5*, 178ra141. [CrossRef]

37. Ciobarca, D.; Catoi, A.F.; Copaescu, C.; Miere, D.; Crisan, G. Bariatric surgery in obesity: Effects on gut microbiota and micronutrient status. *Nutrients* **2020**, *12*, 235. [CrossRef]

38. Michels, N.; Zouiouich, S.; Vanderbauwhede, B.; Vanacker, J.; Indave Ruiz, B.I.; Huybrechts, I. Human microbiome and metabolic health: An overview of systematic reviews. *Obes. Rev.* **2022**, *3*, e13409. [CrossRef]

39. Le Roy, T.; Moens de Hase, E.; Van Hul, M.; Paquot, A.; Pelicaen, R.; Régnier, M.; Depommier, C.; Druart, C.; Everard, A.; Maiter, D.; et al. Dysosmobacter welbionis is a newly isolated human commensal bacterium preventing diet-induced obesity and metabolic disorders in mice. *Gut* **2022**, *71*, 534–543. [CrossRef]

40. Snedeker, S.M.; Hay, A.G. Do interactions between gut ecology and environmental chemicals contribute to obesity and diabetes? *Environ. Health Perspect.* **2012**, *120*, 332–339. [CrossRef] [PubMed]

41. Henao-Mejia, J.; Elinav, E.; Jin, C.; Hao, L.; Mehal, W.Z.; Strowig, T.; Thaiss, C.A.; Kau, A.L.; Eisenbarth, S.C.; Jurczak, M.J.; et al. Inflammasome-mediated dysbiosis regulates progression of NAFLD and obesity. *Nature* **2012**, *482*, 179–185. [CrossRef] [PubMed]

42. Portincasa, P.; Bonfrate, L.; Khalil, M.; Angelis, M.; Calabrese, F.M.; D'Amato, M.; Wang, D.Q.; Di Ciaula, A. Intestinal barrier and permeability in health, obesity and NAFLD. *Biomedicines* **2021**, *10*, 83. [CrossRef] [PubMed]

43. Greenblum, S.; Turnbaugh, P.J.; Borenstein, E. Metagenomic systems biology of the human gut microbiome reveals topological shifts associated with obesity and inflammatory bowel disease. *Proc. Natl. Acad. Sci. USA* **2012**, *109*, 594–599. [CrossRef] [PubMed]

44. Genoni, G.; Prodam, F.; Marolda, A.; Giglione, E.; Demarchi, I.; Bellone, S.; Bona, G. Obesity and infection: Two sides of one coin. *Eur. J. Pediatr.* **2014**, *173*, 25–32. [CrossRef] [PubMed]

45. Green, W.D.; Beck, M.A. Obesity altered T cell metabolism and the response to infection. *Curr. Opin. Immunol.* **2017**, *46*, 1–7. [CrossRef] [PubMed]

46. Frühbeck, G.; Baker, J.L.; Busetto, L.; Dicker, D.; Goossens, G.H.; Halford, J.C.G.; Handjieva-Darlenska, T.; Hassapidou, M.; Holm, J.C.; Lehtinen-Jacks, S.; et al. European Association for the Study of Obesity Position Statement on the Global COVID-19 Pandemic. *Obes. Facts* **2020**, *13*, 292–296. [CrossRef] [PubMed]

47. Gao, M.; Piernas, C.; Astbury, N.M.; Hippisley-Cox, J.; O'Rahilly, S.; Aveyard, P.; Jebb, S.A. Associations between body-mass index and COVID-19 severity in 6·9 million people in England: A prospective, community-based, cohort study. *Lancet Diabetes Endocrinol.* **2021**, *9*, 350–359. [CrossRef]

48. Di Ciaula, A.; Krawczyk, M.; Filipiak, K.J.; Geier, A.; Bonfrate, L.; Portincasa, P. Noncommunicable diseases, climate change and iniquities: What COVID-19 has taught us about syndemic. *Eur. J. Clin. Investig.* **2021**, *51*, e13682. [CrossRef]

49. Dhurandhar, N.V. Infectobesity: Obesity of infectious origin. *J. Nutr.* **2001**, *131*, 2794S–2797S. [CrossRef]

50. Pasarica, M.; Dhurandhar, N.V. Infectobesity: Obesity of infectious origin. *Adv. Food Nutr. Res.* **2007**, *52*, 61–102.

51. van Ginneken, V.; Sitnyakowsky, L.; Jeffery, J.E. Infectobesity: Viral infections (especially with human adenovirus-36: Ad-36) may be a cause of obesity. *Med. Hypotheses* **2009**, *72*, 383–388. [CrossRef]

52. Voss, J.D.; Dhurandhar, N.V. Viral infections and obesity. *Curr. Obes. Rep.* **2017**, *6*, 28–37. [CrossRef] [PubMed]

53. Kanneganti, T.D.; Dixit, V.D. Immunological complications of obesity. *Nat. Immunol.* **2012**, *13*, 707–712. [CrossRef] [PubMed]

54. Dhurandhar, N.V.; Bailey, D.; Thomas, D. Interaction of obesity and infections. *Obes. Rev.* **2015**, *16*, 1017–1029. [CrossRef] [PubMed]

55. Honce, R.; Schultz-Cherry, S. Impact of Obesity on Influenza A Virus Pathogenesis, Immune Response, and Evolution. *Front. Immunol.* **2019**, *10*, 1071. [CrossRef]

56. Sheridan, P.A.; Paich, H.A.; Handy, J.; Karlsson, E.A.; Hudgens, M.G.; Sammon, A.B.; Holland, L.A.; Weir, S.; Noah, T.L.; Beck, M.A. Obesity is associated with impaired immune response to influenza vaccination in humans. *Int. J. Obes.* **2012**, *36*, 1072–1077. [CrossRef] [PubMed]

57. Dicker, D.; Golan, R.; Baker, J.L.; Busetto, L.; Frühbeck, G.; Goossens, G.H.; Halford, J.C.G.; Holm, J.C.; Woodward, E.; Farpour-Lambert, N.J. Vaccinating people with obesity for COVID-19: EASO call for action. *Obes. Facts* **2021**, *14*, 334–335. [CrossRef] [PubMed]

58. Lajunen, T.; Bloigu, A.; Paldanius, M.; Pouta, A.; Laitinen, J.; Ruokonen, A.; Hartikainen, A.L.; Savolainen, M.; Herzig, K.H.; Leinonen, M.; et al. The association of body mass index, waist and hip circumference, and waist-hip ratio with *Chlamydia pneumoniae* IgG antibodies and high-sensitive C-reactive protein at 31 years of age in Northern Finland Birth Cohort 1966. *Int. J. Obes.* **2011**, *35*, 1470–1478. [CrossRef]

59. Rubicz, R.; Leach, C.T.; Kraig, E.; Dhurandhar, N.V.; Grubbs, B.; Blangero, J.; Yolken, R.; Göring, H.H. Seroprevalence of 13 common pathogens in a rapidly growing U.S. minority population: Mexican Americans from San Antonio, TX. *BMC Res. Notes* **2011**, *4*, 433. [CrossRef]

60. Dhurandhar, N.V. A framework for identification of infections that contribute to human obesity. *Lancet Infect. Dis.* **2011**, *11*, 963–969. [CrossRef]

61. Na, H.N.; Nam, J.H. Adenovirus 36 as an obesity agent maintains the obesity state by increasing MCP-1 and inducing inflammation. *J. Infect. Dis.* **2012**, *205*, 914–922. [CrossRef] [PubMed]

62. Na, H.N.; Kim, H.; Nam, J.H. Novel genes and cellular pathways related to infection with adenovirus-36 as an obesity agent in human mesenchymal stem cells. *Int. J. Obes.* **2012**, *36*, 195–200. [CrossRef]

63. Tarantino, G.; Citro, V.; Cataldi, M. Findings from studies are congruent with obesity having a viral origin, but what about obesity-related NAFLD? *Viruses* **2021**, *13*, 1285. [CrossRef] [PubMed]

64. Di Ciaula, A.; Calamita, G.; Shanmugam, H.; Khalil, M.; Bonfrate, L.; Wang, D.Q.; Baffy, G.; Portincasa, P. Mitochondria matter: Systemic aspects of nonalcoholic fatty liver disease (NAFLD) and diagnostic assessment of liver function by stable isotope dynamic breath tests. *Int. J. Mol. Sci.* **2021**, *22*, 7702. [CrossRef] [PubMed]

65. Di Ciaula, A.; Passarella, S.; Shanmugam, H.; Noviello, M.; Bonfrate, L.; Wang, D.Q.; Portincasa, P. Nonalcoholic fatty liver disease (NAFLD). Mitochondria as players and targets of therapies? *Int. J. Mol. Sci.* **2021**, *22*, 5375. [CrossRef] [PubMed]

66. Baldini, F.; Fabbri, R.; Eberhagen, C.; Voci, A.; Portincasa, P.; Zischka, H.; Vergani, L. Adipocyte hypertrophy parallels alterations of mitochondrial status in a cell model for adipose tissue dysfunction in obesity. *Life Sci.* **2021**, *15*, 118812. [CrossRef]

67. Santos, J.P.M.D.; Maio, M.C.; Lemes, M.A.; Laurindo, L.F.; Haber, J.F.D.S.; Bechara, M.D.; Prado, P.S.D.J.; Rauen, E.C.; Costa, F.; Pereira, B.C.A.; et al. Non-alcoholic steatohepatitis (NASH) and organokines: What is now and what will be in the future. *Int. J. Mol. Sci.* **2022**, *23*, 498. [CrossRef]

68. Gallego-Escuredo, J.M.; Gómez-Ambrosi, J.; Catalán, V.; Domingo, P.; Giralt, M.; Frühbeck, G.; Villarroya, F. Opposite alterations in FGF21 and FGF19 levels and disturbed expression of the receptor machinery for endocrine FGFs in obese patients. *Int. J. Obes.* **2015**, *39*, 121–129. [CrossRef]

69. Gomez-Ambrosi, J.; Gallego-Escuredo, J.M.; Catalan, V.; Rodriguez, A.; Domingo, P.; Moncada, R.; Valenti, V.; Salvador, J.; Giralt, M.; Villarroya, F.; et al. FGF19 and FGF21 serum concentrations in human obesity and type 2 diabetes behave differently after diet- or surgically-induced weight loss. *Clin. Nutr.* **2017**, *36*, 861–868. [CrossRef]

70. Basolo, A.; Bechi Genzano, S.; Piaggi, P.; Krakoff, J.; Santini, F. Energy balance and control of body weight: Possible effects of meal timing and circadian rhythm dysregulation. *Nutrients* **2021**, *13*, 3276. [CrossRef]

71. Fatima, N.; Rana, S. Metabolic implications of circadian disruption. *Pflug. Arch.* **2020**, *472*, 513–526. [CrossRef] [PubMed]

72. Karatsoreos, I.N.; Bhagat, S.; Bloss, E.B.; Morrison, J.H.; McEwen, B.S. Disruption of circadian clocks has ramifications for metabolism, brain, and behavior. *Proc. Natl. Acad. Sci. USA* **2011**, *108*, 1657–1662. [CrossRef] [PubMed]

73. Spiegel, K.; Tasali, E.; Leproult, R.; Van Cauter, E. Effects of poor and short sleep on glucose metabolism and obesity risk. *Nat. Rev. Endocrinol.* **2009**, *9*, 253–261. [CrossRef]

74. Tasali, E.; Leproult, R.; Spiegel, K. Reduced sleep duration or quality: Relationships with insulin resistance and type 2 diabetes. *Prog. Cardiovasc. Dis.* **2009**, *51*, 381–391. [CrossRef] [PubMed]

75. Fonken, L.K.; Workman, J.L.; Walton, J.C.; Weil, Z.M.; Morris, J.S.; Haim, A.; Nelson, R.J. Light at night increases body mass by shifting the time of food intake. *Proc. Natl. Acad. Sci. USA* **2010**, *107*, 18664–18669. [CrossRef]

76. Spiegel, K.; Tasali, E.; Leproult, R.; Scherberg, N.; Van Cauter, E. Twenty-four-hour profiles of acylated and total ghrelin: Relationship with glucose levels and impact of time of day and sleep. *J. Clin. Endocrinol. Metab.* **2011**, *96*, 486–493. [CrossRef]

77. Nedeltcheva, A.V.; Kilkus, J.M.; Imperial, J.; Schoeller, D.A.; Penev, P.D. Insufficient sleep undermines dietary efforts to reduce adiposity. *Ann. Intern. Med.* **2010**, *153*, 435–441. [CrossRef]

78. Vgontzas, A.N.; Zoumakis, M.; Bixler, E.O.; Lin, H.M.; Prolo, P.; Vela-Bueno, A.; Kales, A.; Chrousos, G.P. Impaired nighttime sleep in healthy old versus young adults is associated with elevated plasma interleukin-6 and cortisol levels: Physiologic and therapeutic implications. *J. Clin. Endocrinol. Metab.* **2003**, *88*, 2087–2095. [CrossRef]

79. Kritikou, I.; Basta, M.; Tappouni, R.; Pejovic, S.; Fernandez-Mendoza, J.; Nazir, R.; Shaffer, M.; Liao, D.; Bixler, E.; Chrousos, G.; et al. Sleep apnoea and visceral adiposity in middle-aged males and females. *Eur. Respir. J.* **2013**, *41*, 601–609. [CrossRef]

80. Lorton, D.; Lubahn, C.L.; Estus, C.; Millar, B.A.; Carter, J.L.; Wood, C.A.; Bellinger, D.L. Bidirectional communication between the brain and the immune system: Implications for physiological sleep and disorders with disrupted sleep. *Neuroimmunomodulation* **2006**, *13*, 357–374. [CrossRef]

81. van der Valk, E.S.; Savas, M.; van Rossum, E.F.C. Stress and obesity: Are there more susceptible individuals? *Curr. Obes. Rep.* **2018**, *7*, 193–203. [CrossRef] [PubMed]

82. Tasali, E.; Wroblewski, K.; Kahn, E.; Kilkus, J.; Schoeller, D.A. Effect of sleep extension on objectively assessed energy intake among adults with overweight in real-life settings: A randomized clinical trial. *JAMA Intern. Med.* **2022**, *182*, 365–374. [CrossRef] [PubMed]

83. Plagemann, A. Perinatal programming and functional teratogenesis: Impact on body weight regulation and obesity. *Physiol. Behav.* **2005**, *86*, 661–668. [CrossRef] [PubMed]

84. Lindblom, R.; Ververis, K.; Tortorella, S.M.; Karagiannis, T.C. The early life origin theory in the development of cardiovascular disease and type 2 diabetes. *Mol. Biol. Rep.* **2015**, *42*, 791–797. [CrossRef] [PubMed]

85. Heindel, J.J.; Newbold, R.; Schug, T.T. Endocrine disruptors and obesity. *Nat. Rev. Endocrinol.* **2015**, *11*, 653–661. [CrossRef] [PubMed]

86. Azoulay, L.; Bouvattier, C.; Christin-Maitre, S. Impact of intra-uterine life on future health. *Ann. Endocrinol.* **2022**, *83*, 54–58. [CrossRef]

87. Di Ciaula, A.; Portincasa, P. Fat, epigenome and pancreatic diseases. Interplay and common pathways from a toxic and obesogenic environment. *Eur. J. Intern. Med.* **2014**, *25*, 865–873. [CrossRef]

88. Haverinen, E.; Fernandez, M.F.; Mustieles, V.; Tolonen, H. Metabolic syndrome and endocrine disrupting chemicals: An overview of exposure and health effects. *Int. J. Environ. Res. Public Health* **2021**, *18*, 13047. [CrossRef]

89. Heindel, J.J.; Blumberg, B. Environmental obesogens: Mechanisms and controversies. *Annu. Rev. Pharmacol. Toxicol.* **2019**, *59*, 89–106. [CrossRef]

90. Sun, J.; Fang, R.; Wang, H.; Xu, D.X.; Yang, J.; Huang, X.; Cozzolino, D.; Fang, M.; Huang, Y. A review of environmental metabolism disrupting chemicals and effect biomarkers associating disease risks: Where exposomics meets metabolomics. *Environ. Int.* **2022**, *158*, 106941. [CrossRef]

91. Di Ciaula, A.; Portincasa, P. Diet and contaminants: Driving the rise to obesity epidemics? *Curr Med Chem* **2019**, *26*, 3471–3482. [CrossRef] [PubMed]

92. Grun, F.; Blumberg, B. Endocrine disrupters as obesogens. *Mol. Cell. Endocrinol.* **2009**, *304*, 19–29. [CrossRef] [PubMed]

93. Grun, F.; Blumberg, B. Minireview: The case for obesogens. *Mol. Endocrinol.* **2009**, *23*, 1127–1134. [CrossRef] [PubMed]

94. Boudalia, S.; Bousbia, A.; Boumaaza, B.; Oudir, M.; Canivenc Lavier, M.C. Relationship between endocrine disruptors and obesity with a focus on bisphenol A: A narrative review. *Bioimpacts* **2021**, *11*, 289–300. [CrossRef]

95. Elobeid, M.A.; Padilla, M.A.; Brock, D.W.; Ruden, D.M.; Allison, D.B. Endocrine disruptors and obesity: An examination of selected persistent organic pollutants in the NHANES 1999–2002 data. *Int. J. Environ. Res. Public Health* **2010**, *7*, 2988–3005. [CrossRef]

96. Brulport, A.; Le Corre, L.; Chagnon, M.C. Chronic exposure of 2,3,7,8-tetrachlorodibenzo-p-dioxin (TCDD) induces an obesogenic effect in C57BL/6J mice fed a high fat diet. *Toxicology* **2017**, *390*, 43–52. [CrossRef]

97. Guo, J.; Ito, S.; Nguyen, H.T.; Yamamoto, K.; Tanoue, R.; Kunisue, T.; Iwata, H. Effects of prenatal exposure to triclosan on the liver transcriptome in chicken embryos. *Toxicol. Appl. Pharmacol.* **2018**, *347*, 23–32. [CrossRef]

98. Frühbeck, G.; Méndez-Giménez, L.; Fernández-Formoso, J.A.; Fernández, S.; Rodríguez, A. Regulation of adipocyte lipolysis. *Nutr. Res. Rev.* **2014**, *27*, 63–93. [CrossRef]

99. Frühbeck, G.; Gómez Ambrosi, J.; Salvador, J. Leptin-induced lipolysis opposes the tonic inhibition of endogenous adenosine in white adipocytes. *FASEB J.* **2001**, *15*, 333–340. [CrossRef]

100. Frühbeck, G.; Gómez-Ambrosi, J. Modulation of the leptin-induced white adipose tissue lipolysis by nitric oxide. *Cell Signal.* **2001**, *13*, 827–833. [CrossRef]

101. Muruzábal, F.J.; Frühbeck, G.; Gómez-Ambrosi, J.; Archanco, M.; Burrell, M.A. Immunocytochemical detection of leptin in non-mammalian vertebrate stomach. *Gen. Comp. Endocrinol.* **2002**, *128*, 149–152. [CrossRef]

102. Moreno-Navarrete, J.M.; Martínez-Barricarte, R.; Catalán, V.; Sabater, M.; Gómez-Ambrosi, J.; Ortega, F.J.; Ricart, W.; Blüher, M.; Frühbeck, G.; de Córdoba, S.R.; et al. Complement Factor H is expressed in adipose tissue in association with insulin resistance. *Diabetes* **2010**, *59*, 200–209. [CrossRef] [PubMed]

103. Pulido, M.R.; Diaz-Ruiz, A.; Jimenez-Gomez, Y.; Garcia-Navarro, S.; Gracia-Navarro, F.; Tinahones, F.; Lopez-Miranda, J.; Frühbeck, G.; Vazquez-Martinez, R.; Malagon, M.M. Rab18 dynamics in adipocytes in relation to lipogenesis, lipolysis and obesity. *PLoS ONE* **2011**, *6*, e22931. [CrossRef] [PubMed]

104. Catalán, V.; Gómez-Ambrosi, J.; Rodríguez, A.; Ramírez, B.; Rotellar, F.; Valentí, V.; Silva, C.; Gil, M.J.; Fernández-Real, J.M.; Salvador, J.; et al. Increased levels of calprotectin in obesity are related to macrophage content: Impact on inflammation and effect of weight loss. *Mol. Med.* **2011**, *17*, 1157–1167. [CrossRef] [PubMed]

105. Rodríguez, A.; Gómez-Ambrosi, J.; Catalán, V.; Rotellar, F.; Valentí, V.; Silva, C.; Mugueta, C.; Pulido, M.R.; Vázquez, R.; Salvador, J.; et al. The ghrelin O-acyltransferase-ghrelin system reduces TNF-a-induced apoptosis and autophagy in human visceral adipocytes. *Diabetologia* **2012**, *55*, 3038–3050. [CrossRef] [PubMed]

106. Calamita, G.; Delporte, C. Involvement of aquaglyceroporins in energy metabolism in health and disease. *Biochimie* **2021**, *188*, 20–34. [CrossRef] [PubMed]

107. Frühbeck, G. Obesity: Aquaporin enters the picture. *Nature* **2005**, *438*, 436–437. [CrossRef]

108. Francis, C.E.; Allee, L.; Nguyen, H.; Grindstaff, R.D.; Miller, C.N.; Rayalam, S. Endocrine disrupting chemicals: Friend or foe to brown and beige adipose tissue? *Toxicology* **2021**, *463*, 152972. [CrossRef]

109. Ohanyan, H.; Portengen, L.; Huss, A.; Traini, E.; Beulens, J.W.J.; Hoek, G.; Lakerveld, J.; Vermeulen, R. Machine learning approaches to characterize the obesogenic urban exposome. *Environ. Int.* **2022**, *158*, 107015. [CrossRef]

110. Guo, B.; Guo, Y.; Nima, Q.; Feng, Y.; Wang, Z.; Lu, R.; Baimayangji; Ma, Y.; Zhou, J.; Xu, H.; et al. Exposure to air pollution is associated with an increased risk of metabolic dysfunction-associated fatty liver disease. *J. Hepatol.* **2022**, *76*, 518–525. [CrossRef]

111. Koch, C.A.; Sharda, P.; Patel, J.; Gubbi, S.; Bansal, R.; Bartel, M.J. Climate change and obesity. *Horm. Metab. Res.* **2021**, *53*, 575–587. [CrossRef] [PubMed]

112. Cypess, A.M. Reassessing human adipose tissue. *N. Engl. J. Med.* **2022**, *386*, 768–779. [CrossRef] [PubMed]

113. Cypess, A.M.; Lehman, S.; Williams, G.; Tal, I.; Rodman, D.; Goldfine, A.B.; Kuo, F.C.; Palmer, E.L.; Tseng, Y.H.; Doria, A.; et al. Identification and importance of brown adipose tissue in adult humans. *N. Engl. J. Med.* **2009**, *360*, 1509–1517. [CrossRef] [PubMed]

114. Frühbeck, G.; Becerril, S.; Sáinz, N.; Garrastachu, P.; García-Velloso, M.J. BAT: A new target for human obesity? *Trends Pharmacol. Sci.* **2009**, *30*, 387–396. [CrossRef]

115. Tseng, Y.-H.; Cypess, A.M.; Kahn, C.R. Cellular bioenergetics as a target for obesity therapy. *Nat. Rev. Drug. Discov.* **2010**, *9*, 465–481. [CrossRef]

116. Xue, R.; Lynes, M.D.; Dreyfuss, J.M.; Shamsi, F.; Schulz, T.J.; Zhang, H.; Huang, T.L.; Townsend, K.L.; Li, Y.; Takahashi, H.; et al. Clonal analyses and gene profiling identify genetic biomarkers of the thermogenic potential of human brown and white preadipocytes. *Nat. Med.* **2015**, *21*, 760–768. [CrossRef]

117. Shinoda, K.; Luijten, I.H.; Hasegawa, Y.; Hong, H.; Sonne, S.B.; Kim, M.; Xue, R.; Chondronikola, M.; Cypess, A.M.; Tseng, Y.H.; et al. Genetic and functional characterization of clonally derived adult human brown adipocytes. *Nat. Med.* **2015**, *21*, 389–394. [CrossRef]

118. Sakers, A.; De Siqueira, M.K.; Seale, P.; Villanueva, C.J. Adipose-tissue plasticity in health and disease. *Cell* **2022**, *185*, 419–446. [CrossRef]

119. Shamsi, F.; Wang, C.H.; Tseng, Y.H. The evolving view of thermogenic adipocytes—ontogeny, niche and function. *Nat. Rev. Endocrinol.* **2021**, *17*, 726–744. [CrossRef]

120. Symonds, M.E.; Farhat, G.; Aldiss, P.; Pope, M.; Budge, H. Brown adipose tissue and glucose homeostasis—the link between climate change and the global rise in obesity and diabetes. *Adipocyte* **2019**, *8*, 46–50. [CrossRef]

121. Gaspar, R.C.; Pauli, J.R.; Shulman, G.I.; Muñoz, V.R. An update on brown adipose tissue biology: A discussion of recent findings. *Am. J. Physiol. Endocrinol. Metab.* **2021**, *320*, E488–E495. [CrossRef] [PubMed]

122. Yang, F.T.; Stanford, K.I. Batokines: Mediators of inter-tissue communication (a mini-review). *Curr. Obes. Rep.* **2022**. online ahead of print. [CrossRef]

123. Gavaldà-Navarro, A.; Villarroya, J.; Cereijo, R.; Giralt, M.; Villarroya, F. The endocrine role of brown adipose tissue: An update on actors and actions. *Rev. Endocr. Metab. Disord.* **2022**, *23*, 31–41. [CrossRef] [PubMed]

124. Lynes, M.D.; Leiria, L.O.; Lundh, M.; Bartelt, A.; Shamsi, F.; Huang, T.L.; Takahashi, H.; Hirshman, M.F.; Schlein, C.; Lee, A.; et al. The cold-induced lipokine 12,13-diHOME promotes fatty acid transport into brown adipose tissue. *Nat. Med.* **2017**, *23*, 631–637. [CrossRef] [PubMed]

125. Macedo, A.P.A.; Munoz, V.R.; Cintra, D.E.; Pauli, J.R. 12,13-diHOME as a new therapeutic target for metabolic diseases. *Life Sci.* **2022**, *290*, 120229. [CrossRef] [PubMed]

126. Frühbeck, G.; Gómez Ambrosi, J. Rationale for the existence of additional adipostatic hormones. *FASEB J.* **2001**, *15*, 1996–2006. [CrossRef] [PubMed]

127. Landecho, M.F.; Tuero, C.; Valentí, V.; Bilbao, I.; de la Higuera, M.; Frühbeck, G. Relevance of leptin and other adipokines in obesity-associated cardiovascular risk. *Nutrients* **2019**, *11*, 2664. [CrossRef]

128. Klimentidis, Y.C.; Beasley, T.M.; Lin, H.Y.; Murati, G.; Glass, G.E.; Guyton, M.; Newton, W.; Jorgensen, M.; Heymsfield, S.B.; Kemnitz, J.; et al. Canaries in the coal mine: A cross-species analysis of the plurality of obesity epidemics. *Proc. Biol. Sci.* **2011**, *278*, 1626–1632. [CrossRef]
129. Rappaport, S.M.; Smith, M.T. Epidemiology. Environment and disease risks. *Science* **2010**, *330*, 460–461. [CrossRef]
130. Zhang, P.; Carlsten, C.; Chaleckis, R.; Hanhineva, K.; Huang, M.; Isobe, T.; Koistinen, V.M.; Meister, I.; Papazian, S.; Sdougkou, K.; et al. Defining the scope of exposome studies and research needs from a multidisciplinary perspective. *Environ. Sci. Technol. Lett.* **2021**, *8*, 839–852. [CrossRef]
131. Fang, M.; Hu, L.; Chen, D.; Guo, Y.; Liu, J.; Lan, C.; Gong, J.; Wang, B. Exposome in human health: Utopia or wonderland? *Innovation* **2021**, *2*, 100172. [CrossRef]
132. Wild, C.P. Complementing the genome with an "exposome": The outstanding challenge of environmental exposure measurement in molecular epidemiology. *Cancer Epidemiol. Biomarks Prev.* **2005**, *14*, 1847–1850. [CrossRef] [PubMed]
133. Wild, C.P. The exposome: From concept to utility. *Int. J. Epidemiol.* **2012**, *41*, 24–32. [CrossRef] [PubMed]
134. The, N.S.; Suchindran, C.; North, K.E.; Popkin, B.M.; Gordon-Larsen, P. Association of adolescent obesity with risk of severe obesity in adulthood. *JAMA* **2010**, *304*, 2042–2047. [CrossRef] [PubMed]
135. Golding, J.; Gregory, S.; Northstone, K.; Iles-Caven, Y.; Ellis, G.; Pembrey, M. Investigating possible trans/intergenerational associations with obesity in young adults using an exposome approach. *Front. Genet.* **2019**, *10*, 314. [CrossRef] [PubMed]
136. Wu, Y.; Perng, W.; Peterson, K.E. Precision Nutrition and Childhood Obesity: A Scoping Review. *Metabolites* **2020**, *10*, 235. [CrossRef]
137. Mahmoud, A.M. An overview of epigenetics in obesity: The role of lifestyle and therapeutic interventions. *Int. J. Mol. Sci.* **2022**, *23*, 1341. [CrossRef]
138. Rajpurkar, P.; Chen, E.; Banerjee, O.; Topol, E.J. AI in health and medicine. *Nat. Med.* **2022**, *28*, 31–38. [CrossRef]

Perspective

Obesity Subtyping: The Etiology, Prevention, and Management of Acquired versus Inherited Obese Phenotypes

Edward Archer [1],* and Carl J. Lavie [2]

[1] Research & Development, EvolvingFX, LLC., Fort Wayne, IN 46835, USA

[2] Department of Cardiovascular Diseases, John Ochsner Heart & Vascular Institute, Ochsner Clinical School, The University of Queensland School of Medicine, New Orleans, LA 70121, USA; clavie@ochsner.org

* Correspondence: archer.edwardc@gmail.com

Abstract: The etiology of obesity is complex and idiosyncratic—with inherited, behavioral, and environmental factors determining the age and rate at which excessive adiposity develops. Moreover, the etiologic status of an obese phenotype (how and when it developed initially) strongly influences both the short-term response to intervention and long-term health trajectories. Nevertheless, current management strategies tend to be 'one-size-fits-all' protocols that fail to anticipate the heterogeneity of response generated by the etiologic status of each individual's phenotype. As a result, the efficacy of current lifestyle approaches varies from ineffective and potentially detrimental, to clinically successful; therefore, we posit that effective management strategies necessitate a personalized approach that incorporates the subtyping of obese phenotypes. Research shows that there are two broad etiologic subtypes: 'acquired' and 'inherited'. Acquired obesity denotes the development of excessive adiposity after puberty—and because the genesis of this subtype is behavioral, it is amenable to interventions based on diet and exercise. Conversely, inherited obesity subsumes all forms of excessive adiposity that are present at birth and develop prior to pubescence (pediatric and childhood). As the inherited phenotype is engendered in utero, this subtype has irreversible structural (anatomic) and physiologic (metabolic) perturbations that are not susceptible to intervention. As such, the most realizable outcome for many individuals with an inherited subtype will be a 'fit but fat' phenotype. Given that etiologic subtype strongly influences the effects of intervention and successful health management, the purpose of this 'perspective' article is to provide a concise overview of the differential development of acquired versus inherited obesity and offer insight into subtype-specific management.

Keywords: inherited; acquired; obesity; diet; exercise

Citation: Archer, E.; Lavie, C.J. Obesity Subtyping: The Etiology, Prevention, and Management of Acquired versus Inherited Obese Phenotypes. *Nutrients* **2022**, *14*, 2286. https://doi.org/10.3390/nu14112286

Academic Editor: Javier Gómez-Ambrosi

Received: 19 April 2022
Accepted: 27 May 2022
Published: 30 May 2022

Publisher's Note: MDPI stays neutral with regard to jurisdictional claims in published maps and institutional affiliations.

1. Introduction

Obesity is a global health concern [1–5], with the prevalence in the U.S. exceeding 40% in adults and nearly 20% in children and adolescents [6]. Although efforts to stem the increasing prevalence have been unsuccessful, research has led to a clearer understanding of its etiology and how obesity impacts cardiometabolic diseases, such as type-2 diabetes mellitus (T2DM), dyslipidemia, cardiovascular disease, and hypertension [7,8].

Despite these conceptual advances, the development of effective prevention and management protocols has been less successful. Although lifestyle modifications are the cornerstone of obesity management, few individuals achieve long-term benefits with 'one-size-fits-all' diet and exercise approaches [9]. We posit that this lack of success is not due to a deficiency of willpower or adherence by participants and patients but is engendered by the failure to recognize that the obese phenotype is not a single homogenous condition [10]. To be precise, obesity, despite being an increasingly common phenomenon, has a complex, idiosyncratic etiology—with inherited, behavioral, and environmental factors determining the age and rate at which excessive adiposity and cardiometabolic diseases develop.

Thus, because research suggests that that the etiology of an obese phenotype (how and when it developed initially) strongly influences the short-term effectiveness and long-term outcomes of lifestyle interventions [1,3,11,12], successful obesity management necessitates the subtyping of phenotypes. As such, the purpose of this 'perspective' article is to provide a concise overview of the differential development of acquired versus inherited obese phenotypes and offer insight into subtype-specific obesity prevention and management.

2. Etiologic Subtypes of Obesity

Although the defining characteristic of obesity is an excess of bodyfat [2,12], the age and rate at which excessive adiposity develops vary as a result of inherited, behavioral, and environmental factors. Thus, because the obese phenotype may be engendered at any point in an individual's development—from the prenatal period to senescence—research suggests two broad etiologic subtypes: 'acquired' and 'inherited' [1,3].

Acquired obesity, also known as 'adult-onset', denotes the disproportionate development of adiposity after puberty. The genesis of this phenotypic subtype is essentially behavioral, with physical activity (PA) and subsequent hyperphagia (overconsumption) being the major determinants. Stated simply, 'moving too little' leads to 'eating too much' and together these pathologic behaviors lead to acquired obesity and cardiometabolic diseases, such as T2DM [1].

Conversely, inherited obesity—also known as pediatric or childhood—subsumes all forms of excessive adiposity that are present at birth or develop prior to pubescence. Inherited obesity can be further subdivided into 'non-genetic' (common) and 'genetic' (rare) obesity. Common inherited obesity is a ubiquitous, complex, *quantitative* (continuously distributed) phenotype characterized by altered fat, muscle, and pancreatic beta-cell development and function. These structural (anatomic) and physiologic (metabolic) alterations are engendered during prenatal development, and as such, are largely irreversible [3,11,13].

Genetic obesity refers to Mendelian disorders that result in discrete, *qualitative* phenotypes that display excessive adiposity (e.g., leptin deficiency, Prader-Willi syndrome). As explained in detail below, because the genesis of common inherited obese phenotypes differs considerably from the genesis of the genetic inherited phenotype, it is important to distinguish between these subtypes in prevention, diagnosis, and management. As genetic obesity is rare [4], and the role of 'genes' in common obesity is limited (see Section 3 below), in this review, we limit our discussion to common (nongenetic) inherited forms of obesity.

3. Nongenetic versus Genetic Inheritance and the Role of Genes in Obesity

We have written extensively on the role of nongenetic inheritance in the development of obesity and T2DM, and how the conflation of the term 'inherited' with 'genetic' has led to confusion [3,14,15]. Given that a detailed exposition of the conceptual and empirical foundation for our work is beyond the scope of this article, we offer a concise overview below and direct our readers to select publications [1,3,12–16].

Briefly, the functional unit in biology and biological inheritance is the cell, and because each cell's idiosyncratic nature and spaciotemporal context determines gene expression, it is important to distinguish between nongenetic (cellular) inheritance and the two forms of genetic inheritance (nuclear and mitochondrial) [3,17–19]. For example, the fundamental difference between monozygotic (identical) and dizygotic (fraternal) twins is inherent in the nomenclature—identical twins develop from a single cell (a fertilized egg) whereas fraternal twins develop from two different cells (two fertilized eggs). Thus, fraternal twins differ in both cellular and genetic inheritance whereas identical twins do not. Therefore, the greater phenotypic disparity displayed by fraternal twins is due to differences in the genotypic expression induced by different cells in concert with inter-twin differences in genotype. Yet despite the variability in the developmental competence of any given population of eggs, the functional distinction between cellular and genetic inheritance is ignored routinely by those who infer genetic causality from 'twin-studies' and heritability statistics.

To be precise, our work demonstrated that *"there are no 'genes for' quantitative (i.e., nondiscrete) phenotypes, such as common obesity and metabolic diseases (e.g., T2DM)."* [1]. We further detailed the *"fatal flaws of twin studies"* and showed why *"estimates of genetic heritability are misused, and often meaningless statistical abstractions derived from attempts to impose an artificial and false dichotomy (i.e., nature vs. nurture (genes vs. environment)) on demonstrably non-dichotomous biologic processes"* [1].

These conclusions—which form the basis for our perspective on the limited role of 'genes' in obesity—are most clearly supported by the nonlinear processes that lead to 'one-to-many', 'many-to-one', and 'many-to-many' genotype–phenotype relations. These processes include reaction norms, phenotypic accommodation, alternative splicing, RNA editing, chimeric transcripts, protein multifunctionality, epistatic variance, maternal effects, the metabolic regulation of transcription, and post-translational modifications.

Thus, a 'great deal of biology'—both established and undiscovered—links an individual's genotype, the cellular expression of that genotype, and the development of specific phenotypes; therefore, as Felder and Lewontin wrote, there is *"a vast loss of information in going from a complex machine* [an organism] *to a few descriptive parameters* [heritability estimates]" [20]. Moreover, because estimates of genetic heritability are mere statistical associations, they cannot be used to quantify the relative contributions of presumed etiologic factors outside of highly controlled animal and plant breeding operations [1,21]. In other words, 'correlation does not equal causation'—especially when the relations are nonlinear, and the fundamental constructs are inherently flawed or misconstrued.

In sum, the emerging field of non-genetic inheritance [17–19] and our work suggests that genes are *"tools of the cell"*, and as such, *"are merely a necessary but not sufficient component for the development of obesity/T2DM phenotypes"* [14]; therefore, an understanding of the etiology, prevention, management, and treatment of these phenotypes *"will not be found in the genome"* [3].

As detailed below, because the etiologies of acquired and 'non-genetic' (common) inherited obese phenotypes differ markedly, strategies for their prevention and management must be subtype-specific.

4. Acquired Obesity: Its Etiology and Response to Intervention

Although the etiology of acquired obesity is often contested [1], there is strong evidence dating from the mid-20th century that reductions in PA, high physical inactivity (PI), and excessive sedentary behavior (SB) are strong determinants of this phenotype in both human and non-human animals [1,12,22–33]. To summarize briefly, first, PA is the major *modifiable* determinant of caloric consumption [27,28,31,34–37]. Second, when individuals reduce their PA, their consumption declines more slowly than their caloric expenditure [27,28,31,34–36]. This leads to relative hyperphagia (overconsumption) and positive energy balance—with individuals consuming more calories than they expend.

Third, as PA declines, the energetic demands of skeletal muscle decline. This reduces the number of calories partitioned to skeletal-muscle and increases the number of calories available for storage in fat-cells (adipogenic partitioning). Fourth, PI decreases skeletal muscle insulin-sensitivity, which induces hyperinsulinemia (higher levels of insulin) during and after each meal with concomitant increments in adipogenic partitioning and decrements in lipolysis. The confluence of PI-induced hyperphagia and hyperinsulinemia causes a greater percentage of the calories consumed at each meal to be stored and sequestered in fat cells (reduced lipid turnover) with concomitant increments in body and fat mass.

When PI and excessive SB become habitual, the attendant metabolic perturbations [33] lead to acquired obesity via increments in fat-cell size, number (hypertrophy and hyperplasia, respectively), and ectopic development (fat-cell intrusions into non-adipose tissue). If the increased demands for insulin production and caloric storage cannot be met by parallel increments in pancreatic beta-cell functioning and fat-cell plasticity, the declining skeletal muscle insulin-sensitivity progresses to whole-body insulin-resistance, and over time, to overt T2DM [38–41]. Evidence for these phenomena was established decades ago, with the loss of skeletal muscle insulin sensitivity being the initial and primary metabolic insult in cardiometabolic diseases [38–41]. Thus, PI, high levels of SB, and concomitant hyperphagia are the major etiologic factors leading to acquired obesity and cardiometabolic diseases [1,12,16,30,33,38].

Nevertheless, despite the strong influence of PA on the development of acquired obesity and T2DM, the management of these metabolic maladies *must* include dietary interventions because exercise-only interventions have trivial impacts on body mass and weight loss, despite clinically important impacts on body composition, and blood glucose and insulin levels.

5. The Prevention and Management of the Acquired Obese Phenotype

As the genesis and maintenance of the acquired obese phenotype are largely behavioral (moving less and eating more), prevention entails adequate levels of daily PA and relative caloric consumption from childhood to senescence. To be precise, 30–60 min of daily PA and a physical activity level (PAL) reaching 1.7–1.8, are necessary for the primary prevention of acquired obesity and the maintenance of a reduced (post-obese) phenotype [31,42,43]. Furthermore, in the early stages of development, the acquired subtype is extremely amenable to interventions emphasizing diet, PA, and exercise; however, it is important to note that despite the demonstrated impact on metabolic (glycemic and lipidemic) control, exercise-only interventions have a limited impact on weight-loss and body mass [44,45]. Thus, dietary and caloric restriction *must* play a dominant role if body mass is to be reduced.

This may be particularly important in patients with obesity and cardiometabolic diseases, such as dyslipidemia, especially hypertriglyceridemia, hypertension, and elevated blood glucose levels, including metabolic syndrome and T2DM. These patients require increased PA and exercise in concert with reductions in caloric intakes, particularly simple and complex carbohydrates and alcohol—even more so than reductions in fat intake—to improve both weight and metabolic control [46–48].

It is important to note, however, that if the chronic positive energy balance and metabolic perturbations induced by PI and excessive SB continue over time, the growth in the number of fat-cells (hyperplasia), in concert with the degradation of pancreatic beta-cell function and insulin sensitivity eventually lead to diminished health and responsiveness to lifestyle interventions. As such, long-standing acquired obesity will resemble the common inherited obese phenotype in its response to intervention [1,3,12].

6. Inherited Obesity: Its Etiology and Response to Intervention

In contrast to the behavioral genesis of the acquired (adult-onset) phenotype, the common inherited phenotype is engendered during in utero (prenatal) development. As briefly explained below, and detailed elsewhere [3,11–15], this subtype exhibits irreversible structural (anatomic) and physiologic (metabolic) perturbations engendered by the mother's behavioral and metabolic phenotypes (e.g., PA levels, adiposity, glycemic control).

Briefly, it is well-established that during pregnancy, a mother's cells compete for calories with those of her fetus [3,12,13,16]. Thus, to ensure that the fetus receives the number of calories it needs for development, pregnancy leads to hormonal changes that induce insulin-resistance in maternal skeletal muscle. This naturally developing insulin-resistance increases caloric consumption while decreasing the number of calories partitioned to maternal skeletal muscle. This leads to increased maternal serum lipid and glucose levels with concomitant increments in maternal body and fat mass, and caloric transfer to the fetus [3].

For comparison, stunting and common inherited obesity represent opposing ends of the maternal–fetal competitive continuum and they impact at least three generations: the mother, the fetus, and the germline of female fetuses. Stunting develops when a mother's diet and body-fat stores cannot keep pace with the competitive demands of her cells and fetal development. This causes fewer fetal muscle, fat, bone, and pancreatic beta-cells to be created, and permanently alters the offspring's structural (anatomic) and physiologic (metabolic) phenotypes (e.g., shorter height and impaired glucose and lipid metabolism). These changes are irreversible and substantially increase the risk of cardiometabolic diseases [49–51].

Conversely, common inherited obesity is engendered by insufficient maternal PA and metabolic control which reduces the competition for calories between mother and fetus. More specifically, when the naturally occurring insulin-resistance of pregnancy acts in concert with the pathological insulin-resistance induced by maternal PI and excessive SB, the escalation in insulin-resistance exponentially increases caloric consumption, while decreasing the number of calories partitioned to maternal skeletal muscle. This causes an excessive number of calories to be transferred to the fetus—which stimulates a disproportionate increment in fat-cell size and number, fetal insulin production, and dysfunctional skeletal muscle development (more structural and less contractile elements) [3,11–15].

These pathologic 'maternal-effects' (non-genetic mechanisms of inheritance) are irreversible and produce children who are predisposed to *'eating more and moving less'*, independent of genotype [3,11–15]. Infants and children with this subtype will consume more calories than those with normal phenotypes because their excessive fat-cell hyperplasia, reduced skeletal muscle function, and hyperinsulinemia, increase the number of calories stored and sequestered in fat-cells after each meal—both in adipose tissue and ectopically. Over time, this adipogenic partitioning causes increments in body and fat mass, and concomitant obesity [1,12,16]. These 'maternal-effects' offer a comprehensive explanation for the inheritance of compromised metabolic phenotypes in both human and nonhuman animals [3,12–15].

Thus, increments in childhood obesity and adolescent T2DM are most plausibly explained by the substantial decline in PA and increments in SB over the past 50 years by young women and mothers [52–54]. As the PI-driven maternal-effects escalated from one generation to the next, the prevalence of both obesity and T2DM increased markedly [3,11–15]. Our research suggests that these pathological maternal effects also explain the increased prevalence of obesity and cardiometabolic maladies in nonhuman mammals inclusive of dogs, cats, laboratory mice, monkeys, and feral moose [1,12].

7. The Prevention and Management of the Inherited Obese Phenotype

The inherited obese phenotype represents a continuum of metabolic perturbations instantiated during prenatal development. Thus, unlike acquired obesity, the structural (anatomic) and physiologic (metabolic) perturbations are not a behavioral manifestation, but are inherent to the phenotype, and therefore, are irreversible. This means that the prevention of common inherited obesity must begin with the current generation of female children and adolescents (future mothers). Sufficient increments in pre-pubertal, pubertal, pre-conception, and prenatal PA will ameliorate or prevent the pathologic maternal effects that lead to this phenotypic subtype. More specifically, as with the prevention of acquired obesity, future mothers must perform at least 30–60 min of daily PA and reach a PAL of 1.7–1.8 to prevent the development of common inherited obesity in future generations.

Nevertheless, once instantiated in utero, the structural and physiologic perturbations engendered by accumulative maternal effects are irreversible. To be precise, the inherited phenotype exhibits both hypertrophic and hyperplastic obesity (greater fat-cell size and number) in concert with dysfunctional pancreatic-cell function and reduced muscle-cell contractility. No behavioral interventions can reduce the number of fat-cells, nor wholly overcome the reduced muscle-cell function; therefore, individuals with this subtype will always find it more difficult to 'move more and eat less' than individuals with normal or acquired obese phenotypes.

Importantly, as detailed in the following section, the amount of PA and caloric restriction necessary to induce and maintain weight loss may be beyond many individuals' physical and/or psychological capacity for exercise and caloric deprivation. Thus, the inherited obese phenotype is less amenable to interventions than the acquired subtype and in many cases, the best health trajectory achievable will be a 'fit but fat' phenotype [7,55–59].

8. The *'Metabolic Tipping-Point'* and Its Effect on Intervention

There is a large body of observational and experimental research dating from the 1950s showing that although body mass and composition and concomitant basal energy expenditure are the major determinants of caloric consumption [60–64], PA is the major *modifiable* determinant of consumption, expenditure, and storage [16,27,28,31,34–37,65–68]. Thus, because PA plays an essential role in all aspects of metabolism, we previously coined the term *'Metabolic Tipping-point'* to denote the amount of PA necessary to prevent overconsumption and weight gain [1,12]. As briefly explained below, and in detail elsewhere [1,12], this concept offers a concise framework for understanding the heterogeneity of response of caloric consumption and body and fat mass to altered levels of PA.

As depicted in Figure 1, PA, body mass, and caloric consumption have complex, nonlinear relations [16]. When an individual's PA declines below their lower metabolic tipping-point (the left side of Figure 1), caloric intake declines more slowly than energy expenditure (a nonlinear relation). This leads to increments in body fat and mass, and decrements in skeletal muscle insulin-sensitivity. If habitual, these individuals will develop acquired obesity and T2DM—dependent on fat-cell plasticity and pancreatic beta-cell function. Nevertheless, any intervention that increases their PA above their lower metabolic tipping-point will reduce hyperphagia, positive energy balance, and prevent further gains in body and fat mass. Nevertheless, as explained in a previous section, *caloric restriction is essential* if the excess body mass is to be reduced because interventions that rely exclusively on PA and exercise have trivial effects on body mass.

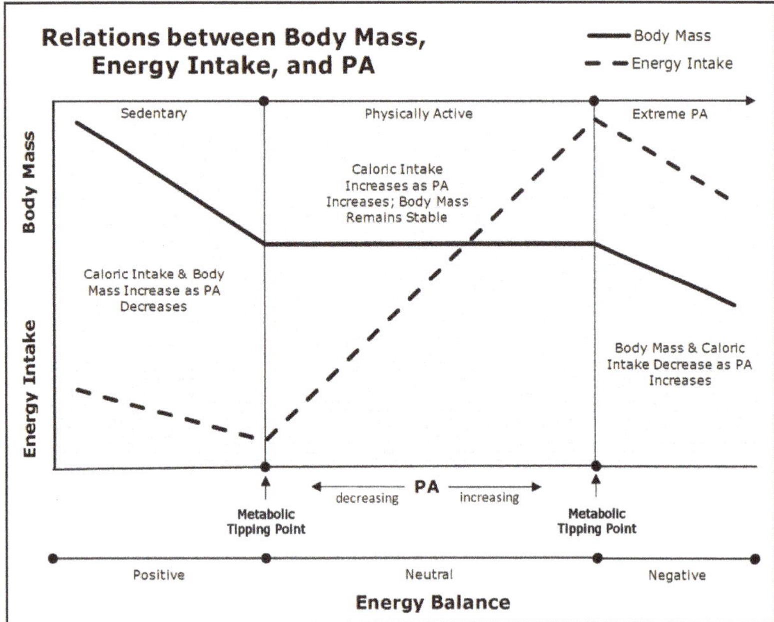

Figure 1. Relations between PA, Body Mass, and Energy Intake (adapted from [69]). As PA declines below the lower metabolic tipping-point into the 'Sedentary' range (left panel), energy intake and energy expenditure become dissociated due to insufficient PA. Body mass begins to increase as energy balance becomes positive and insulin sensitivity is diminished.

When individuals maintain PA levels between the upper and lower metabolic tipping-points (the center portion of Figure 1), their body and fat mass remain stable, regardless of increments and decrements in PA within this range. This occurs because of a linear relation between caloric consumption and expenditure at moderate levels of PA. Thus, as

PA increases, caloric consumption increases in parallel. To be precise, the *nonlinear* relations below the lower tipping-point explains why decrements in PA lead to increments in body and fat mass in highly sedentary individuals, whereas the *linear* relation between PA and consumption in the range between the upper and lower metabolic tipping-points explains why increased PA and exercise have little or no effect on body mass in individuals who are already moderately active.

Conversely, when individuals increase their PA above their upper metabolic tipping-point (the right side of Figure 1), they experience declines in caloric consumption, basal energy metabolism, energy expenditure, and body and lean mass. This level of PA is not sustainable and leads to incomplete recovery, reduced physical performance, injury, and exhaustion [70].

In summary, the left side of Figure 1 depicts the *nonlinear* relations between caloric consumption and expenditure, and the concomitant development of acquired obesity and T2DM. The center panel depicts the *linear* relations between PA and caloric consumption and explains why exercise interventions without caloric restriction will not reduce body and fat mass. The right side of Figure 1 depicts unsustainable levels of PA that lead to the loss of body and lean mass. Therefore, it is the transition from a nonlinear to a linear relation between caloric consumption and expenditure as PA increases from a sedentary to an active lifestyle that explains the heterogeneity of response to diet and exercise in individuals with varied levels of baseline PA.

Nevertheless, what Figure 1 does not depict is how an individual's obese subtype impacts their metabolic tipping-points. Because the acquired obese phenotype is essentially a behavioral phenomenon, any intervention that increases PA and reduces caloric consumption will be successful in the early and mid-stages of phenotypic development; however, the longer the physical inactivity-induced metabolic perturbations continue, the less amenable to intervention the acquired subtype becomes. In this respect, long-standing acquired obesity will mimic the inherited subtype in its response.

Conversely, individuals with an inherited subtype represent a continuum of irreversible structural (anatomic) and physiologic (metabolic) perturbations that are inherent to their phenotype. As such, the amount of PA and caloric restriction necessary to reduce body mass and maintain weight loss depends on where they fall in the continuum of perturbations—from mild to extreme. The more extreme an individual's inherited obese phenotype, the higher their metabolic tipping points, and the greater the amount of PA and caloric restriction required to prevent overconsumption and achieve and maintain a healthy weight.

Nevertheless, the physical and psychological burdens induced by large amounts of PA and severe caloric restriction are beyond the perseverative capacity of most humans. As such, the long-term maintenance of weight loss becomes an increasingly unachievable goal as the structural and physiologic perturbations become more severe. Therefore, the management objective for individuals with inherited subtypes should be along the continuum of 'fit but fat'. The refusal to appreciate this reality has led to unrealistic expectations, management 'failure', and the stigmatization of individuals with an inherited obese phenotype [71,72].

9. Assumptions and Limitations

Our 'perspective' is based on several assumptions that may limit our conclusions. The most critical is that obesity and cardiometabolic diseases are wholly anatomical (structural) and physiological (metabolic) disorders. Thus, we posit that if psychological, social, economic, or other non-physiologic phenomena influence obese or diabetic phenotypes, they must act through cellular mechanisms that cause increments in skeletal muscle-cell insulin-resistance and its sequelae (e.g., hyperphagia, adipogenic caloric partitioning, and increased fat-cell mass and number).

Although a large body of experimental evidence demonstrating the causal effects of PI on skeletal muscle-cell insulin resistance and its sequelae exists, the only support for speculations regarding the effects of psychological, social, and economic phenomena is correlational.

Moreover, we assert that distinguishing between etiology and treatment is critical for discussions revolving around the roles of PA, genes, diet, and exercise. For example, we contend that although specific macro-nutrients are not causal to obesity and other disease states, except as a source of calories (for details please see [69,73–76]), we argue strongly that caloric restriction with further reductions in carbohydrates are essential protocols for reducing body and fat mass and the treatment of acquired obesity and T2DM.

Finally, although our work on nongenetic inheritance and the developmental origins of disease is rigorous, consilient, and supported by voluminous research across species (please see [12] for details), our theories are novel and may therefore appear controversial to those unfamiliar with this emerging area of research and science (for reviews see [3,17–19]). Nevertheless, it remains to be seen if our conclusions withstand the 'test of time'.

10. Summary and Conclusions

The age and rate at which an individual's obese phenotype develops is a strong determinant of its response to intervention. Thus, the development of effective management strategies necessitates a personalized approach that incorporates the subtyping of obese phenotypes by etiologic status (acquired or inherited). The acquired phenotype denotes the development of excessive adiposity after puberty and is essentially a behavioral phenomenon induced by low levels of PA and concomitant hyperphagia (overconsumption). Thus, effective prevention and treatment strategies can be based on diet and exercise [32,47,77]. Although this subtype is amenable to lifestyle interventions in the early stages of development, the longer the PI, excessive SB, and overconsumption continue, the less amenable to intervention this subtype becomes.

In contrast, inherited obesity subsumes all forms of excessive adiposity that develop prior to pubescence (pediatric and childhood). The prevention of non-genetic inherited obese phenotypes in the next generation necessitates adequate levels of PA by the current generation of young females, potential mothers, and pregnant women. Nevertheless, once instantiated during the prenatal period, this subtype has irreversible structural (anatomic) and physiologic (metabolic) perturbations that are not amenable to intervention because no amount of diet and exercise can reduce the excessive number of fat cells and adipogenic partitioning, or significantly improve skeletal muscle function. Therefore, the objective in the management of inherited subtypes is the development of a 'fit but fat' phenotype. Importantly, because the amount of PA and caloric restriction necessary for the maintenance of weight loss with an inherited subtype may be beyond the physical and psychological capabilities of most individuals, it should not be the goal.

In closing, clinicians and investigators must recognize that despite its ubiquity, obesity is not a homogenous condition. Moreover, because obesity is a complex and idiosyncratic phenotype determined by inherited, behavioral, and environmental factors, a personalized approach based on etiologic subtype is essential for successful health management.

Author Contributions: Conceptualization, E.A. and C.J.L.; writing—original draft preparation, E.A.; writing—review and editing, E.A. and C.J.L. All authors have read and agreed to the published version of the manuscript.

Funding: This research received no external funding.

References

1. Archer, E.; Lavie, C.J.; Hill, J.O. The Contributions of 'Diet', 'Genes', and Physical Activity to the Etiology of Obesity: Contrary Evidence and Consilience. *Prog. Cardiovasc. Dis.* **2018**, *61*, 89–102. [CrossRef] [PubMed]
2. Schwartz, M.W.; Randy, J.; Seeley, R.J.; Zeltser, L.M.; Drewnowski, A.; Ravussin, E.; Redman, L.M.; Leibel, R.L. Obesity Pathogenesis: An Endocrine Society Scientific Statement. *Endocr. Rev.* **2017**, *38*, 267–296. [CrossRef] [PubMed]

3.   Archer, E. The Childhood Obesity Epidemic as a Result of Nongenetic Evolution: The Maternal Resources Hypothesis. *Mayo Clin. Proc.* **2015**, *90*, 77–92. [CrossRef] [PubMed]

4.   Heymsfield, S.B.; Wadden, T.A. Mechanisms, Pathophysiology, and Management of Obesity. *N. Engl. J. Med.* **2017**, *376*, 254–266. [CrossRef] [PubMed]

5.   Blair, S.N.; Sallis, R.E.; Hutber, A.; Archer, E. Exercise therapy—The public health message. *Scand. J. Med. Sci. Sports* **2012**, *22*, e24–e28. [CrossRef]

6.   Fryar, C.D.; Carroll, M.D.; Ogden, C.L. *Prevalence of Overweight, Obesity, and Severe Obesity among Children and Adolescents Aged 2–19 Years: United States, 1963–1965 through 2015–2016*; Centers for Disease Control and Prevention, 2018. Available online: https://www.cdc.gov/nchs/data/hestat/obesity_adult_15_16/obesity_adult_15_16.htm (accessed on 28 May 2022).

7.   Lavie, C.J.; Parto, P.; Archer, E. Obesity, fitness, hypertension, and prognosis: Is physical activity the common denominator? *JAMA Intern. Med.* **2016**, *176*, 217–218. [CrossRef]

8.   Lavie, C.J.; De Schutter, A.; Parto, P.; Jahangir, E.; Kokkinos, P.; Ortega, F.B.; Arena, R.; Milani, R.V. Obesity and Prevalence of Cardiovascular Diseases and Prognosis-The Obesity Paradox Updated. *Prog. Cardiovasc. Dis.* **2016**, *58*, 537–547. [CrossRef]

9.   Gadde, K.M.; Martin, C.K.; Berthoud, H.-R.; Heymsfield, S.B. Obesity. *J. Am. Coll. Cardiol.* **2018**, *71*, 69–84. [CrossRef]

10.   Field, A.E.; Camargo, C.A.; Ogino, S. The merits of subtyping obesity: One size does not fit all. *JAMA* **2013**, *310*, 2147–2148. [CrossRef]

11.   Archer, E. The Mother of All Problems. *New Sci.* **2015**, *225*, 32–33. [CrossRef]

12.   Archer, E.; Pavela, G.; McDonald, S.; Lavie, C.J.; Hill, J.O. Cell-Specific "Competition for Calories" Drives Asymmetric Nutrient-Energy Partitioning, Obesity, and Metabolic Diseases in Human and Non-human Animals. *Front. Physiol.* **2018**, *9*, 1053. [CrossRef] [PubMed]

13.   Archer, E.; McDonald, S.M. The Maternal Resources Hypothesis and Childhood Obesity. In *Fetal and Early Postnatal Programming and Its Influence on Adult Health*; Patel, M.S., Nielsen, J.S., Eds.; CRC Press: Boca Raton, FL, USA; Taylor and Francis Group: New York, NY, USA, 2017; pp. 17–32.

14.   Archer, E. In reply-Maternal, paternal, and societal efforts are needed to "cure" childhood obesity. *Mayo Clin. Proc.* **2015**, *90*, 555–557. [CrossRef] [PubMed]

15.   Archer, E. In reply—epigenetics and childhood obesity. *Mayo Clin. Proc.* **2015**, *90*, 693–695. [CrossRef] [PubMed]

16.   Archer, E.; Hill, J.O. Body and Fat Mass are not Regulated, Controlled, or Defended: An Introduction to the 'Invisible Hand' and 'Competition Model of Metabolism'. *Prog. Cardiovasc. Dis.* **2022**, *in press*.

17.   Bonduriansky, R.; Day, T. Nongenetic Inheritance and Its Evolutionary Implications. *Annu. Rev. Ecol. Evol. Syst.* **2009**, *40*, 103–125. [CrossRef]

18.   Bonduriansky, R.; Day, T. *Extended Heredity: A New Understanding of Inheritance and Evolution*; Princeton University Press: Princeton, NJ, USA, 2018.

19.   Day, T.; Bonduriansky, R. A unified approach to the evolutionary consequences of genetic and nongenetic inheritance. *Am. Nat.* **2011**, *178*, E18–E36. [CrossRef]

20.   Feldman, M.W.; Lewontin, R.C. The heritability hang-up. *Science* **1975**, *190*, 1163–1168. [CrossRef]

21.   Lewontin, R.C. The analysis of variance and the analysis of causes. *Am. J. Hum. Genet.* **1974**, *26*, 400–411. [CrossRef]

22.   Weinsier, R.L.; Hunter, G.R.; Heini, A.F.; Goran, M.I.; Sell, S.M. The etiology of obesity: Relative contribution of metabolic factors, diet, and physical activity. *Am. J. Med.* **1998**, *105*, 145–150. [CrossRef]

23.   Jebb, S.A.; Moore, M.S. Contribution of a sedentary lifestyle and inactivity to the etiology of overweight and obesity: Current evidence and research issues. *Med. Sci. Sports Exerc.* **1999**, *31*, S534–S541. [CrossRef]

24.   LaMonte, M.J.; Blair, S.N.; Church, T.S. Physical activity and diabetes prevention. *J. Appl. Physiol.* **2005**, *99*, 1205–1213. [CrossRef] [PubMed]

25.   Blair, S.N.; Church, T.S. The fitness, obesity, and health equation: Is physical activity the common denominator? *JAMA* **2004**, *292*, 1232–1234. [CrossRef] [PubMed]

26.   Blair, S.N.; Archer, E.; Hand, G.A. Commentary: Luke and Cooper are wrong: Physical activity has a crucial role in weight management and determinants of obesity. *Int. J. Epidemiol.* **2013**, *42*, 1836–1838. [CrossRef] [PubMed]

27.   Mayer, J.; Marshall, N.B.; Vitale, J.J.; Christensen, J.H.; Mashayekhi, M.B.; Stare, F.J. Exercise, food intake and body weight in normal rats and genetically obese adult mice. *Am. J. Physiol.* **1954**, *177*, 544–548. [CrossRef] [PubMed]

28.   Mayer, J. Decreased activity and energy balance in the hereditary obesity-diabetes syndrome of mice. *Science* **1953**, *117*, 504–505. [CrossRef]

29.   Chirico, A.M.; Stunkard, A.J. Physical activity and human obesity. *N. Engl. J. Med.* **1960**, *263*, 935–940. [CrossRef]

30.   Archer, E.; Blair, S.N. Physical activity and the prevention of cardiovascular disease: From evolution to epidemiology. *Prog. Cardiovasc. Dis.* **2011**, *53*, 387–396. [CrossRef]

31.   Melzer, K.; Kayser, B.; Saris, W.H.; Pichard, C. Effects of physical activity on food intake. *Clin. Nutr.* **2005**, *24*, 885–895. [CrossRef]

32.   Lavie, C.J.; Ozemek, C.; Carbone, S.; Katzmarzyk, P.T.; Blair, S.N. Sedentary behavior, exercise, and cardiovascular health. *Circ. Res.* **2019**, *124*, 799–815. [CrossRef]

33.   Archer, E.; Artero, E.G.; Blair, S.N. Sedentary Behavior and Cardiovascular Disease. In *Sedentary Behavior and Health: Concepts, Assessment & Intervention—Human Kinetics*; Zhu, W., Owen, N., Eds.; Human Kinetics: Champaign, IL, USA, 2017; pp. 203–225.

34. Stubbs, R.J.; Hughes, D.A.; Johnstone, A.M.; Horgan, G.W.; King, N.; Blundell, J.E. A decrease in physical activity affects appetite, energy, and nutrient balance in lean men feeding ad libitum. *Am. J. Clin. Nutr.* **2004**, *79*, 62–69. [CrossRef]

35. Bosy-Westphal, A.; Hägele, F.A.; Müller, M.J. Impact of Energy Turnover on the Regulation of Energy and Macronutrient Balance. *Obesity* **2021**, *29*, 1114–1119. [CrossRef] [PubMed]

36. Shook, R.P.; Hand, G.A.; Drenowatz, C.; Hebert, J.R.; Paluch, A.E.; Blundell, J.E.; Hill, J.O.; Katzmarzyk, P.T.; Church, T.S.; Blair, S.N. Low levels of physical activity are associated with dysregulation of energy intake and fat mass gain over 1 year. *Am. J. Clin. Nutr.* **2015**, *102*, 1332–1338. [CrossRef] [PubMed]

37. Westerterp, K.R.; Meijer, G.A.; Janssen, E.M.; Saris, W.H.; Ten Hoor, F. Long-term effect of physical activity on energy balance and body composition. *Br. J. Nutr.* **1992**, *68*, 21–30. [CrossRef] [PubMed]

38. DeFronzo, R.A.; Tripathy, D. Skeletal Muscle Insulin Resistance Is the Primary Defect in Type 2 Diabetes. *Diabetes Care* **2009**, *32*, S157–S163. [CrossRef]

39. DeFronzo, R.A. Lilly lecture 1987. The triumvirate: Beta-cell, muscle, liver. A collusion responsible for NIDDM. *Diabetes* **1988**, *37*, 667–687. [CrossRef] [PubMed]

40. DeFronzo, R.A.; Bonadonna, R.C.; Ferrannini, E. Pathogenesis of NIDDM. A balanced overview. *Diabetes Care* **1992**, *15*, 318–368. [CrossRef]

41. Taylor, R. Pathogenesis of type 2 diabetes: Tracing the reverse route from cure to cause. *Diabetologia* **2008**, *51*, 1781–1789. [CrossRef]

42. Saris, W.; Blair, S.; Van Baak, M.; Eaton, S.; Davies, P.; Di Pietro, L.; Fogelholm, M.; Rissanen, A.; Schoeller, D.; Swinburn, B. How much physical activity is enough to prevent unhealthy weight gain? Outcome of the IASO 1st Stock Conference and consensus statement. *Obes. Rev.* **2003**, *4*, 101–114. [CrossRef]

43. Piercy, K.L.; Troiano, R.P.; Ballard, R.M.; Carlson, S.A.; Fulton, J.E.; Galuska, D.A.; George, S.M.; Olson, R.D. The physical activity guidelines for Americans. *JAMA* **2018**, *320*, 2020–2028. [CrossRef]

44. Boulé, N.G.; Haddad, E.; Kenny, G.P.; Wells, G.A.; Sigal, R.J. Effects of Exercise on Glycemic Control and Body Mass in Type 2 Diabetes MellitusA Meta-analysis of Controlled Clinical Trials. *JAMA* **2001**, *286*, 1218–1227. [CrossRef]

45. Umpierre, D.; Ribeiro, P.A.B.; Kramer, C.K.; Leitão, C.B.; Zucatti, A.T.N.; Azevedo, M.J.; Gross, J.L.; Ribeiro, J.P.; Schaan, B.D. Physical Activity Advice Only or Structured Exercise Training and Association with HbA1c Levels in Type 2 Diabetes: A Systematic Review and Meta-analysis. *JAMA* **2011**, *305*, 1790–1799. [CrossRef] [PubMed]

46. Lavie, C.J. Sugar Wars-Commentary from the Editor. *Prog. Cardiovasc. Dis.* **2018**, *61*, 382–383. [CrossRef] [PubMed]

47. Lavie, C.J.; Laddu, D.; Arena, R.; Ortega, F.B.; Alpert, M.A.; Kushner, R.F. Healthy weight and obesity prevention: JACC health promotion series. *J. Am. Coll. Cardiol.* **2018**, *72*, 1506–1531. [CrossRef] [PubMed]

48. Lavie, C.J.; Milani, R.V.; O'Keefe, J.H. Dyslipidemia intervention in metabolic syndrome: Emphasis on improving lipids and clinical event reduction. *Am. J. Med. Sci.* **2011**, *341*, 388–393. [CrossRef]

49. Hales, C.N.; Barker, D.J.; Clark, P.M.; Cox, L.J.; Fall, C.; Osmond, C.; Winter, P.D. Fetal and infant growth and impaired glucose tolerance at age 64. *BMJ* **1991**, *303*, 1019–1022. [CrossRef]

50. Barker, D.J.P.; Osmond, C. Infant Mortality, Childhood Nutrition, And Ischaemic Heart Disease in England And Wales. *Lancet* **1986**, *327*, 1077–1081. [CrossRef]

51. Hales, C.N.; Barker, D.J. Type 2 (non-insulin-dependent) diabetes mellitus: The thrifty phenotype hypothesis. *Diabetologia* **1992**, *35*, 595–601. [CrossRef]

52. Archer, E.; Lavie, C.J.; McDonald, S.M.; Thomas, D.M.; Hebert, J.R.; Taverno Ross, S.E.; McIver, K.L.; Malina, R.M.; Blair, S.N. Maternal inactivity: 45-year trends in mothers' use of time. *Mayo Clin. Proc.* **2013**, *88*, 1368–1377. [CrossRef]

53. Archer, E.; Shook, R.P.; Thomas, D.M.; Church, T.S.; Katzmarzyk, P.T.; Hebert, J.R.; McIver, K.L.; Hand, G.A.; Lavie, C.J.; Blair, S.N. 45-Year trends in women's use of time and household management energy expenditure. *PLoS ONE* **2013**, *8*, e56620. [CrossRef]

54. Church, T.S.; Thomas, D.M.; Tudor-Locke, C.; Katzmarzyk, P.T.; Earnest, C.P.; Rodarte, R.Q.; Martin, C.K.; Blair, S.N.; Bouchard, C. Trends over 5 decades in U.S. occupation-related physical activity and their associations with obesity. *PLoS ONE* **2011**, *6*, e19657. [CrossRef]

55. Kennedy, A.; Lavie, C.J.; Blair, S.N. Fitness or fatness: Which is more important? *JAMA* **2018**, *319*, 231–232. [CrossRef]

56. Ortega, F.B.; Ruiz, J.R.; Labayen, I.; Lavie, C.J.; Blair, S.N. The Fat but Fit paradox: What we know and don't know about it. *Br. J. Sports Med.* **2018**, *52*, 151–153. [CrossRef] [PubMed]

57. Ortega, F.B.; Lavie, C.J.; Blair, S.N. Obesity and cardiovascular disease. *Circ. Res.* **2016**, *118*, 1752–1770. [CrossRef] [PubMed]

58. Elagizi, A.; Kachur, S.; Lavie, C.J.; Carbone, S.; Pandey, A.; Ortega, F.B.; Milani, R.V. An overview and update on obesity and the obesity paradox in cardiovascular diseases. *Prog. Cardiovasc. Dis.* **2018**, *61*, 142–150. [CrossRef] [PubMed]

59. Myers, J.; McAuley, P.; Lavie, C.J.; Despres, J.-P.; Arena, R.; Kokkinos, P. Physical activity and cardiorespiratory fitness as major markers of cardiovascular risk: Their independent and interwoven importance to health status. *Prog. Cardiovasc. Dis.* **2015**, *57*, 306–314. [CrossRef]

60. Blundell, J.E.; Caudwell, P.; Gibbons, C.; Hopkins, M.; Naslund, E.; King, N.; Finlayson, G. Role of resting metabolic rate and energy expenditure in hunger and appetite control: A new formulation. *Dis. Models Mech.* **2012**, *5*, 608–613. [CrossRef]

61. Blundell, J.E.; Caudwell, P.; Gibbons, C.; Hopkins, M.; Naslund, E.; King, N.A.; Finlayson, G. Body composition and appetite: Fat-free mass (but not fat mass or BMI) is positively associated with self-determined meal size and daily energy intake in humans. *Br. J. Nutr.* **2012**, *107*, 445–449. [CrossRef]

62. Caudwell, P.; Finlayson, G.; Gibbons, C.; Hopkins, M.; King, N.; Naslund, E.; Blundell, J.E. Resting metabolic rate is associated with hunger, self-determined meal size, and daily energy intake and may represent a marker for appetite. *Am. J. Clin. Nutr.* **2013**, *97*, 7–14. [CrossRef]

63. Blundell, J.E.; Finlayson, G.; Gibbons, C.; Caudwell, P.; Hopkins, M. The biology of appetite control: Do resting metabolic rate and fat-free mass drive energy intake? *Physiol. Behav.* **2015**, *152*, 473–478. [CrossRef]

64. Blundell, J.E.; Gibbons, C.; Beaulieu, K.; Casanova, N.; Duarte, C.; Finlayson, G.; Stubbs, R.J.; Hopkins, M. The drive to eat in homo sapiens: Energy expenditure drives energy intake. *Physiol. Behav.* **2020**, *219*, 112846. [CrossRef]

65. Mayer, J.; Vitale, J.; Bates, M.W. Mechanism of the regulation of food intake. *Nature* **1951**, *167*, 562–563. [CrossRef] [PubMed]

66. Beaudoin, R.; Mayer, J. Food intakes of obese and non-obese women. *J. Am. Diet. Assoc.* **1953**, *29*, 29–33. [CrossRef]

67. Mayer, J.; Roy, P.; Mitra, K.P. Relation between Caloric Intake, Body Weight, and Physical Work: Studies in an Industrial Male Population in West Bengal. *Am. J. Clin. Nutr.* **1956**, *4*, 169–175. [CrossRef] [PubMed]

68. Mayer, J. Correlation between metabolism and feeding behavior and multiple etiology of obesity. *Bull. N. Y. Acad. Med.* **1957**, *33*, 744–761. [PubMed]

69. Archer, E. In Defense of Sugar: A Critique of Diet-Centrism. *Prog. Cardiovasc. Dis.* **2018**, *61*, 10–19. [CrossRef] [PubMed]

70. Lee, D.C.; Lavie, C.J.; Vedanthan, R. Optimal dose of running for longevity: Is more better or worse? *J. Am. Coll. Cardiol.* **2015**, *65*, 420–422. [CrossRef]

71. Friedman, J.M. Modern science versus the stigma of obesity. *Nat. Med.* **2004**, *10*, 563–569. [CrossRef]

72. Hartlev, M. Stigmatisation as a Public Health Tool against Obesity—A Health and Human Rights Perspective. *Eur. J. Health Law* **2014**, *21*, 365–386. [CrossRef]

73. Archer, E.; Lavie, C.J.; Hill, J.O. The Failure to Measure Dietary Intake Engendered a Fictional Discourse on Diet-Disease Relations. *Front. Nutr.* **2018**, *5*, 105. [CrossRef]

74. Archer, E. The Demonization of 'Diet' Is Nothing New. *Prog. Cardiovasc. Dis.* **2018**, *61*, 386–387. [CrossRef]

75. Archer, E.; Arjmandi, B. Falsehoods and facts about dietary sugars: A call for evidence-based policy. *Crit. Rev. Food Sci. Nutr.* **2020**, *61*, 3725–3739. [CrossRef] [PubMed]

76. Dobersek, U.; Wy, G.; Adkins, J.; Altmeyer, S.; Krout, K.; Lavie, C.J.; Archer, E. Meat and mental health: A systematic review of meat abstention and depression, anxiety, and related phenomena. *Crit. Rev. Food Sci. Nutr.* **2020**, *61*, 622–635. [CrossRef] [PubMed]

77. Fletcher, G.F.; Landolfo, C.; Niebauer, J.; Ozemek, C.; Arena, R.; Lavie, C.J. Promoting physical activity and exercise: JACC health promotion series. *J. Am. Coll. Cardiol.* **2018**, *72*, 1622–1639. [CrossRef] [PubMed]

Article

Is There a Link between Obesity Indices and Skin Autofluorescence? A Response from the ILERVAS Project

Enric Sánchez [1,†], Marta Sánchez [1,†], Carolina López-Cano [1], Marcelino Bermúdez-López [2], José Manuel Valdivielso [2], Cristina Farràs-Sallés [3], Reinald Pamplona [4], Gerard Torres [5,6], Dídac Mauricio [7,8], Eva Castro [2], Elvira Fernández [2] and Albert Lecube [1,8,*]

[1] Endocrinology and Nutrition Department, University Hospital Arnau de Vilanova, Obesity, Diabetes and Metabolism (ODIM) Research Group, IRBLleida, University of Lleida, 25198 Lleida, Spain
[2] Vascular and Renal Translational Research Group, IRBLleida, Red de Investigación Renal, Instituto de Salud Carlos III (RedinRen-ISCIII), 25198 Lleida, Spain
[3] Centre d'Atenció Primària Cappont, Gerència Territorial de Lleida, Institut Català de la Salut, Research Support Unit Lleida, Fundació Institut Universitari per a la Recerca a l'Atenció Primària de Salut Jordi Gol i Gorina (IDIAPJGol), 08007 Barcelona, Spain
[4] Department of Experimental Medicine, IRBLleida, University of Lleida, 25003 Lleida, Spain
[5] Respiratory Medicine Department, University Hospital Arnau de Vilanova and Santa María, Group of Translational Research in Respiratory Medicine, IRBLleida, University of Lleida, 25198 Lleida, Spain
[6] Centro de Investigación Biomédica en Red de Enfermedades Respiratorias (CIBERES), Instituto de Salud Carlos III (ISCIII), 28029 Madrid, Spain
[7] Department of Endocrinology and Nutrition, Hospital de la Santa Creu i Sant Pau, Sant Pau Biomedical Research Institute (IIB Sant Pau), 08025 Barcelona, Spain
[8] Centro de Investigación Biomédica en Red de Diabetes y Enfermedades Metabólicas Asociadas (CIBERDEM), Instituto de Salud Carlos III (ISCIII), 28029 Madrid, Spain
*   Correspondence: alecube@gmail.com; Tel.: +34-973-70-51-83; Fax: +34-973-70-51-89
†   These authors contributed equally to this work.

Citation: Sánchez, E.; Sánchez, M.; López-Cano, C.; Bermúdez-López, M.; Valdivielso, J.M.; Farràs-Sallés, C.; Pamplona, R.; Torres, G.; Mauricio, D.; Castro, E.; et al. Is There a Link between Obesity Indices and Skin Autofluorescence? A Response from the ILERVAS Project. *Nutrients* **2023**, *15*, 203. https://doi.org/10.3390/nu15010203

Academic Editor: Javier Gómez-Ambrosi

Received: 17 December 2022
Revised: 26 December 2022
Accepted: 27 December 2022
Published: 31 December 2022

Abstract: There is controversial information about the accumulation of advanced glycation end-products (AGEs) in obesity. We assessed the impact of total and abdominal adiposity on AGE levels via a cross-sectional investigation with 4254 middle-aged subjects from the ILERVAS project. Skin autofluorescence (SAF), a non-invasive assessment of subcutaneous AGEs, was measured. Total adiposity indices (BMI and Clínica Universidad de Navarra-Body Adiposity Estimator (CUN-BAE)) and abdominal adiposity (waist circumference and body roundness index (BRI)) were assessed. Lean mass was estimated using the Hume index. The area under the receiver operating characteristic (ROC) curve was evaluated for each index. Different cardiovascular risk factors (smoking, prediabetes, hypertension and dyslipidemia) were evaluated. In the study population, 26.2% showed elevated SAF values. No differences in total body fat, visceral adiposity and lean body mass were detected between patients with normal and high SAF values. SAF levels showed a very slight but positive correlation with total body fat percentage (estimated by the CUN-BAE formula) and abdominal adiposity (estimated by the BRI). However, none of them had sufficient power to identify patients with high SAF levels (area under the ROC curve <0.52 in all cases). Finally, a progressive increase in SAF levels was observed in parallel with cardiovascular risk factors in the entire population and when patients with normal weight, overweight and obesity were evaluated separately. In conclusion, total obesity and visceral adiposity are not associated with a greater deposit of AGE. The elevation of AGE in obesity is related to the presence of cardiometabolic risk.

Keywords: adipose tissue; advanced glycation end-products; body composition; cardiometabolic risk; cardiovascular risk factors; novel targets; obesity; skin autofluorescence

1. Introduction

Obesity is a multifactorial chronic disease that can shorten the quality of life and life expectancy of patients due to its high morbidity and mortality [1]. While this is clearly established, population studies have revealed that more than 30% of patients with obesity do not present associated metabolic pathology, which has given rise to the concept of "metabolically healthy obesity" [2]. However, these patients have shown a higher risk of both diabetes and cardiometabolic disease in the medium–long term. It has been hypothesized that this is a possible initial phase prior to the development of comorbidities [3–5].

To date, the trigger for the development of these comorbidities is unknown. However, among the different hypotheses, the possible role of inflammatory adaptation against tissue hypoxia produced by the expansion of white adipose tissue is becoming increasingly relevant [6–8]. This continuous hypoxia facilitates a change towards a proinflammatory profile that enhances the secretion of cytokines such as tumor necrosis factor alpha, interleukin-6 or hypoxia inducible factor type 1 with the consequent increase in acute phase indicators such as C-reactive protein and fibrinogen [6,9,10]. These factors are related to the appearance of both local and systemic insulin resistance, endothelial dysfunction and arteriosclerosis, as well as a higher rate of cardiovascular events [11,12].

Hypoxia, along with a proinflammatory pattern and oxidative stress, are common features of obesity, all of which have been associated with increased protein glycation [5]. Taken together, increased advanced glycation end-products (AGEs) have been related to the formation of atherosclerotic plaques and increased cardiovascular risk [13,14]. Lifeline cohort studies have shown that increased AGEs are independently related to BMI, age and HbA1c level [15]. Others have shown its increase in patients with visceral obesity, related to an increased prevalence of metabolic syndrome [16]. Similarly, our group, has previously published that the increase in AGE concentration in patients with severe obesity is clearly at the expense of those with metabolic syndrome, suggesting its determination as a way of identifying those patients with "metabolically diseased obesity" [17]. However, we are missing a study specifically designed to assess the impact of obesity, as measured by both BMI and body fat, on AGE levels. With this objective, and to verify if the accumulation of AGEs could help us to identify early and easily those people with a higher risk of metabolic syndrome, we have analyzed the population of the ILERVAS project. This large cohort included subjects with one or more cardiometabolic risk factors and different weight ranges.

2. Materials and Methods

2.1. Study Design

In this work, we analyze the information collected in the ILERVAS project (ClinTrials.gov Identifier: NCT03228459), a prospective study whose main goal was learning the prevalence of non-clinical atheromatous disease and occult kidney disease in a cohort with moderate cardiovascular risk [18,19]. Data were analyzed from 4254 people recruited between 2015 and 2018. Patients were recruited aged 45 to 70 years, with no previous cardiovascular event but at least one cardiometabolic risk factor (obesity, hypertension, dyslipidemia, smoking or first-degree relative with prematurity (<55 years in men, <65 in women) cardiovascular disease (myocardial infarction, stroke and peripheral arterial disease)). Those with diabetes, chronic kidney disease, active neoplasia, a life expectancy of less than 18 months and/or pregnancy were excluded.

The ILERVAS project protocol was approved by the ethics committee of the Arnau de Vilanova University Hospital (CEIC-1410) and written informed consent was acquired from all subjects. The ethical guidelines of the Declaration of Helsinki and Spanish legislation on the protection of personal data were also followed.

2.2. Definition of Cardiovascular Risk Factors

The diagnosis of dyslipidemia was obtained from patients who had an assigned code for disorders of lipoprotein metabolism and other lipidemias by means of the International Classification of Diseases (ICD-10) codes, namely E78.0–78.9 (pure hypercholesterolemia, pure hyperglyceridemia, mixed hyperlipidemia, hyperchylomicronemia, other hyperlipidemia, unspecified hyperlipidemia, lipoprotein deficiency, other disorders of lipoprotein metabolism and unspecified disorders of lipoprotein metabolism). A diagnosis of hypertension was obtained from patients coded for hypertensive diseases using ICD-10 codes, i.e., I10–I13 (essential hypertension, hypertensive heart disease, hypertensive renal disease and hypertensive heart and renal disease) and I15 (secondary hypertension).

Prediabetes was defined as a glycated hemoglobin (HbA1c) level between 39 to 47 mmol/mol (5.7 to 6.4%), and normal glucose metabolism as HbA1c <39 mmol/mol (<5.7%), agreeing with the American Diabetes Association guidelines. Smoking habits (never, former or current smoker) were also considered. Smokers who quit smoking a year or more before the visit were considered ex-smokers. As patients with diabetes were excluded from the ILERVAS project, this diagnosis was not considered a cardiovascular risk factor in our study.

The antihypertensive and lipid-lowering treatments that were prescribed in the ILERVAS population have been taken from the prescription and billing databases provided by CatSalut (Catalan Health Service), which were incorporated annually into the SIDIAP database. Antihypertensive medications include angiotensin-converting enzyme inhibitors, diuretics, type II aldosterone receptor antagonists, beta-blockers, calcium channel blockers and other antihypertensives. Lipid-lowering drugs included statins, fibrates, ezetimibe and omega-3 fatty acids.

2.3. Anthropometric Measures

Both weight and height were analyzed almost without clothing and without shoes with a precision of 0.5 kg and 1.0 cm, respectively [20]. Waist circumference was measured between the iliac crest and the lower rib in the horizontal plane with the subject standing and with a non-elastic tape to a precision of 0.1 cm [21]. To decrease interobserver and device variability, all anthropometric measures were performed by trained nurses under standardized conditions. The relative technical error of intra-rater measurement was less than 1% for height, weight and waist and circumferences.

BMI was obtained by weight (kg) divided by the square of body height (m), and obesity was classified according to clinical guidelines as BMI $\geq$30 kg/m^2. The percentage of total body fat was estimated using the Body Adiposity Estimator of the Clínica Universidad de Navarra (CUN-BAE) using the formula: $-44.988 + (0.503 \times \text{age}) + (10.689 \times \text{sex}) + (3.172 \times \text{BMI}) - (0.026 \times \text{BMI}^2) + (0.181 \times \text{BMI}) \times \text{sex} - (0.02 \times \text{BMI} \times \text{age}) - (0.005 \times \text{BMI}^2 \times \text{sex}) + (0.00021 \times \text{BMI2} \times \text{age})$, where sex is 1 for women and 0 for men and age is in years [22].

For the estimation of central adiposity, in addition to waist circumference, the body roundness index was included. This index, suggested by Thomas et al., is based on a geometric model defined to quantify body circularity. Those with abdominal fat look like a perfect circle, compared to those with more linear figures. It was calculated as: WC $(\text{m})/(\text{BMI}^{2/3} \times \text{height (m)})^{1/2}$ [23]. In addition, we evaluated the Hume index for the amount of lean mass based on the analysis of the body composition of the antipyrine dilution space through the formula: $(0.29569 \times \text{weight}) + (0.41813 \times \text{height}) - 43.2933$ [24].

2.4. Skin Autofluorescence

SAF was assessed using the AGE Reader™ device (DiagnOptics Technologies, Groningen, The Netherlands), a computerized non-invasive tool that quantifies AGE deposits in the forearm via the ultraviolet spectrum [25]. A device calibrated according to the manufacturer's recommendations was used. Three analyses were carried out in areas free of tattoos, cosmetics or with a concentration of freckles or superficial vessels, and their mean value (arbitrary units: AU) was taken. Measurements made on the same day showed an overall Altman error rate of 5.03%, and intra-individual seasonal deviation showed an Altman error rate of 5.87% [25]. Since AGEs accumulate progressively with aging, there is a normal sum of AGEs at each age. When this number is higher than expected, the software classifies the patient as a "high AGE" individual. Therefore, participants in the ILERVAS project were classified as a group with "normal" and "high" SAF levels.

2.5. Statistical Methods

The Shapiro–Wilk test was used to estimate the normal distribution of the sample. Quantitative baseline characteristics were analyzed using the Mann–Whitney U test or Kruskal–Wallis test, and categorical characteristics using Pearson's chi-squared test. Spearman's correlation was used to assess the relationship between AGE levels and anthropometric data. Data are expressed as median and interquartile range or n (percentage). Patients were differentiated based on their elevated and normal SAF results. In addition, patients were also categorized according to the number of cardiovascular risk factors.

The evaluation of the diagnostic performance of the anthropometric formulas was carried out by analyzing the area under the receiver operating characteristic (ROC) curves and the Youden J statistic. The results of the area under the ROC curve were interpreted following the guidelines stipulated by the scientific community: excellent, between 0.9 and 1.0; good, between 0.8 and 0.9; fair, between 0.7 and 0.8; poor, between 0.6 and 0.7; and not useful, between 0.5 and 0.6. SSPS software (IBM SPSS Statistics for Windows, version 20.0., Armonk, NY, USA) was used for statistical analysis. Statistical significance was determined with a p value < 0.05.

3. Results

The main clinical and metabolic data according to the presence of SAF levels are shown in Table 1. The ILERVAS cohort consisted of 1115 (26.2%) individuals with elevated SAF values. This group of individuals were mainly smokers with a characteristic cardiovascular risk profile better than participants with normal SAF values. This high SAF group also received significant undertreatment with antihypertensive and lipid-lowering medications. However, no differences in the prevalence of obesity, according to BMI, were observed between groups (30.0 vs. 27.9%, $p = 0.189$). Similarly, no differences in total body fat percentage or estimated visceral adiposity and lean body mass were detected between the groups (Table 2).

Table 1. Central clinical and metabolic data in the ILERVAS cohort according to skin autofluorescence values.

	Normal SAF (n = 3139)	High SAF (n = 1115)	p-Value
Women, n (%)	1576 (50.2)	548 (49.1)	0.543
Age (years)	57 (52–63)	57 (54–61)	0.003
Smoking habits (current and former), n (%)	1840 (58.6)	856 (76.8)	<0.001
Obesity diagnosis, n (%)	941 (30.0)	311 (27.9)	0.189
Blood hypertension diagnosis, n (%)	1296 (41.3)	390 (35.0)	<0.001
Antihypertensive drugs, n (%)	1036 (33.0)	313 (28.1)	0.002
Dyslipidemia diagnosis, n (%)	1635 (52.1)	495 (44.4)	<0.001
Lipid-lowering agents, n (%)	555 (17.7)	157 (14.1)	0.006
Prediabetes diagnosis, n (%)	1081 (34.4)	363 (32.6)	0.311

Data are expressed as a median (interquartile range) or n (percentage). Antihypertensive drugs include angiotensin-converting enzyme (ACE) inhibitors, diuretics, angiotensin-II receptor antagonists (ARA II), beta-blockers, calcium antagonists and other antihypertensives. Lipid-lowering treatments involve statins, fibrates, ezetimibe and omega-3 fatty acids.

Table 2. Data of the anthropometric indices in all individuals according to skin autofluorescence values.

	Normal SAF	**High SAF**	*p***-Value**
Total adiposity			
BMI (kg/m^2)	28.4 (25.7–31.6)	28.1 (24.7–31.8)	0.104
CUN-BAE (%)	35.9 (30.1–42.3)	35.3 (29.1–42.1)	0.132
Visceral adipose tissue			
Waist circumference (cm)	101 (94–108)	100 (94–108)	0.19
Body roundness index	5.68 (4.73–6.83)	5.58 (4.50–6.94)	0.127
Lean body mass			
Hume index (kg)	49.5 (43.0–56.2)	49.5 (43.4–55.5)	0.693

Data are expressed as a median (interquartile range). BMI: body mass index; CUN-BAE: Clínica Universidad de Navarra-Body Adiposity Estimator.

Regarding the bivariate analysis, the SAF levels showed a very slight but positive correlation with the percentage of total body fat (estimated by the CUN-BAE formula) and abdominal adiposity (estimated by body roundness index) (Table 3). These correlations disappeared when anthropometric formulas such as BMI and waist circumference were used. In addition, a negative correlation with lean body mass was also observed. In the same way, the measures related to obesity and body composition had no power to identify the patients with higher levels of SAF, being in all cases areas under the ROC curve <0.52 (Figure 1).

Table 3. Bivariate correlations of SAF with anthropometric formulas in the ILERVAS population.

	r	*p*
BMI (kg/m^2)	−0.019	0.221
CUN-BAE (%)	0.080	<0.001
Waist circumference (cm)	0.005	0.746
Body roundness index	0.065	<0.001
Lean body mass (kg)	−0.122	<0.001

BMI: body mass index; CUN-BAE: Clínica Universidad de Navarra-Body Adiposity Estimator.

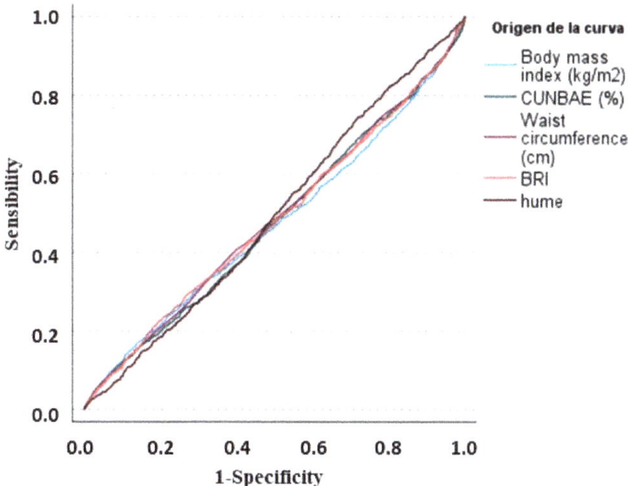

Figure 1. Receiver operating characteristic (ROC) curve analysis in the ILERVAS population to assess the diagnostic accuracy of obesity indices to identify patients with higher AGEs from those with normal AGEs.

We also analyzed SAF levels according to a number of cardiometabolic risk factors (dyslipidemia, hypertension, history of smoking and prediabetes) in both the total study population and according the degree of obesity. A progressive increase in SAF levels was observed in parallel with cardiovascular risk factors in the entire population (Figure 2). There were also significant differences in terms of skin autofluorescence values according to the number of cardiometabolic risk factors in subjects with normal weight, overweight and obesity (Figure 3).

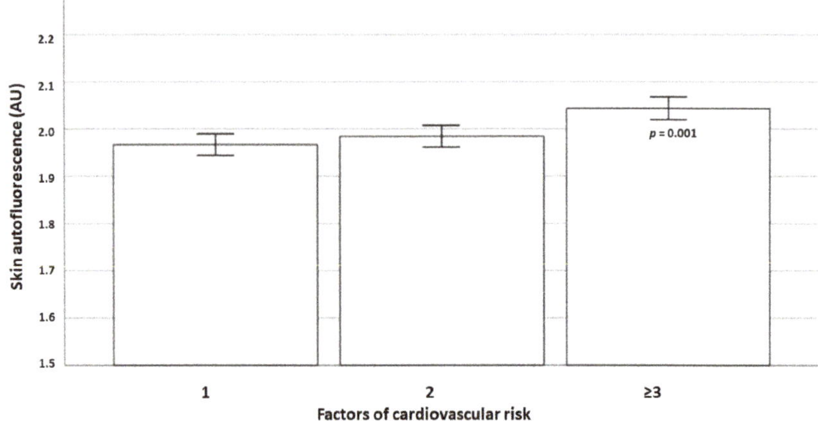

Figure 2. Results of the skin autofluorescence values in the entire population of the ILERVAS project according to the number of cardiovascular risk factors (a history of smoking habits, hypertension, dyslipidemia and prediabetes).

Finally, the multivariable logistic regression model in patients with obesity according to their BMI showed that male sex, degree of obesity and the presence of three or more cardiovascular risk factors (prediabetes, smoking, hypertension and dyslipidemia) were independently associated with AGE levels (Table 4).

Table 4. The multivariable logistic regression model for high AGEs in subjects with obesity.

	Odds Ratio (95% CI)	*p*
Sex Women	Ref.	
Men	0.94 (0.79 to 1.11)	0.435
Age (years)	1.00 (0.99 to 1.01)	0.992
Body mass index (kg/m^2)	1.03 (1.01 to 1.05)	0.006
Smoking status Never	Ref.	
Current or former	1.09 (0.092 to 1.28)	0.314
Cardiovascular risk factors 1	Ref.	
2	1.15 (0.98 to 1.36)	0.087
≥3	1.28 (1.04 to 1.58)	0.019
Hosmer–Lemeshow test of fit		0.492
Area under the ROC curve	0.74 (0.72 to 0.77)	<0.001

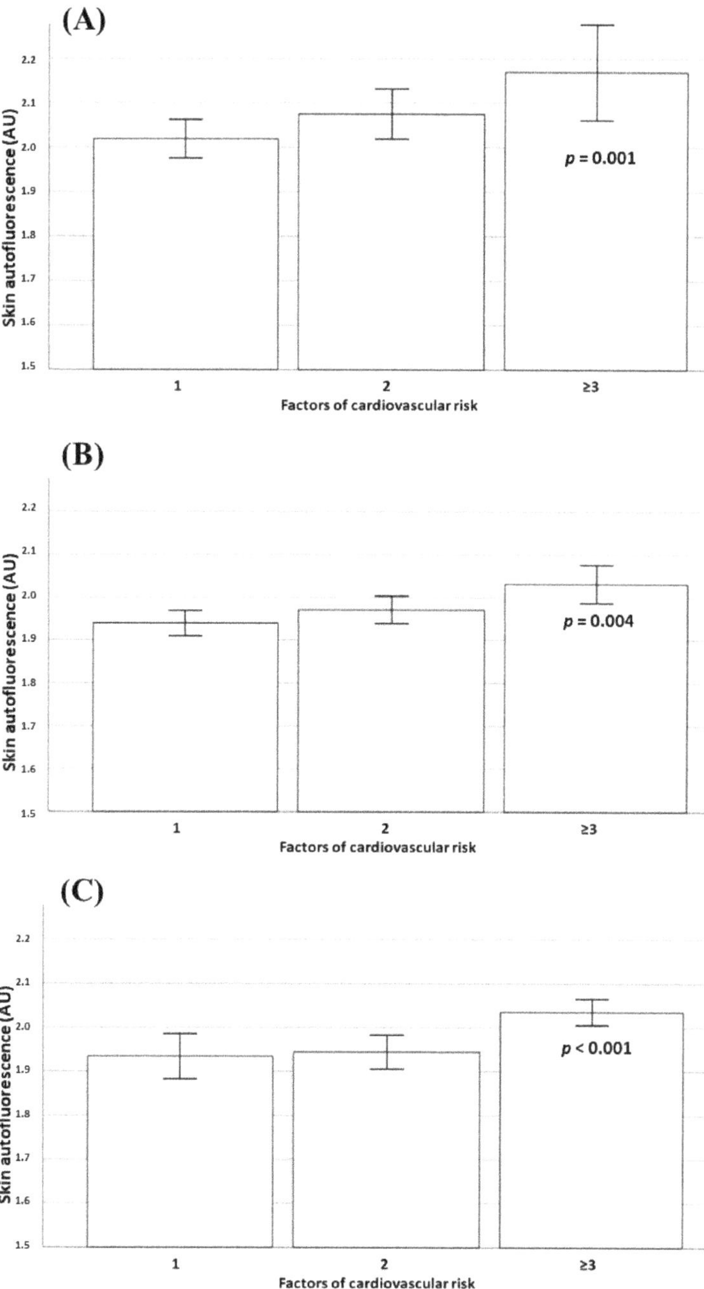

Figure 3. Results of the skin autofluorescence values in subjects with normal weight (**A**), overweight (**B**) and obesity (**C**) according to the number of cardiovascular risk factors (history of smoking habits hypertension, dyslipidemia and prediabetes).

4. Discussion

In our middle-aged Caucasian population cohort, no significant increase in AGEs (measured as skin autofluorescence) was observed with respect to overall obesity or abdominal obesity. However, the subcutaneous deposition of AGE seems to be positively related to the prevalence of cardiometabolic risk factors, both in patients with and without obesity. Until now, when the relationship between obesity and AGE deposition has been evaluated, controversial data have been shown. For example, in a study of the child population, Lentferink et al. found a correlation of AGEs with the highest standard deviation of BMI, which disappeared when adjusting for skin type [26]. Likewise, in their study, Gogas et al. observed a positive correlation of AGEs with BMI, being higher in those with type 2 diabetes [27].

Visceral adipose tissue has been shown to have a greater inflammatory capacity, so it would be expected that this confers greater oxidative conditions favoring the formation of AGEs [28]. Despite this, Den Engelsen et al. did not observe significant differences in AGEs in those with or without central obesity measured by waist circumference. However, they demonstrated a progression: from those with healthy normal weight (1.63 $\pm$ 0.37 AU), increasing in those with abdominal obesity (1.74 $\pm$ 0.44 AU) and being even higher in subjects with abdominal obesity and comorbidities (1.87 $\pm$ 0.43 AU; $p < 0.001$) [16]. In the same study, after a medium follow-up period of 3 years after bariatric surgery, the SAF values did not change, although there was a marked reduction in weight and remission of comorbidities.

Elevated AGE levels have been linked to increased cardiometabolic risk, coronary artery disease and cardiovascular mortality [14,29]. Similarly, increased subcutaneous AGE content has previously been associated with increased atheromatous plaque burden in the ILERVAS project [13]. In the present study, we found differences when we assessed patients according to their cardiovascular risk, with SAF values that progressively increased according to the accumulation of cardiometabolic risk factors. Our results are in line with those of Koyama et al. who found a significant relationship between AGE receptors and metabolic syndrome, blood pressure, hypertriglyceridemia, and subclinical atheromatosis in both patients with and without diabetes [30]. Reinforcing the role of metabolic control in AGE deposition, plasma AGEs were higher in patients with type 2 diabetes and atherosclerotic disease than in patient with atherosclerotic disease without type 2 diabetes, especially in those with higher HbA1c levels in recent years (r = 0.46, $p < 0.001$) [31].

Other cardiovascular risk factors not evaluated in our study, such as chronic kidney disease or adherence to the Mediterranean diet, also cause an increase in AGE concentration [32]. In fact, studies based on dietary surveys have associated a low intake of exogenous AGEs with lower insulin resistance, TNF alpha levels, peripheral cell mononuclear cells and leptin concentration, as well as higher adiponectin, which ultimately means less proinflammatory activity [33]. Thus, adherence to the Mediterranean diet, an eating pattern associated with lower proinflammatory state, has been independently associated with AGEs, especially in those with a high consumption of vegetables, fruits and low sugar [34]. In our investigation it is also interesting to note that for the first time the negative but statistically significant link between the levels of SAF and lean body mass is shown.

Our research has some limitations. First, we do not use a precise measure of body composition to correlate with AGEs. However, anthropometric formulas have been validated with other gold standard tests such as dual-energy X-ray absorptiometry or magnetic resonance imaging. Second, we used an indirect test based on skin fluorescence to measure AGEs instead of a direct plasma test, but there is an extensive literature demonstrating the accuracy of this test compared to skin biopsy or plasma measurements. Third, an intrinsic characteristic of the ILERVAS study population is that participants have one or more cardiometabolic risk factors, so care must be taken when generalizing our results to the general population.

5. Conclusions

In conclusion, total obesity and visceral adiposity are not associated with a higher AGE deposit. The elevated levels of AGEs detected in subjects with obesity seem more related to the presence of cardiometabolic risk factors than to the percentage of body fat. With all this evidence, the measurement of SAF is a non-invasive test that can be helpful to identify those patients with unhealthy obesity, which opens the door to a new management of obesity in clinical practice.

Author Contributions: Conceptualization, E.S., M.S. and A.L.; Data curation, E.S., M.S. and M.B.-L.; Formal analysis, E.S., M.S., C.L.-C. and A.L.; Funding acquisition, J.M.V., E.F. and M.B.-L.; Investigation, E.C. and J.M.V.; Methodology, C.F.-S., G.T. and M.B.-L., Software, C.F.-S., E.C.; Project administration, M.B.-L., J.M.V. and A.L.; Resources, M.B.-L., E.F. and A.L.; Supervision, R.P., and D.M.; Visualization, M.B.-L. and J.M.V.; Writing—original draft, E.S., M.S. and C.L.-C.; Writing—review & editing, A.L. All authors have read and agreed to the published version of the manuscript.

Funding: This investigation was sustained by grants from the Generalitat de Catalunya and Diputació de Lleida (SLT0021600250 and 2017SGR696). CIBER de Diabetes y Enfermedades Metabólicas Asociadas and CIBER de Enfermedades Respiratorias are initiatives of the Instituto de Salud Carlos III.

Institutional Review Board Statement: The research was conducted agreeing to the guidelines of the Declaration of Helsinki and accepted by Arnau de Vilanova University Hospital Ethics Committee (CEIC-1410, 19 December 2014).

Informed Consent Statement: Informed consent was obtained from all subjects involved in the study.

Data Availability Statement: The evidence presented in this investigation are accessible on request from the corresponding author. The data are not publicly presented due to the signed consent agreements around data sharing, which only allow access to the researchers of the ILERVAS project following the project purposes. Requestors wishing to access the data used in this work can make a demand to A.L. and M.B.-L. The request will be subjected to authorization and formal agreements regarding confidentiality and secure data storage being signed the data would be the provided.

Acknowledgments: The authors would like to thank to ILERVAS Project collaborators: Marta Hernández, Ferran Rius, Ferrán Barbé, Pere Godoy, Manuel Portero-Otin, Mariona Jové, Marta Ortega, Eva Miquel, Montserrat Martínez-Alonso, Jordi de Batlle, Silvia Barril, Manuel Sánchez-de-la-Torre, Josep Franch-Nadal and Esmeralda Castellblanco. We would like also to thank to more than a few nurses and staff of the project, Fundació Renal Jaume Arnó, and the Primary Care teams of the province of Lleida for enrolling individuals and their energies in the correct progress of the study. The funders had no role in work design, the collection, analysis and interpretation of data, report writing, or the decision to submit the article for publication.

Conflicts of Interest: The authors declare no conflict of interest.

References

1. Manson, J.E.; Willett, W.C.; Stampfer, M.J.; Colditz, G.A.; Hunter, D.J.; Hankinson, S.E.; Hennekens, C.H.; Speizer, F.E. Body weight and mortality among women. *N. Engl. J. Med.* **1995**, *333*, 677–685. [CrossRef] [PubMed]
2. Lin, H.; Zhang, L.; Zheng, R.; Zheng, Y. The prevalence, metabolic risk and effects of lifestyle intervention for metabolically healthy obesity: A systematic review and meta-analysis: A PRISMA-compliant article. *Medicine* **2017**, *96*, e8838. [CrossRef] [PubMed]
3. Aung, K.; Lorenzo, C.; Hinojosa, M.A.; Haffner, S.M. Risk of developing diabetes and cardiovascular disease in metabolically unhealthy normal-weight and metabolically healthy obese individuals. *J. Clin. Endocrinol. Metab.* **2014**, *99*, 462–468. [CrossRef]
4. Stefan, N.; Fritsche, A.; Schick, F.; Häring, H.U. Phenotypes of prediabetes and stratification of cardiometabolic risk. *Lancet Diabetes Endocrinol.* **2016**, *4*, 789–798. [CrossRef] [PubMed]
5. Jung, C.H.; Lee, M.J.; Kang, Y.M.; Jang, J.E.; Leem, J.; Hwang, J.Y.; Kim, E.H.; Park, J.Y.; Kim, H.K.; Lee, W.J. The risk of incident type 2 diabetes in a Korean metabolically healthy obese population: The role of systemic inflammation. *J. Clin. Endocrinol. Metab.* **2015**, *100*, 934–941. [CrossRef]
6. Wu, H.; Ballantyne, C.M. Metabolic Inflammation and Insulin Resistance in Obesity. *Circ. Res.* **2020**, *126*, 1549–1564. [CrossRef]
7. Saltiel, A.R.; Olefsky, J.M. Inflammatory mechanisms linking obesity and metabolic disease. *J. Clin. Investig.* **2017**, *127*, 1–4. [CrossRef]
8. Scherer, P.E. The many secret lives of adipocytes: Implications for diabetes. *Diabetologia* **2019**, *62*, 223–232. [CrossRef]
9. Sun, K.; Kusminski, C.M.; Scherer, P.E. Adipose tissue remodeling and obesity. *J. Clin. Investig.* **2011**, *121*, 2094–2101. [CrossRef]
10. Hwang, I.; Kim, J.B. Two Faces of White Adipose Tissue with Heterogeneous Adipogenic Progenitors. *Diabetes Metab J.* **2019**, *43*, 752–762. [CrossRef]

11. Dragoljevic, D.; Westerterp, M.; Veiga, C.B.; Nagareddy, P.; Murphy, A.J. Disordered haematopoiesis and cardiovascular disease: A focus on myelopoiesis. *Clin. Sci.* **2018**, *132*, 1889–1899. [CrossRef] [PubMed]

12. Engström, G.; Hedblad, B.; Stavenow, L.; Jonsson, S.; Lind, P.; Janzon, L.; Lindgärde, F. Incidence of obesity-associated cardiovascular disease is related to inflammation-sensitive plasma proteins: A population-based cohort study. *Arterioscler. Thromb. Vasc. Biol.* **2004**, *24*, 1498–1502. [CrossRef] [PubMed]

13. Sánchez, E.; Betriu, À.; Yeramian, A.; Fernández, E.; Purroy, F.; Sánchez-de-la-Torre, M.; Pamplona, R.; Miquel, E.; Kerkeni, M.; Hernández, C.; et al. Skin Autofluorescence Measurement in Subclinical Atheromatous Disease: Results from the ILERVAS Project. *J. Atheroscler. Thromb.* **2019**, *26*, 879–889. [CrossRef] [PubMed]

14. Fishman, S.L.; Sonmez, H.; Basman, C.; Singh, V.; Poretsky, L. The role of advanced glycation end-products in the development of coronary artery disease in patients with and without diabetes mellitus: A review. *Mol. Med.* **2018**, *24*, 59. [CrossRef] [PubMed]

15. van Waateringe, R.P.; Slagter, S.N.; van der Klauw, M.M.; van Vliet-Ostaptchouk, J.V.; Graaff, R.; Paterson, A.D.; Lutgers, H.L.; Wolffenbuttel, B.H.R. Lifestyle and clinical determinants of skin autofluorescence in a population-based cohort study. *Eur. J. Clin. Investig.* **2016**, *46*, 481–490. [CrossRef]

16. den Engelsen, C.; van den Donk, M.; Gorter, K.J.; Salomé, P.L.; Rutten, G.E. Advanced glycation end products measured by skin autofluorescence in a population with central obesity. *Derm. Endocrinol.* **2012**, *4*, 33–38. [CrossRef]

17. Sánchez, E.; Baena-Fustegueras, J.A.; de la Fuente, M.C.; Gutiérrez, L.; Bueno, M.; Ros, S.; Lecube, A. Advanced glycation end-products in morbid obesity and after bariatric surgery: When glycemic memory starts to fail. *Endocrinol. Diabetes Nutr.* **2017**, *64*, 4–10. [CrossRef]

18. Betriu, À.; Farràs, C.; Abajo, M.; Martinez-Alonso, M.; Arroyo, D.; Barbé, F.; Buti, M.; Lecube, A.; Portero, M.; Purroy, F.; et al. Randomised intervention study to assess the prevalence of subclinical vascular disease and hidden kidney disease and its impact on morbidity and mortality: The ILERVAS project. *Nefrologia* **2016**, *36*, 389–396. [CrossRef]

19. Sánchez, E.; Sánchez, M.; Betriu, À.; Rius, F.; Torres, G.; Purroy, F.; Pamplona, R.; Ortega, M.; López-Cano, C.; Hernández, M.; et al. Are Obesity Indices Useful for Detecting Subclinical Atheromatosis in a Middle-Aged Population? *Obes. Facts* **2020**, *13*, 29–39. [CrossRef]

20. WHO Consultation on Obesity. *Obesity: Preventing and Managing the Global Epidemic: Report of a WHO Consultation*; World Health Organization: Geneva, Switzerland, 2000; pp. 1–253.

21. Ma, W.Y.; Yang, C.Y.; Shih, S.R.; Hsieh, H.J.; Hung, C.S.; Chiu, F.C.; Lin, M.S.; Liu, P.H.; Hua, C.H.; Hsein, Y.C.; et al. Measurement of Waist Circumference: Midabdominal or iliac crest? *Diabetes Care* **2013**, *36*, 1660–1666. [CrossRef]

22. Gómez-Ambrosi, J.; Silva, C.; Catalán, V.; Rodríguez, A.; Galofré, J.C.; Escalada, J.; Valentí, V.; Rotellar, F.; Romero, S.; Ramírez, B.; et al. Clinical usefulness of a new equation for estimating body fat. *Diabetes Care* **2012**, *35*, 383–388. [CrossRef] [PubMed]

23. Thomas, D.M.; Bredlau, C.; Bosy-Westphal, A.; Mueller, M.; Shen, W.; Gallagher, D.; Maeda, Y.; McDougall, A.; Peterson, C.M.; Ravussin, E.; et al. Relationships between body roundness with body fat and visceral adipose tissue emerging from a new geometrical model. *Obesity* **2013**, *21*, 2264–2271. [CrossRef] [PubMed]

24. Hume, R. Prediction of lean body mass from height and weight. *J. Clin. Pathol.* **1966**, *19*, 389–391. [CrossRef] [PubMed]

25. Meerwaldt, R.; Graaff, R.; Oomen, P.H.N.; Links, T.P.; Jager, J.J.; Alderson, N.L.; Thorpe, S.R.; Baynes, J.W.; Gans, R.O.B.; Smit, A.J. Simple non-invasive assessment of advanced glycation endproduct accumulation. *Diabetologia* **2004**, *47*, 1324–1330. [CrossRef] [PubMed]

26. Lentferink, Y.E.; van Teeseling, L.; Knibbe, C.A.J.; van der Vorst, M.M.J. Skin autofluorescence in children with and without obesity. *J. Pediatr. Endocrinol. Metab.* **2019**, *32*, 41–47. [CrossRef] [PubMed]

27. Gogas Yavuz, D.; Apaydin, T.; Imre, E.; Uygur, M.M.; Yazici, D. Skin Autofluorescence and Carotid Intima-Media Thickness Evaluation Following Bariatric Surgery in Patients with Severe Obesity. *Obes. Surg.* **2021**, *31*, 1055–1061. [CrossRef] [PubMed]

28. Son, K.H.; Son, M.; Ahn, H.; Oh, S.; Yum, Y.; Choi, C.H.; Park, K.Y.; Byun, K. Age-related accumulation of advanced glycation end-products-albumin, S100β, and the expressions of advanced glycation end product receptor differ in visceral and subcutaneous fat. *Biochem. Biophys. Res. Commun.* **2016**, *477*, 271–276. [CrossRef]

29. Semba, R.D.; Ferrucci, L.; Sun, K.; Beck, J.; Dalal, M.; Varadhan, R.; Walston, J.; Guralnik, J.M.; Fried, L.P. Advanced glycation end products and their circulating receptors predict cardiovascular disease mortality in older community-dwelling women. *Aging. Clin. Exp. Res.* **2009**, *21*, 182–190. [CrossRef]

30. Koyama, H.; Shoji, T.; Yokoyama, H.; Motoyama, K.; Mori, K.; Fukumoto, S.; Emoto, M.; Shoji, T.; Tamei, H.; Matsuki, H.; et al. Plasma level of endogenous secretory RAGE is associated with components of the metabolic syndrome and atherosclerosis. *Arterioscler. Thromb. Vasc. Biol.* **2005**, *25*, 2587–2593. [CrossRef]

31. Kiuchi, K.; Nejima, J.; Takano, T.; Ohta, M.; Hashimoto, H. Increased serum concentrations of advanced glycation end products: A marker of coronary artery disease activity in type 2 diabetic patients. *Heart* **2001**, *85*, 87–91. [CrossRef]

32. Sánchez, E.; Betriu, À.; Salas-Salvadó, J.; Pamplona, R.; Barbé, F.; Purroy, F.; Farràs, C.; Fernández, E.; López-Cano, C.; Mizab, C.; et al. Mediterranean diet, physical activity and subcutaneous advanced glycation end-products' accumulation: A cross-sectional analysis in the ILERVAS project. *Eur. J. Nutr.* **2020**, *59*, 1233–1242. [CrossRef] [PubMed]

33. Baye, E.; Kiriakova, V.; Uribarri, J.; Moran, L.J.; de Courten, B. Consumption of diets with low advanced glycation end products improves cardiometabolic parameters: Meta-analysis of randomised controlled trials. *Sci. Rep.* **2017**, *7*, 2266. [CrossRef] [PubMed]
34. Snelson, M.; Coughlan, M.T. Dietary Advanced Glycation End Products: Digestion, Metabolism and Modulation of Gut Microbial Ecology. *Nutrients* **2019**, *11*, 215. [CrossRef] [PubMed]

Review

The Potential of the Mediterranean Diet to Improve Mitochondrial Function in Experimental Models of Obesity and Metabolic Syndrome

Mohamad Khalil [1,2], Harshitha Shanmugam [1], Hala Abdallah [1], Jerlin Stephy John Britto [1], Ilaria Galerati [1], Javier Gómez-Ambrosi [3,4,5], Gema Frühbeck [3,4,5,6] and Piero Portincasa [1,*]

[1] Clinica Medica "A. Murri", Department of Biomedical Sciences & Human Oncology, University of Bari Medical School, Piazza Giulio Cesare 11, 70124 Bari, Italy; mohamad.khalil@uniba.it (M.K.); harshitha.shanmugam@uniba.it (H.S.); hala.abdallah@uniba.it (H.A.); jerlin.johnbritto@uniba.it (J.S.J.B.); ilariagalerati@live.it (I.G.)

[2] Department of Soil, Plant and Food Sciences, University of Bari Aldo Moro, Via Amendola 165/a, 70126 Bari, Italy

[3] Metabolic Research Laboratory, Clínica Universidad de Navarra, 31008 Pamplona, Spain; jagomez@unav.es (J.G.-A.); gfruhbeck@unav.es (G.F.)

[4] CIBER Fisiopatología de la Obesidad y Nutrición (CIBEROBN), ISCIII, 28029 Pamplona, Spain

[5] Obesity and Adipobiology Group, Instituto de Investigación Sanitaria de Navarra (IdiSNA), 31008 Pamplona, Spain

[6] Department of Endocrinology & Nutrition, Clínica Universidad de Navarra, 31008 Pamplona, Spain

* Correspondence: piero.portincasa@uniba.it; Tel.: +39-328-4687215

Citation: Khalil, M.; Shanmugam, H.; Abdallah, H.; John Britto, J.S.; Galerati, I.; Gómez-Ambrosi, J.; Frühbeck, G.; Portincasa, P. The Potential of the Mediterranean Diet to Improve Mitochondrial Function in Experimental Models of Obesity and Metabolic Syndrome. *Nutrients* 2022, *14*, 3112. https://doi.org/10.3390/nu14153112

Received: 5 June 2022
Accepted: 25 July 2022
Published: 28 July 2022

Publisher's Note: MDPI stays neutral with regard to jurisdictional claims in published maps and institutional affiliations.

Abstract: The abnormal expansion of body fat paves the way for several metabolic abnormalities including overweight, obesity, and diabetes, which ultimately cluster under the umbrella of metabolic syndrome (MetS). Patients with MetS are at an increased risk of cardiovascular disease, morbidity, and mortality. The coexistence of distinct metabolic abnormalities is associated with the release of pro-inflammatory adipocytokines, as components of low-to-medium grade systemic inflammation and increased oxidative stress. Adopting healthy lifestyles, by using appropriate dietary regimens, contributes to the prevention and treatment of MetS. Metabolic abnormalities can influence the function and energetic capacity of mitochondria, as observed in many obesity-related cardio-metabolic disorders. There are preclinical studies both in cellular and animal models, as well as clinical studies, dealing with distinct nutrients of the Mediterranean diet (MD) and dysfunctional mitochondria in obesity and MetS. The term *"Mitochondria nutrients"* has been adopted in recent years, and it depicts the adequate nutrients to keep proper mitochondrial function. Different experimental models show that components of the MD, including polyphenols, plant-derived compounds, and polyunsaturated fatty acids, can improve mitochondrial metabolism, biogenesis, and antioxidant capacity. Such effects are valuable to counteract the mitochondrial dysfunction associated with obesity-related abnormalities and can represent the beneficial feature of polyphenols-enriched olive oil, vegetables, nuts, fish, and plant-based foods, as the main components of the MD. Thus, developing mitochondria-targeting nutrients and natural agents for MetS treatment and/or prevention is a logical strategy to decrease the burden of disease and medications at a later stage. In this comprehensive review, we discuss the effects of the MD and its bioactive components on improving mitochondrial structure and activity.

Keywords: obesity; mitochondria; Mediterranean diet; metabolic syndrome; plant-based foods; polyphenols; polyunsaturated fatty acids

1. Introduction

Trends for obesity and metabolic syndrome (MetS) are dramatically increasing worldwide and represent the "malnutrition" burden of the disease [1]. Obesity is characterized by excessive accumulation of adipose tissue combined with adipocytokine-mediated chronic inflammation, mitochondrial dysfunction, and the inhibition of antioxidant defenses [2]. Obesity is typically linked to metabolic disorders such as hypertension, dyslipidemia, and insulin resistance predisposing to type 2 diabetes (T2DM). Such metabolic abnormalities tend to cluster within MetS [3–5].

Mitochondria contribute to the pathogenesis of obesity-related metabolic disorders. Mitochondria are essential for cellular energy metabolism, as they generate adenosine triphosphate (ATP) by oxidizing carbohydrates, lipids, and proteins [6–8]. Mitochondria produce and eliminate the reactive oxygen species (ROS) [9]. The inability of mitochondria to produce and maintain sufficient levels of ATP is known as "mitochondrial dysfunction", which is the result of an imbalance in nutrient signal input, energy production, and oxidative respiration [8,10]. Several studies suggest that an excessive intake of nutrients influences mitochondrial function [11], and that obesity predisposes to mitochondrial dysfunction [12–15].

Basic, translational, clinical research, epidemiological studies, and society guidelines find that the adoption of a healthy diet and lifestyle has beneficial preventive and therapeutic effects on obesity and MetS. Among all dietary patterns, the typical Mediterranean Diet (MD) is high in monounsaturated fatty acids, fiber, antioxidants, and glutathione [16,17]. Since adherence to the MD has been associated with a lower risk of obesity, T2DM, MetS, coronary heart disease, and cardiovascular mortality [18–23], the MD is considered a potential remedy for the prevention of obesity-related diseases [24].

In this scenario, the term "*mitochondrial nutrients*" refers to specific nutrients that can preserve mitochondrial function. Cellular and animal models, as well as clinical studies, have investigated the effects of components of the MD on dysfunctional mitochondria in obesity and MetS. Thus, mitochondria represent a promising target for novel, natural supplements or functional foods designed for the prevention and treatment of obesity-related MetS. This is a reasonable strategy to decrease the impact of medications at a later stage of the disease.

In this review, we will discuss the main features of obesity and MetS with respect to mitochondrial function, as well as the effects of the MD and its bioactive components on improving mitochondrial structure and activity [25].

2. Obesity

2.1. Definition

From a physiological perspective, body fat consists of brown and white adipose tissue. By location, fat is found at the subcutaneous and visceral levels. According to the World Health Organization (WHO), obesity is defined as the excessive accumulation of fat in the body [26], as a result of sustained positive-energy balance where energy intake exceeds energy expenditure [27]. Obesity is considered a disease of body-weight regulation [28]. Expanded visceral adipocytes act as an endocrine organ, releasing adipocytokines actively involved in metabolic control, inflammation, and tissue repair [29,30], as well as tumorigenesis [31,32]. Excessive visceral adipose tissue is associated with increased efflux of long-chain fatty acids from adipocytes resulting in ectopic-fat deposition in the liver, skeletal muscle, pancreas, and heart. These changes are associated with insulin resistance and systemic gluco-lipidic toxicity. In a clinical context, obesity is associated with higher cardiovascular risk, mortality, and morbidity [33–36].

Obesity is typically assessed by the calculation of body mass index (BMI), expressed as body weight in kilograms divided by the square of height in meters (kg/m^2) [37]. Specific reference standards exist for children by age and sex between the ages of 2 and 20 years. In adults, BMI is independent of age and sex and is a surrogate marker of fat in the body [36]. In adults, BMI is classified into the following categories: underweight ($<18.5 \ kg/m^2$), normal weight ($18.5–24.9 \ kg/m^2$), overweight ($25–29.9 \ kg/m^2$), and obese ($BMI \geq 30 \ kg/m^2$). Obesity is further classified as class I (BMI $30–34.9 \ kg/m^2$), class II (BMI $35–39.9 \ kg/m^2$), and class III obesity (BMI $> 40 \ kg/m^2$), also known as severe obesity [38,39].

Although simple to obtain, the classification based on BMI does not take into account several subtypes of obesity and the interaction between body composition and cardiometabolic risk [36,40–45]. For example, the concept of metabolically healthy obesity (MHO) describes a subtype of obese subjects with limited or no features of cardiometabolic abnormalities. Conversely, some normal-weight subjects can display an elevated risk of cardiometabolic disorders, termed "metabolically unhealthy normal weight" [46,47]. The MHO phenotype displays a normal lipid and pro-inflammatory cytokine profile and insulin sensitivity [48]. These patients have low visceral adiposity, high cardiorespiratory fitness, and minimal or absent intima media thickness. Caution is required when classifying MHO for several reasons. Longitudinal studies show that MHO can evolve into the metabolically altered obesity (MAO) phenotype [49]. Nearly one-third of MHO patients, according to fasting glycemia, exhibit impaired glucose tolerance or T2DM following an oral-glucose tolerance test. Individuals with MHO and MAO have similar patterns of inflammatory biomarkers such as C reactive protein, fibrinogen, uric acid, leukocyte count, serum amyloid A and hepatic enzymes, as well as adipokines such as adiponectin, resistin, leptin, and angiotensin II. In addition, typical inflammatory gene expression in adipose tissue and the liver shows comparable patterns in MHO and MAO individuals [45,50–52]. Notably, the MHO phenotype is associated with accelerated age-related declines in functional ability and jeopardizes the independence in older age [53].

Another phenotype of obesity is sarcopenic obesity (SO), a condition characterized by the combination of low skeletal-muscle mass and decreased strength, i.e., "dynapenic" abdominal obesity [54]. In the obesogenic environment, this condition is becoming more important when considering the aging population [55,56]. A dangerous link exists between obesity and sarcopenia, characterized by a mismatch between muscle mass and fat mass with a negative impact on energy balance. This pathway, in turn, paves the way for weight gain. In addition, obesity-associated chronic inflammation has a catabolic effect on muscle mass, facilitating the loss of lean muscle combined with an increased risk for developing metabolic alterations, cardiovascular disease (CVD), and mortality, at a much higher rate than sarcopenia or obesity alone [57–59].

2.2. Epidemiology of Overweight and Obesity

Overweight and obesity are chronic non-communicable diseases, and since 1980 their prevalence has doubled worldwide. Over one-third of the population worldwide is now classified as overweight or obese. By 2030, nearly 38% of the adult population will be overweight and another 20% will be obese worldwide [60,61]. In 2015, evidence estimated that obesity affected around 604 million adults and 108 million children worldwide [62]. In 2015, the prevalence of obesity had become higher among women than men, for all age groups and at all socio-economic levels. From 1980 to 2015, the most pronounced increase in the prevalence of obesity (11.1% to 38.3%) was observed in men aged 25 to 29 years in low-to-middle income countries. Continuous increasing trends of severe types of obesity is an area of concern. For instance, between the years 2007 and 2018, the age-adjusted prevalence of class III obesity (BMI $\geq$ 40 kg/m^2) increased from 5.7% to 9.2% [63,64]. As will be discussed in the next sub-section, obesity is the key component of MetS [65–67], and for this reason overweight and obese [68–71] populations are at elevated risk of several metabolic disorders, including insulin resistance, dyslipidemia, hyperglycemia, CVD, and many specific cancers [37,72–76].

2.3. Metabolic Syndrome

MetS is characterized by specific criteria defined by the National Cholesterol Education Program Adult Treatment Panel (ATP) III [77] and the International Diabetes Federation (IDF) [78] (Table 1). The classification is based on the combination of at least three out of the five following factors: visceral adiposity, increased serum triglycerides, low HDL cholesterol, arterial hypertension, and elevated serum glucose (Figure 1).

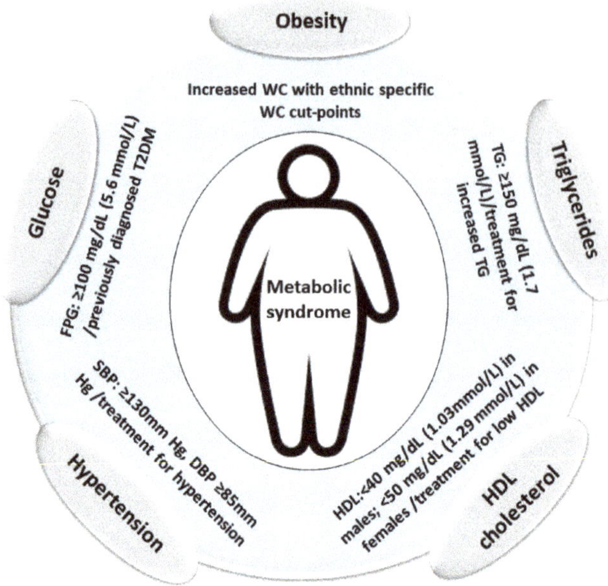

Figure 1. Criteria for the definition of metabolic syndrome. WC: waist circumference, TG: triglycerides, HDL: high-density lipoprotein, SBP: systolic blood pressure, DSP: diastolic blood pressure, FPG: fasting plasma glucose.

The estimated prevalence of MetS according to IDF definition is higher than the prevalence of MetS, according to the ATP III definition [79]. MetS is gaining increasing epidemiologic relevance [80–82]. According to the Third National Health and Nutrition Examination Survey, the overall prevalence of MetS was 22%. In 2002 [83], another study reported a worldwide prevalence of 10–30%, including children and adolescents [84]. An age-dependent increase was observed from 6.7% to 43.5% to 42.0%, for ages 20 to 29, 60 to 69, and over 70 years, respectively. Ethnic differences exist, with the highest age-adjusted prevalence among Mexican Americans (31.9%). Among Black Americans and Mexican Americans, the prevalence of MetS was 57% and 26%, which was higher in women than men.

Table 1. Criteria for the definition of metabolic syndrome.

	National Cholesterol Education Program ATP III [77]	**International Diabetes Federation (IDF)** [78]
	Any three of the following five abnormalities:	*Central obesity plus any two of the following four factors:*
Obesity	Abdominal obesity is defined as a waist circumference ≥102 cm in men and ≥88 cm in females	Increased waist circumference, with ethnic-specific waist-circumference cut-off points *
Triglycerides	Serum triglycerides ≥ 1.7 mmol/L or drug treatment for elevated triglycerides	Triglycerides ≥ 1.7 mmol/L or drug treatment for elevated triglycerides
HDL cholesterol	Serum high-density lipoprotein (HDL) cholesterol <1 mmol/L in males and <1.3 mmol/L in females or drug treatment for low HDL cholesterol	HDL cholesterol < 1.03 mmol/L in men or <1.29 mmol/L in females or drug treatment for low HDL cholesterol
Hypertension	Systolic blood pressure ≥ 130mm Hg, diastolic blood pressure ≥ 85 mm Hg or drug treatment for elevated blood pressure	Systolic blood pressure ≥ 130 mm Hg, diastolic blood pressure ≥ 85 mm Hg, or treatment for hypertension
Glucose	Fasting plasma glucose (FPG) ≥ 100 mg/dL (5.6 mmol/L) or drug treatment for elevated blood glucose	FPG ≥ 100 mg/dL (5.6 mmol/L) or previously diagnosed type 2 diabetes; an oral glucose tolerance test is recommended for patients with an elevated FPG, but it is not required

* Europid populations: males ≥ 94 cm; females ≥ 80 cm; South Asian populations, Chinese populations, and Japanese populations: males ≥ 90 cm; females ≥ 80 cm; South and Central American populations: use South Asian recommendations until more specific data are available; Sub-Saharan African, Eastern Mediterranean, and Middle Eastern populations: use European data until more specific data are available [85].

The cause of MetS is complex, and the major etiological components are considered to be genetic, environmental, and lifestyle factors [67,86–90].

Defective cell metabolism is an important contributing factor for MetS because of an imbalance between nutrient intake and utilization for energy. Diminished fatty-acid oxidation accelerates elevation in the intracellular aggregation of fatty acyl-CoAs as well as other fat-derived molecules in the liver, skeletal muscle, and adipose tissue [86]. Patients with MetS display early evidence of insulin resistance, with initial elevated serum-insulin levels. At a later stage, and if not properly treated, this condition can progress to T2DM. Associated conditions with MetS include cholesterol cholelithiasis and liver steatosis [91,92]. Cholesterol cholelithiasis originates from excessive secretion of hepatic cholesterol, which makes bile supersaturated and prone to the precipitation and aggregation of monohydrate cholesterol crystals, which grow into stones in the gallbladder [93–97]. The second condition is non-alcoholic fatty liver disease (NAFLD), recently renamed metabolic dysfunction-associated fatty liver disease (MAFLD) [98]. NAFLD/MAFLD originates from excessive intrahepatic influx of circulating long-chain fatty acids, with accumulation of triglycerides and toxic metabolites in the hepatocytes [69,99–103].

In this metabolically unhealthy scenario, several determining metabolic pathways converge in mitochondria, suggesting that obesity and MetS are associated with mitochondrial dysfunction, and pointing to a type of metabolic mitochondrial disease [68,69,102,104,105]. In MetS, mitochondrial dysfunction has been identified in various target organs such as the liver, heart, and skeletal muscle, as well as in tissue and cells such as adipocyte and pancreatic islet beta cells [106]. Nevertheless, it is still unclear if mitochondrial dysfunction is the primary cause or a secondary effect of MetS.

3. Mitochondria, Bioenergetics, Obesity, and MetS

3.1. Mitochondria and Bioenergetics

Mitochondria are small intracellular organelles with a double membrane structure, i.e., the outer mitochondrial membrane (OMM) and inner mitochondrial membrane (IMM), separated by the intermembranous space [102,107]. Mitochondria are the "powerhouse of the cell" and the main sites for ATP production. Using beta-oxidation and the citric acid cycle, mitochondria oxidize the long-chain fatty acids and glucose derived from foods [108]. Starting from chemical bonds in foods, high-energy electrons are produced and captured by nicotinamide adenine dinucleotide (NAD) and flavin adenine dinucleotide (FAD) and later reduced to NADH and $FADH_2$ [109]. High-energy electrons are donated to the electron transport chain (ETC) by NADH and $FADH_2$. The ETC is based in IMM and consists of five complexes [110,111]. NADH donates electrons to complex I, $FADH_2$ donates electrons to complex II, and both complexes I and II donate electrons to coenzyme Q (CoQ) [68–71,112].

CoQ is freely diffusible through IMM and provides electrons to complex III and reduces cytochrome c. Complex IV oxidizes cytochrome c and transfers electrons to oxygen to produce water. The movement of electrons along the transport chain releases free energy that is used to pump protons at complex I, III, and IV from the mitochondrial matrix into the intermembranous space, generating a proton gradient [110,113]. Protons diffuse along its concentration gradient at complex V, releasing energy that is used to create ATP from ADP [114]. This process is also known as oxidative phosphorylation (OXPHOS) [115]. Over 90% of the total cellular ATP is generated in the mitochondria, and this pathway is at the center of energy metabolism [102] and can become dysfunctional in MetS.

3.2. Mitochondria and Reactive Oxygen Species (ROS)

Mitochondria play a key role in ATP production but are also an important source of physiological levels of intracellular ROS [116]. As electrons pass through the ETC, a small fraction escape and prematurely react with molecular oxygen, generating superoxide radicals that are spontaneously or enzymatically converted into hydrogen peroxide [117,118]. Furthermore, by undergoing the Fenton reaction, hydrogen peroxide can produce hydroxyl radicals that are harmful and highly reactive molecules [119,120], which can cause cell death by damaging membranes, proteins, DNA, and enzymes [121]. Mitochondria host a very well-structured antioxidant mechanism that includes the homotetrameric Manganese (Mn) superoxide dismutase (MnSOD = SOD2), which in mammals is found solely in the mitochondria at the level of the matrix and intermembrane space [122]. SOD2 converts superoxide radicals to hydrogen peroxide and molecular oxygen. In the presence of reduced glutathione, hydrogen peroxide is converted to water by the enzyme glutathione peroxidase, minimizing the production of hydroxyl radical [123]. This process is very highly efficient and scavenges most of the ROS produced locally in mitochondria. Mitochondria also play a key role in ROS scavenging from other cellular sources, and mitochondrial dysregulation can lead to unrestricted ROS generation and cell injury [124–127]. Excessive production of ROS exceeding cellular antioxidant defense causes cellular macromolecule damage and affects cellular viability and functions, a process called oxidative stress [128]. Oxidative stress is widely recognized as one of the deciding mechanisms for several disease processes including MetS [129]. An increase in hydrogen peroxide and superoxide in cells modifies intracellular signaling and can lead to metabolic reprogramming resulting in increased fat synthesis and storage [130]. Therefore, increased ROS production with subsequent oxidative stress may add to the pathogenesis of MetS.

3.3. Mitochondrial Biogenesis

Mitochondria have their own DNA, which encodes for only 22 mitochondrial t-RNA and some components of the ETC [131]. The key master regulator and transcriptional activator of mitochondrial biogenesis is the peroxisome proliferator–activated receptor gamma coactivator-1 α (PGC-1α) [132,133]. Furthermore, by activating various other transcription factors, PGC-1α stimulates the process of mitochondrial biogenesis involved in nuclear and mitochondrial gene expression [134]. The induction of mitochondrial transcription factor A (TFAM) is led by the activation of nuclear respiratory factors 1 and 2 (NRF-1 and NRF-2), transcription factors, and estrogen-related receptors (ERRs) [135,136]. TFAM interacts directly with mitochondrial transcription factor B2 (TFB2M) along with the mitochondrial genome, to induce mitochondrial gene transcription [137]. Mitochondrial biogenesis is the physiological response to increased energy demand by AMP-activated kinase (AMPK) to monitor cellular-energy status [138]. The AMPK system responds to rises in the AMP:ATP ratio rather than to rises in AMP alone [139]. Increased AMP mediated by AMPK and elevated NAD+ mediated by Sirtuin-1 pathways can cause PGC1α activation, which in turn decreases cellular oxidative stress by enhancing the expression of mitochondrial antioxidant enzymes [140,141]. Hence, PGC1α has become a significant therapeutic target for MetS [142–146]. Therapeutic approaches focusing on enhanced mitochondrial biogenesis not only improve mitochondrial efficacy for substrate handling but also decrease oxidative stress, providing multifactorial benefits [147–149].

3.4. Mitochondrial Dysfunction in Obesity

The pathological expansion of body fat is associated with a chronic status of low-to-medium grade inflammation, oxidative stress, and insulin resistance [30,150]. These changes can be paralleled by dysregulation of mitochondrial function and biogenesis.

An excessive intake of nutrients, especially lipids and carbohydrates, can promote mitochondrial dysfunction. Due to high-calorie intakes, the metabolism is shifted towards the lipid reservoir, reduced mitochondrial function, and biogenesis, with subsequent production of ROS and the progression of insulin resistance in the liver, muscle, and adipose tissue [8]. By using hypertrophic adipocytes as an experimental in vitro cellular model of obesity, we showed that lipid accumulation and oxidative stress are associated with impaired mitochondrial oxygen consumption and alteration of mitochondrial complexes [105]. In addition, lipolysis, adipogenesis, and adipocyte-derived adiponectin production were abnormal in adipocytes along with deranged insulin sensitivity [151]. In skeletal muscle obtained from rodents and humans, the obesity-induced status by a high-fat diet increased the H_2O_2-emitting potential of mitochondria, shifting the cellular-redox environment to a more oxidized state and decreasing the redox-buffering capacity. These events occurred in the absence of any changes in mitochondrial respiratory function. Notably, the authors reported that attenuating mitochondrial H_2O_2 emission by either treating rats with an antioxidant-targeting mitochondrial or by genetically engineering the overexpression of catalase in mitochondria of muscle cells in mice, preserved insulin sensitivity despite the high-fat diet [152].

The relationship between mitochondrial dysfunction and obesity has been also investigated in animal models. More precisely, *db/db* mice and mice on a high-fat diet under diabetes/obesity conditions displayed a reduction in mitochondrial ATP production and alterations of mitochondrial structure [153]. In high-fat-diet, obese mice, mitochondrial dysfunction also occurs in the liver, mediated by a decrease in the expression of carnitine palmitoyltransferase-1 (CPT-1), citrate synthase, nuclear respiratory factor-1 (NRF-1), and mitochondrial transcription factor A (TFAM) [154].

The role of obesity in mitochondrial dysfunction has been also investigated in humans [155]. Adipocytes collected from omental and/or abdominal subcutaneous adipose samples of obese patients showed a reduction in mitochondrial oxygen-consumption rates and citrate synthase activity, compared to non-obese subjects [13]. Mitochondrial biogenesis, mitochondrial oxidative phosphorylation, and oxidative metabolic pathways in subcutaneous adipose tissue are downregulated in obese subjects, when compared to lean subjects [15]. These effects were accompanied by a reduction in the amount of mtDNA and the mtDNA-dependent translation system. At the molecular level, obese subjects showed reduced peroxisome proliferator–activated receptor-α (PGC1-α) expression, as a marker of altered mitochondrial biogenesis [156].

3.5. Mitochondrial Dysfunction in MetS

Mitochondrial dysfunction is a cardinal hallmark of MetS [69]. In the liver, mitochondria are involved in the metabolic pathways of lipids, proteins, carbohydrates, and xenobiotics [157,158]. Mitochondrial dysfunction is documented in NAFLD/MAFLD, the most common chronic liver disease [69,102,104,105,159]. In the early stages of NAFLD/MAFLD the increased intrahepatic influx of circulating FFAs causes early mitochondrial biogenesis mediated by the activation of PGC1-α and increased β-oxidation rates [160,161]. The high rate of FFA oxidation and ATP synthesis cause the uncontrolled increase in ROS level and changes in mitochondrial structure/function such as swelling, alteration in the mitochondrial electron transporter chain, mitochondrial DNA (mtDNA) damage, and sirtuin alteration. Despite the endogenous mitochondrial antioxidant system works to counteract the oxidative stress, the mitochondrial dysfunction occurs with imbalance between ROS production and mitochondrial defense mechanisms [162].

As NAFLD/MAFLD progresses, the increased levels of ROS severely impairs mtDNA function [39] and mitochondrial ATP synthesis promoting further hepatic dysfunction [163–166], and inflammation [167–170]. At the structural level, the mitochondrial electron transfer chain seems to be altered as consequence of the excessive accumulation of toxic lipids and mitochondrial ROS (mtROS) with a direct impact on the permeability of the inner mitochondrial membrane and increased oxidative damage [171]. At the molecular level, mitochondrial cytochrome P450 2E1 (CYP2E1), which is responsible for long-chain fatty acid metabolism, is directly involved in mitochondrial ROS production, and is considered a fundamental player in NAFLD/MAFLD pathophysiology [172]. Indeed, experimental studies on non-alcoholic steatohepatitis (NASH) in animal models and in humans, showed an increased activity of CYP2E1 [173,174]. Besides, mitochondrial enzymatic oxidative defense mechanisms resulted also impaired in NAFLD and NASH with progressive mitochondrial dysfunction. Furthermore, alteration of the expression PGC-1α are associated with NAFLD pathogenesis and to NASH-hepatocellular carcinoma progression [175].

The relationship between insulin resistance and mitochondrial dysfunction is not fully understood. Increased production of mtROS has been associated with a high glucose intake and FFAs accumulation, the two principal factors of insulin resistance. Despite the established role of genetic and environmental factors associated with T2DM pathophysiology, different metabolic abnormalities are directly implicated in the etiology of T2DM.With ongoing insulin resistance and pancreatic β-cell dysfunction, mitochondrial dysfunction has been indicated as a principal contributor. The reduction in insulin sensitivity in adipocytes, hepatocytes, and skeletal muscles is related to other complications such as the increased production of ROS and an accumulation of FFAs, both of which are associated with mitochondrial dysfunction and impaired mitochondrial biogenesis in diabetic patients [176–178].

In overweight/obesity and MetS, high fat and carbohydrate intake leads to lipid deposition resulting in the expansion of visceral adipocytes and an excessive influx of circulating FFAs [36]. The involvement of metabolic abnormalities (e.g., visceral fat accumulation, insulin resistance, and inflammation) in obesity is closely related to mitochondrial dysfunction and vice versa [179,180]. In experimental animal models, the accumulation of fat and formation of adipose tissue in obesity are correlated with increased ROS production. In addition, obese mice showed a mitochondrial dysfunction phenotype indicated by increasing NADPH oxidase expression and reducing antioxidative enzymes [181,182]. In obese mice, Choo et al. [183] showed that the number of mitochondria and mtDNA are reduced in adipocytes. Dysfunctional fatty acid oxidation and mitochondrial respiration were also observed. Similarly, mitochondrial biogenesis was strongly suppressed in the adipocytes of obese mice. Mitochondrial ATP production occurred with molecular (PGC-1α/β, estrogen-related receptor alpha, and PPAR-α) and structural (outer and inner membrane translocases and mitochondrial ribosomal proteins) alteration in adipose tissue [153].

Two main mechanisms of damage include ATP depletion and excessive ROS production. Mitochondrial dysfunction in association with adipose tissue dysfunction, plays a role in aging [184,185]. Thus, based on the pivotal role of mitochondria in the pathogenesis of MetS, targeting mitochondrial dysfunction for the treatment of MetS is of great interest. There is growing evidence from animal and human models that sheds light on the beneficial effects of nutrition-based intervention targeting mitochondria in MetS. Diets rich in polyphenols such as the MD could represent one of the healthiest approaches for nutritional intervention for the prevention and/or treatment of MetS.

4. Diet, Features, and Effects

The proper maintenance of metabolic homeostasis is closely related to food and nutrient intake. Both epidemiological and clinical evidence suggests that dietary patterns are closely related to the incidence and complications of MetS [186,187]. The Western diet, characterized by the high intake of refined grains, red meat, and fried foods, is associated with a greater risk of developing one or more components of MetS [188]. Low-fat diets such as the vegan diet, characterized by the absence of all animal-based products, if well-balanced, can promote health and reduce the risk of MetS [189]. A well-balanced diet, such as the MD is associated with lower incidence and risk of MetS (Table 2).

Table 2. Principal features of Western diet, Vegan diet, and Mediterranean diet.

	Western Diet	Vegan Diet	Mediterranean Diet
Characteristics	High fat and sugar	High vegetable Low fat No meat	Low meat High vegetable and olive oil High plant-based foods
Main components	Red meat (Saturated fat and cholesterol) Refined grains Fructose beverage	Fiber Grain Cereals	Fiber Antioxidants Unsaturated fats Whole grain
Health consequences	Obesity Insulin resistance NAFLD Diabetes CVD	Healthy (if balanced) Deficiency of essential macro and micronutrients (if unbalanced)	Healthy

Table 2. *Cont.*

	Western Diet	Vegan Diet	Mediterranean Diet
Mechanisms	↑ Adipose tissue ↑ Circulating FFAs ↑ Hepatic lipid accumulation ↑ Triglycerides ↑ Cholesterol ↑ Fasting glucose ↑ De novo lipogenesis ↑ VLDL ↑ ER stress ↑ Lysosomal permeabilization ↓ Insulin sensitivity	↓ Circulating FFAs ↓ Hepatic steatosis ↓ Lipolysis ↓ De novo lipogenesis ↑ Insulin sensitivity	↓ Circulating FFAs ↓ Hepatic steatosis ↓ Triglycerides ↓ Cholesterol ↓ Inflammation ↓ Lipolysis ↓ De novo lipogenesis ↓ ROS ↓ CRP ↑ Insulin sensitivity ↓ Inflammatory markers
Effect on Mitochondria	↑ mtROS ↓ mitochondrial biogenesis ↓ mitochondrial respiration	↓ mtROS ↑ mitochondrial biogenesis ↑ mitochondrial respiration	↓ mtROS ↑ mitochondrial biogenesis ↑ mitochondrial respiration
References	[188,190,191]	[189,192,193]	[194–199]

Abbreviation: **NAFLD**: non-alcoholic fatty liver disease, **CVD**: cardiovascular disease, **FFAs**: free fatty acids, **ROS**: reactive oxygen species, **CRP**: C-reactive protein, **mtROS**: mitochondrial reactive oxygen species, **ER**: endoplasmic reticulum, ↑: increased, ↓: decreased.

Mediterranean Diet and Beneficial Effects

The Mediterranean dietary pattern is particularly popular among people living in the Mediterranean Sea basin. The MD is mainly characterized by a high intake of vegetables, fruits, nuts, cereals, and whole grains, a moderate intake of white meat such as fish and poultry, and low intake of dairy products, sweets, red meat, processed meat, and red wine. Extra virgin olive oil becomes the principal source of fat [200–203] (Figure 2). The promotion of a healthy lifestyle is an effective strategy is to decrease the risk of MetS onset by promoting healthy lifestyle. Evidence suggests that the MD possesses antioxidant and anti-inflammatory properties [25] with protective effects in regard to the disorders associated with MetS and in the prevention of cardiovascular disease (CVD) [201].

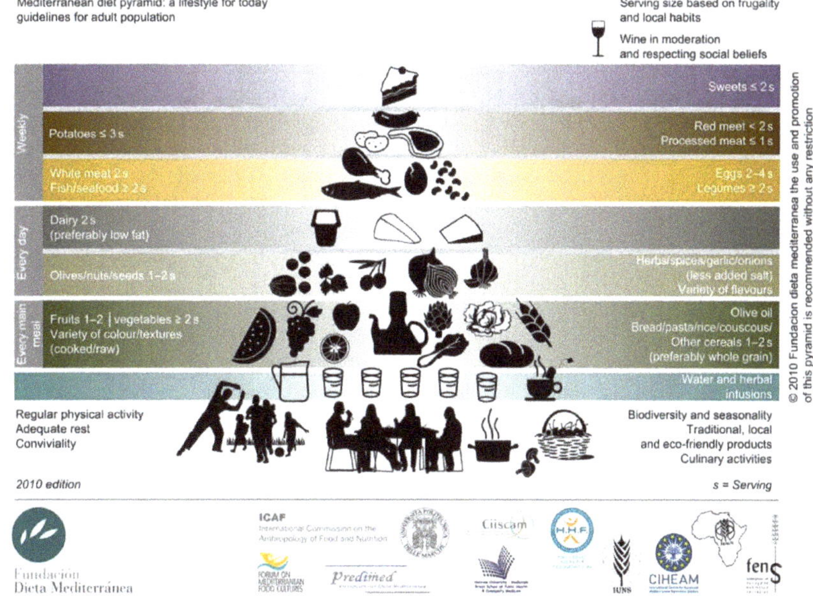

Figure 2. The concept of the healthy food pyramid is based on differences across countries which include food quality and quantity, social and cultural context, and economical aspects encountered in the Mediterranean basin. The graphical abstracts provide information about the type of seasonal food, weekly intake in relation to standard portions, and the role of macro- and micro-nutrients. The

idea is that of promoting healthy lifestyles among different populations. The importance of regular physical activity and social relationships is also indicated. The final design of the MD pyramid today and a brief complementary text for the general public have been developed by the Mediterranean Diet Foundation Expert Group that includes the Mediterranean Diet Foundation's International Scientific Committee expertise, the in situ discussions by a representative group of members that met within the Barcelona VIII International Congress on the Mediterranean diet, and several other experts who provided support on the design, editing, and translation to 10 different languages (English, French, Italian, Spanish, Catalan, Basque, Galician, Greek, Portuguese, and Arabic). With permission from Cambridge University Press, 2022 [200]. Website http://dietamediterranea.com/en/ (accessed on 4 June 2022).

Studies show that adherence to the MD has protective effects against obesity, stroke, CVD, hypertension, diabetes, some types of cancer, allergic diseases, and Alzheimer and Parkinson's disease [204–215]. The American Diabetes Association and the American Heart Association both recommend the MD in order to decrease cardiovascular risk factors in T2DM and to improve glycemic control [216]. In younger subjects, low adherence to the MD can trigger functional gastrointestinal symptoms, as component of the irritable bowel syndrome and functional dyspepsia, mainly in younger subjects [217]. Unprocessed plant-based food such as fruits, vegetables, legumes, seeds, spices, and nuts, rich in polyphenols, are the principal aspect of the MD with a wide range of biological and pharmacological effects [218–221]. The polyphenols undergo biotransformation process by gut microbiota before reaching the liver by the portal vein with beneficial effects either locally (i.e., intestine) and systemically (i.e., liver, brain) [222]. In contrast, the Western diet, high in calories, is characterized by the high intake of processed macronutrients (cholesterol, fat, protein, and sugars) and salt (sodium chloride), trans fats, and the low intake of fiber and magnesium. In the long term, this diet predisposes to obesity, insulin resistance, T2DM, CVD, and MetS. At the molecular level, the Western diet stimulates oxidative stress and inflammation by inducing mitochondrial dysfunction, decreasing the activity of antioxidant enzymes such as catalase, dismutase, and glutathione peroxidase, and peroxisomal oxidation of fatty acids [223]. As discussed earlier, the protective effects of the MD are the result of the diet as a whole, rather than individual components, reinforcing the idea that the interaction of various dietary components can have a beneficial synergistic effect [222]. However, several scientific-based evidence about the beneficial effects of individual components of the MD have been documented. For example, olive oil exerts antidiabetic, cardioprotective, neuroprotective, and nephroprotective effects due to the presence of tyrosol, oleocanthal, and hydroxytyrosol [224]. The long-term consumption of olive oil counteracts inflammation, promotes blood vessels' relaxation, protects against T2DM, reduces blood pressure, and increases insulin circulation [225]. The MD contains sea foods and fish rich in by fatty acids such as docosahexaenoic acid (DHA) and eicosapentaenoic acids (EPAs), which are metabolized producing 5-series leukotrienes and resolvins (RvE1 and RvE2). These metabolites possess anti-inflammatory effects in vivo [226]. Red grapes and wine found in the MD contain the polyphenol resveratrol (3,40,5-trihydroxystilbene), which not only has cardioprotective, antiaging, and anticarcinogenic effects but also promotes neuroprotective activities leading to anti-inflammatory, antioxidant, and gene-modulating effects. Resveratrol in patients with T2DM modulates the genes that influence mitochondrial function, such as PGC-1α, which is a key regulator of mitochondrial biogenesis and leads to elevation of mitochondrial content [227]. Furthermore, resveratrol indirectly activates AMP-activated protein kinase (AMPK), leading to increased mitochondrial biogenesis, improved glucose tolerance, insulin sensitivity, physical endurance, and a reduction in fat accumulation [228]. Moreover, due to its structural similarity to the synthetic estrogen diethylstilbesterol, resveratrol interacts with estrogen receptors inducing favorable cardiovascular effects. Several studies have demonstrated that the overall pattern of the MD produces beneficial effects by reducing the risk of obesity, hypertension, dyslipidemia, glucose metabolism, and CVD in T2DM patients [229,230]. The MD includes a high consumption of green vegetables rich in magnesium, which is a main constituent of chlorophyll. The magnesium present in chlorophyll plays a crucial role in the metabolism of insulin and glucose by translocating the phosphate from ATP to protein through its influence on tyrosine kinase activity of the insulin receptor. Magnesium is one of the cofactors of more than 300 enzymic reactions and it is important for ATP metabolism. It is also necessary for the regulation of blood pressure, insulin metabolism, muscle contraction, cardiac excitability, neuromuscular conduction, and vasomotor tone. The deficiency of magnesium is known to be associated with the onset of T2DM, while its consumption reduces the intensity of diabetes by sensitizing insulin [231,232].

The MD also has significant protective effects in MetS [19,233]. Scientific-based evidence suggest that many component of the MD display anti-inflammatory effects by reducing the activation of NF-κB signaling pathway and the expression of chemokine and proinflammatory cytokines such as TNF-α, IL-1β, and IL-6 [234]. The decreased expression of cytokines reduces oxidative stress, low-grade inflammation, and apoptotic cell death in brain and visceral tissues [235]. Another biomarker for inflammation is C-reactive protein (CRP), and prolonged intake of the MD diminishes CRP and unusual quantity of cytokines and adipokines irrespective of weight loss increase [235–237]. Furthermore, the MD is associated with lower mortality and a decreased incidence of common chronic diseases such as CVD [238], several cancers [239], T2DM [240], fatty liver disease [241], and some types of allergies [242] as a result of the inhibition of oxidative stress, reduction in inflammation, and improved lipid profiles [218,219,243–245]. Along with improving physical health, long-term adherence to the MD also improves the quality of life and longevity [246,247].

5. MD and Mitochondrial Activity

5.1. Preclinical Studies

The MD is characterized by the high intake of several ingredients with beneficial, nutraceutical and pharmaceutical properties, involved in the prevention and recovery of metabolic diseases. This is achieved through different pathways, including the attenuation of mitochondrial dysfunction (Table 3). Despite the difference in some components of the MD between different countries, most essential ingredients are the same, such as olive oil, PUFA (Omega-3), fruits, and polyphenol-rich plants and vegetables.

Table 3. Summary of in vitro and in vivo studies about effects of Mediterranean diet on metabolic diseases targeting mitochondria.

Compound	Study	Model	Effects	Reference
Chlorogenic Acid (CGA)	In vitro OxLDL-treated HUVECs	Oxidative Damage/Mitochondrial Dysfunction	↑ SIRT1 expression ↓ OxLDL-impaired SIRT1 Level ↓ ROS ↑ SIRT1, AMPK, and PGC-1α pathway	[248]
Delphinidin	In vitro VEGF-treated HUVECs	Post-ischemic neovascularization	↑ NRF1, Tfb2m, Tfam and PolG ↓ Abnormal increase in mitochondrial respiration, mtDNA content, and complex IV activity	[249]
Lycopene (LYC)	In vivo LPS-treated mice	Inflammation	↑ SIRT1 ↑ PGC1α ↑ Cox5b, Cox7a1, Cox8b, and Cycs ↑ Complexes I, II, III, and IV	[250]
Lycopene (LYC)	In vitro H_2O_2-treated SH-SY5Y	Oxidative stress /Apoptosis	↑ Depolarization ↑ Bcl2 ↓ Bax	[251]
5-Heptadecylresorcinol (AR-C17)	In vitro H_2O_2-treated PC-12	Apoptosis/Mitochondrial dysfunction	↓ ROS ↑ Mitochondrial respiration ↑ ATP ↑ SIRT-3 ↑ FOXO3a ↓ H_2O_2-cell apoptosis	[252]

Table 3. *Cont.*

Compound	Study	Model	Effects	Reference
Resveratrol	In vivo Mice	Insulin resistance/Obesity	↑ SIRT1 activity ↑ PGC-1α activity ↑ Mitochondrial activity ↑ Aerobic capacity	[253]
Resveratrol	In vitro solubilized complex I In vivo Mice	Aging	↑ Complex I activity in vitro ↑ Complex I activity in young mice ↑ Oxidative stress in old mice	[227]
Resveratrol	In vivo HFD mice	Obesity/Ageing	↑ SIRT1 enzymatic activity ↑ PGC-1α deacetylation and activity	[254]
Butyric acid	In vivo HFD mice	Metabolic syndrome	↑ PGC-1α ↑ CPT1b ↑ COX-I ↑ PPAR-δ ↑ Fatty acid oxidation	[255]
Butyrate and its synthetic derivative FBA	In vivo mice In vitro HepG2 cells	Insulin resistance/Obesity	↑ Oxygen consumption ↑ Citrate synthase activity ↓ H_2O_2 ↑ Aconitase activity ↑ *Mfn1, Mfn2, Opa1* ↓ Drp1 and Fis1	[256]
Ginger extract (GE)/ 6-gingerol	In vivo mice In vitro HepG2 cells	-	↑ mtDNA ↑ OXPHOS ↑ATP ↑ Complex I and IV activity ↑ AMPK-PGC-1α signaling	[257]
Ferulic acid (FA)	In vivo HFD mice In vitro PBMC and EPC	Cardiovascular disease	↑ Mitochondrial biogenesis markers ↓ Oxidative stress ↓ PBMC apoptosis ↑ PGC-1α	[258]
Different ω-3/ω-6 PUFAs ratios	In vivo mice	Metabolic syndrome	↓ Metabolic risk factors ↓ p-mTOR ↑ Mitochondrial electron transport chain ↑ Tricarboxylic acid cycle ↑ Mitochondrial activities ↓ Fumaric acid ↓ Oxidative stress	[259]
Extra virgin olive oil (EVOO) and nitrite (NO_3)	In vivo mice	NAFLD	↑ HO-1 expression ↑ Complexes II and V ↑ NO_2-OA ↓ Cholesterol ↓ LDL ↓ Endothelial dysfunction ↓ Blood pressure ↓ Thrombosis ↓ Hyperglycemia	[260]
Hydroxytyrosol (HT)	In vivo mice	Metabolic syndrome	↓ Drp1 ↑ Complex I and II ↓ Complex V ↓ PARP	[261]
Hydroxytyrosol (HT)	In vivo HFD-Megalobrama amblycephala fish In vitro hepatocytes	Hepatic fat deposition	↑ Citrate synthase activity ↑ ATP content ↑ Mitochondria number ↑ PGC-1α, PGC-1β, NRF1 and TFAM	[262]
Ellagic acid (EA)	In vivo chronic arsenic-rats	Diabetes/Cancer	↓ ROS ↓ Mitochondrial damage ↑ Dehydrogenase complex II-associated activity	[263].
Apigenin (APG)	In vivo multiwall CNT (MWCNT)-exposed rats	Kidney toxicity	↑ Succinate dehydrogenase ↓ ROS ↑ Mitochondrial membrane potential ↓ Mitochondrial swelling ↓ Release cytochrome	[264]
Apigenin (APG)	In vivo aged Mice	Muscle Atrophy	↑ Basal oxygen consumption ↑ Complexes I, II, and IV activity ↑ ATP content ↑ PGC-1α, TFAM, and NRF-1 ↓ Cyt-C release to cytosol	[265]

Table 3. *Cont.*

Compound	Study	Model	Effects	Reference
Cocoa Flavanols	In vivo mice	Healthy and SIRT3-/-mice	↑ Mitochondrial respiration ↑ AMPK phosphorylation ↑ Mitochondria mass ↑ NAD+/NADH ↑ Complex I and IV activity	[266]

Abbreviations: Drp1: mitochondrial fission-related protein, **Bak, Bax, and Bad**: proapoptotic Bcl-2 members, **Bcl-2 and Bcl-XL**: antiapoptotic Bcl-2 proteins, **PARP**: poly(ADP-ribose) polymerase, **HFD**: high-fat diet, **EVOO**: extra virgin olive oil, **HO-1**: heme oxygenase-1, **NO2-OA**: nitro-fatty acids, **LDL**: low-density lipoprotein, **LYC**: lycopene, **SH-SY5Y**: human neuroblastoma cells **LPS**: lipopolysaccharides, **SIRT1**: sirtuin 1, **PGC1α**: peroxisome proliferator–activated receptor gamma coactivator-1α, **Cox**: cyclooxygenase, **PBMC**: peripheral blood mononuclear cell, **EPC**: endothelial progenitor cells, **ROS**: reactive oxygen species, **HUVECs**: human umbilical vein endothelial cells, **OxLDL**: oxidized low-density lipoprotein, **FOXO3a**: forkhead box O3 (transcription factors), **HepG2**: human liver cancer cell line, **OXPHOS**: oxidative phosphorylation, **NAD+**: nicotinamide adenine dinucleotide, **CPT1b**: carnitine palmitoyltransferase 1B, **COX-1**: cytochrome c oxidase I, **PPAR-δ**: peroxisome proliferator–activated receptor-δ, **FBA**: *N*-(1-carbamoyl-2-phenyl-ethyl) butyramide, **MetS**: metabolic syndrome, ↑: increased, ↓: decreased.

In vitro studies evaluated the beneficial effects of polyphenol-rich foods on MetS mediating mitochondrial modulation. In detail, the protective role of chlorogenic acid (CGA) found in coffee beans and apples against ox-LDL-induced endothelial cells dysfunction as cellular model of atherosclerosis was evaluated using human endothelial cells HUVECs. CGA displayed mitochondria-mediated effects by enhancing SIRT1 activity and up-regulating AMPK and PGC-1 expression to maintain mitochondrial biogenesis. In addition, CGA treatment exhibited a cytoprotective effect by reducing ROS production in endothelial cells. [248]. Similarly, in endothelial cells with VEGF-induced mitochondrial dysfunction, delphinidin (a flavonoid present in red wine and berries) restored the elevated level of mitochondrial respiration, mtDNA content, and complex IV activity. In addition, delphinidin increased the expression of NRF1, Tfb2m, Tfam, and PolG, all of which are involved in the regulation of mitochondrial biogenesis [249]. Lycopene (LYC), a member of the carotene phytochemical family, present in tomatoes and grapefruits, exerted an anti-inflammatory effect on mice exposed to LPS through improving mitochondrial dysfunction. In detail, LYC upregulated the expression of SIRT1, PGC1α, Cox5b, Cox7a1, Cox8b, and Cycs. In addition, a partial effect of LYC was proved in regulating the expression of many complexes in the respiratory chain [250]. Another in vitro study using human neuroblastoma cells SH-SY5Y showed a protective effect of lycopene against H_2O_2-induced depolarization of the mitochondrial membrane [251]. LYC increased the expression of Bcl2 and decreased Bax expression [251]. Whole grains also represent an important category in the MD, with a beneficial impact on metabolic diseases. Especially, 5-heptadecylresorcinol, a biomarker of whole grain rye consumption, protects against H_2O_2-induced oxidative stress in rat pheochromocytoma (PC-12) by activating the SIRT3-FOXO3a signaling pathway. In addition, it reduced mitochondrial ROS levels and maintained the mitochondrial respiration and membrane potential, which leads to an increase in ATP production and cell respiration [252].

Another study found that the antioxidant effect of resveratrol found in grapes, berries, and cacao is dose- and age-dependent [253]. This polyphenol competes with NAD+ in a solubilized complex of mitochondria to improve their activity [253]. In addition, resveratrol prevents metabolic diseases (obesity and insulin resistance) in mice through improving mitochondrial function via PGC-1α and SIRT-1 activation [227]. These results have also been confirmed by Baur et al. [254] using a high-calorie-diet mice model which demonstrated a SIRT-1-dependent effect of resveratrol on the activation of PGC-1α resulting in the improvement of mitochondrial biogenesis.

By monitoring the increase in CO_2 level in skeletal muscle tissue and L6 muscle cells in butyrate-treated mice, an increase in PGC-1α level accompanied by an increase in CPT1b and COX-I genes expression was observed. Moreover, the levels of peroxisome proliferator–activated receptors (PPARs) were also increased in the treated group. Overall, these data suggest that butyrate (found in legumes, fruits, and nuts) promotes fatty acid oxidation and improves mitochondrial function [255]. In addition, butyrate increased citrate synthase activity, aconitase activity, and oxygen consumption in butyrate-treated mice and FBA-treated human HepG2 cells, with a decrease in H_2O_2 yield. In mitochondrial dynamics, butyrate and FBA upregulated the expression of fusion genes (*Mfn1*, *Mfn2*, and *Opa1*) and decreased the expression of fission-related genes (*Drp1* and *Fis1*) [256].

Ginger extract and its bioactive compound 6-gingerol promote mitochondrial biogenesis and function through improving AMPK-PGC1α signaling in vivo (skeletal muscle, liver, and BAT) and in vitro (HepG2 cells). Furthermore, 6-gingerol enhanced p-AMPKα, PGC-1α, NRF1, and TFAM protein expression and stimulated the subunits of OXPHOS complexes in HepG2 cells [257]. Further study links the role of ferulic acid (FA), the main active phenolic acid in rice bran, with the improvement of mitochondrial biogenesis and dynamic by increasing the expression of *Pgc-1α*, *Pgc-1β*, *Nrf-1*, *Mfn1*, *Mfn2*, *Fis1*, and *Beclin-1*. In addition, the rice bran enzymatic extract (RBEE) diet upregulated AMPK activity with enhanced PGC-1α expression in mice. The latter was also observed in peripheral blood mononuclear cell (PBMC) and endothelial progenitor cells (EPC), in addition to an increase in fusion MFN1 [258].

The effect of a high omega-3 to omega-6 ratio (ω-3/ω-6) on metabolic syndrome was investigated in vivo using high-fat-diet mice. A high ω-3/ω-6 ratio significantly decreased the insulin index, body weight, atherosclerosis markers, and accumulation of hepatic lipid. These effects were mediated by a reduction in p-mTOR expression, accompanied by an upregulation of the mitochondrial electron-transport chain and tricarboxylic acid-cycle pathway, when compared to a diet with a low or moderate ω-3/ω-6 ratio. Therefore, a diet with a high ω-3/ω-6 ratio displayed an enhancement of mitochondrial complexes activities, accompanied by an alleviation of fumaric acid and oxidative stress [259]. The Mediterranean diet also contains a variety of vegetables rich in NO_2- and nitrate (NO_3). Sánchez-Calvo et al. analyzed the involvement of nitro-fatty acids (NO_2-FA) on the beneficial effects of extra virgin olive oil (EVOO) consumption on an NAFLD experimental animal model. EVOO and nitrite supplementation improved the function of liver mitochondrial complexes II and V and exerted antioxidant and anti-inflammatory effects. The authors concluded that EVOO-NO_2- consumption may promote additional nutraceutical effects in NAFLD patients [260].

Hydroxytyrosol (HT), a polyphenol from olive oil, was effective in the regulation of multiple HFD-induced MetS, especially those related to mitochondrial dysfunction, through the modulation of mitochondrial apoptotic pathway in the liver and skeletal muscles. Moreover, HT treatment normalized the down-expression of Complex I and II and the up-expression of complex V, while Drp1 and PARP were decreased after treatment [261]. HT may also improve mitochondrial biogenesis (increase mtDNA and number of mitochondria) through the AMPK pathway, by enhancing the expression of involved genes (PGC-1α, NRF-1, and TFAM). ATP content and citrate synthase activity were also shown to increase after HT treatment of HFD (high-fat diet) and LFD (low-fat diet) groups [262]. Another phenolic acid, ellagic acid, which is found in strawberries and walnuts, prevents metabolic disorders by targeting the mitochondria via two ways: directly, by decreasing the ROS amounts and mitochondrial damage, or indirectly, by restoring the total dehydrogenase activity in mitochondria through complex II maintenance [263]. Apigenin (APG), a flavonoid found in many fruits and vegetables, increased the respiratory complex II succinate dehydrogenase (SDH) activity on carbon-nanotubes-induced mitochondrial damage. APG acts as antioxidant by decreasing ROS generation in kidney, which leads to a decrease in MMP collapse [264]. Results were confirmed in another study in old mice by Wang et al. [265]. In addition, APG improves mitochondrial biogenesis (by increasing mtDNA, PGC-1α, TFAM, and NRF-1), and the activity of complexes I, II, and IV and ATP synthesis [265]. Cocoa flavanol supplementation boosted the NAD metabolism, which stimulates sirtuins metabolism and improved mitochondrial function. These results suggest that flavanols likely contributed to the observed whole-body metabolism adaptation, with a greater ability to use carbohydrates, at least partially through Sirt3 [266].

5.2. Clinical Studies

In subjects with NASH, a six-month treatment with omega-3 showed a regulation of lipogenesis, ER stress, and mitochondrial function [267]. These effects were mediated by an overexpression of FABPL and PRDX6, with a reduction in PGRMC1 level. Meanwhile, an up-representation of PEBP1 and ApoJ was detected after the oral consumption of omega-3, confirming its role in the modulation of insulin resistance. In addition, FASTKD2, mitochondrial proteins related to aerobic-cell respiration, was overexpressed in this situation [267]. A study in patients on 2–3 weeks of a PUFA diet before elective cardiac surgery confirmed that omega-3 fatty acids from fish oil upregulated the nuclear transactivation of peroxisome proliferator–activated receptor-γ (PPARγ). This effect improved the mitochondrial oxidation of fatty acid and enhanced the antioxidant effect in the human atrial myocardium [268]. An EPA+DHA diet increases the expression of mitochondrial uncoupling protein 3 (UCP3) and ubiquinol cytochrome c reductase (UQCRC1) genes, which reduces ROS production. In addition, this diet improved oxidative phosphorylation activity and the extracellular matrix (ECM)-related pathways [269]. DHA, an omega 3 fatty acid, present in marine foods, increases the expression of the genes responsible for integrating fatty acid into mitochondria, as a new source of energy. In addition, DHA-enriched food consumption enhanced mitochondrial antioxidant capabilities and decreased mitochondrial ROS production [270]. The effect of resveratrol was studied in overweight/T2DM patients. Resveratrol improves the mitochondrial function through increasing state 3 respiration, while decreasing complex IV [271]. Resveratrol can stimulate the ENDOG gene to further stimulate the PGC-1α activity in biogenesis and to increase the number of mitochondria [272]. Resveratrol combined with epigallocatechin-3-gallate (EGCG) increases complexes III and V and improves the electron transport chain capacity, in addition to the upregulation of the citric acid cycle and fat oxidation in muscles during fasting [273]. Furthermore, a mixture of ancient peat and apple extract exerts a beneficial effect on mitochondrial function and ATP production, accompanied with a decrease in ROS production and oxidative stress in resistance-trained [274] (Table 4).

Table 4. Clinical studies in metabolic syndrome assessing the efficacy of MD components on mitochondria.

Authors	Year	Sample Size	Gender M/F (Age)	Participants	Format, Dose	Duration of Study	Main Findings
Anderson et al. [268]	2014	24	16/8 (63.1 ± 8.4; 65.8 ±9.9)	Elective cardiac surgery for patients	Oral consumption of EPA and DHA capsule, 3.4 g/day	2–3 weeks	↑ PPARγ ↑ Mitochondrial fatty acid oxidation ↑ TxnRd2 enzyme
Capo et al. [270]	2014	15	15/0 (20.4 ± 0.5)	Exercise-induced oxidative stress	Beverage enriched with DHA	2 months	↑ Antioxidant activity ↓ ROS production ↑ DHA
Yoshino et al. [269]	2016	20	60 to 85	Large hypertrophic response	Consumption of 4 pills (1.86g EPA+ 1.50 g DHA)	6 months	↑ Respiratory electron transport activity ↑ Oxidative phosphorylation ↑ ECM organization ↑ UCP3 and UQCRC1
Most et al. [273]	2016	38	18/20 (38 ± 2)	Subjects with obesity	Consumption of 282 mg EGCG + 80 mg RES	12 weeks	↑ Complexes III and V ↑ Citric acid cycle ↑ Respiratory electron-transport chain ↑ Fat oxidation
Joy et al. [274]	2016	25	25/0 (28 ± 5)	Resistance-trained subjects	Consumption of 150 mg (ancient peat and apple extract (TRT))	12 weeks	↑ Mitochondrial ATP production ↓ ROS ↓ Oxidative stress

Table 4. *Cont.*

Authors	Year	Sample Size	Gender M/F (Age)	Participants	Format, Dose	Duration of Study	Main Findings
Pollack et al. [272]	2017	30	19/11 (67 ± 7)	Older glucose-intolerant patients	Treated with 2–3 g Resveratrol/day	6 weeks	↑ Mitochondrial number ↑ Oxidative phosphorylation ↑ ENDOG ↑ P GC1α
Samara et al. [267]	2018	60	18–75 years	Patients with NASH	Oral consumption of n − 3 PUFA capsules, 0.945 g/day	6 months	↑ ALA, EPA, glycerophospholipids ↓ Arachidonic acid ↑ FABPL ↑ PRDX6 ↓ PGRMC1 ↑ PEBP1, ApoJ ↑ FASTKD2
de Ligt et al. [271]	2018	13	13/0 (59.2 to 67.6)	Patients with overweight/T2DM	Consumption of 150 mg Resveratrol/day	6 months	↑ State 3 respiration ↓ Complex IV ↑ Mitochondrial function

Abbreviations: **ALA**: alpha-linolenic acid, **EPA**: eicosapenteanoic acid, **FABPL**: fatty acid binding protein—liver type, **PRDX6**: peroxiredoxin 6, **PEBP1**: phosphatidylethanolamine-binding protein 1, **ApoJ**: apolipoprotein J, **FASTKD2**: FAST kinase domain-containing protein 2, **PGRMC1**: progesterone receptor membrane component 1 protein, **DHA**: doxosahexaenoic acid, **ECM**: extracellular matrix, **UCP3**: 1 uncoupling protein 3, **UQCRC1**: ubiquinol cytochrome c reductase, **EGCG**: epigallocatechin-3-gallate, **RES**: resveratrol, **ENDOG**: endonuclease G, ↑: increased, ↓: decreased.

6. Summary

Mitochondrial dysfunction can occur along with many diseases, and the dysfunction is associated with changes in gene expression reflecting on both cell morphology and function. Key events include disrupted mitochondrial ATP production, impaired metabolism, and regulation of apoptosis. Altered metabolic homeostasis will also influence the physiological mitochondrial dynamics. [275]. In the last decade, so-called *"mitochondrial medicine"* and *"mitochondrial nutrients"* have attracted the attention of researchers, with the idea that improving mitochondrial structure and function is a plausible strategy for MetS prevention and treatment. It is important to note that the MD is rich in polyphenols and other naturally derived compounds that have substantial antioxidant properties, the capacity to scavenge free radicals, and the ability to modulate endogenous antioxidant defense mechanisms. These effects involve mitochondrial antioxidant enzymes. Due to their antioxidant properties, polyphenols can reduce the inflammation and mitochondrial dysfunction characteristic of MetS. Hereby, we discussed the effects of the main nutrients and polyphenols in the MD on mitochondrial dysfunction in MetS. Preclinical studies (in vitro cellular and in vivo animal studies) show that the nutrients and polyphenols present in the MD, such as chlorogenic acid, resveratrol, hydroxytyrosol, and apigenin, exert a vast range of beneficial effects on mitochondrial dysfunction. Figure 3 summarizes the possible mechanisms, including the effects on key regulators of mitochondrial function and biogenesis such as SIRT-1, AMPK, and PGC-1α. In addition, the antioxidant properties of the polyphenols present in the MD reduced mtROS production and ameliorated mitochondrial damage and apoptosis in different experimental studies. Several studies reported the health-promoting effects of the MD due to its high fiber content.

Short-chain fatty acids are the end products of the fermentation of insoluble fiber by the gut microbiota. Evidence suggests SCFAs can modulate several metabolic disorders such as obesity, insulin resistance, and T2DM [276]. Butyrate, an SCFA present in the MD, promotes fatty acid oxidation and improves mitochondrial function. The vegetables, nuts, and fish characteristics of the MD contain significant amounts of PUFA. The correlation between PUFA intake (especially ω-3) and decreased cardiometabolic risk has been well-documented [277]. Additionally, dietary n-3 PUFAs have shown substantial positive effects on mitochondrial function and structure [278]. These effects seem to be mediated by a reduction in the expression of p-mTOR, accompanied by the upregulation of the mitochondrial electron-transport chain and tricarboxylic acid cycle. Several studies in humans have demonstrated the beneficial effect of the bioactive compounds present in the MD on MetS, suggesting advanced health-promoting effects through the targeting of mitochondria. This could be used to promote additional pharmacological and nutraceutical effects, especially on the gastrointestinal system and muscle strength.

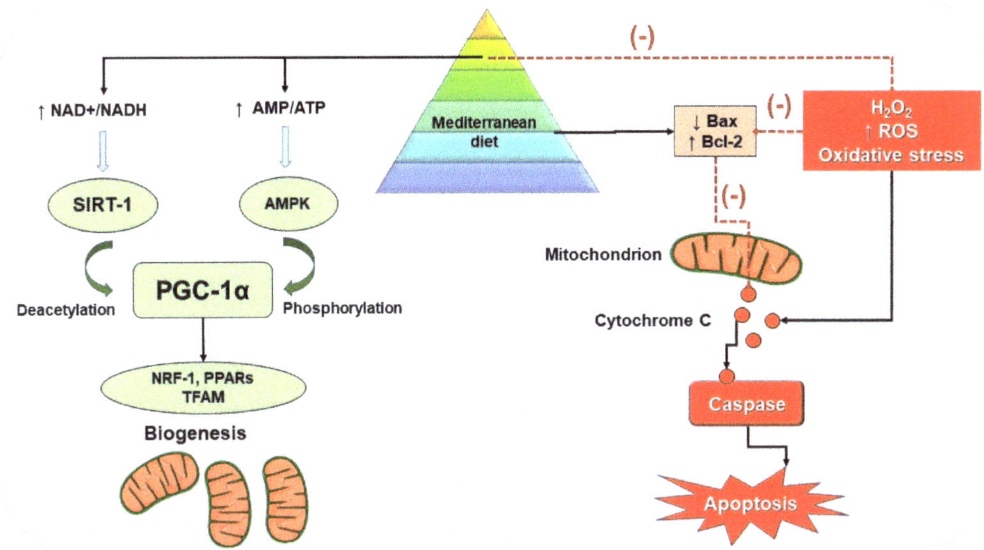

Figure 3. Potential molecular mechanisms of MD on mitochondrial dysfunction in MetS. The dotted red line represents inhibitory pathways. Abbreviations: AMPK: AMP-activated protein kinase; BAX: bcl2-like protein 4; Bcl−2: B-cell lymphoma 2; NRF−1: nuclear respiratory factor 1; PPARs: peroxisome proliferator–activated receptors; PGC−1α: peroxisome proliferator–activated receptor-gamma coactivator-1α; SIRT−1: sirtuin 1; TFAM: transcription factor A, mitochondrial.

7. Conclusions

Obesity is closely linked to metabolic disorders that pave the way for organ, tissue, cellular, and sub-cellular dysfunction. Mitochondria are dynamic cell organelles, which are essential for energy metabolism and represent cardinal players in obesity and metabolic disease. Cumulative evidence from pre-clinical studies indicates that the MD is rich in polyphenols, essential oils, and fiber and plays a beneficial role by stimulating mitochondrial biogenesis and exerting an antioxidant effect.

Despite the substantial positive effects reported for the MD and its components in obesity and MetS, the bioactive mechanisms of the MD on mitochondrial dysfunction are not fully understood. Therefore, further animal and human studies are necessary to elucidate the translational aspects of "mitochondrial nutrition" and to fully characterize its role in the prevention and treatment of obesity-related MetS.

Author Contributions: Conceptualization, H.S. and P.P.; writing—original draft preparation, H.S., H.A., J.S.J.B. and I.G.; revision, J.G.-A., G.F., M.K. and P.P. All authors have read and agreed to the published version of the manuscript.

Funding: This paper has been partly supported by funding from the European Union's Horizon 2020 Research and Innovation program under the Marie Skłodowska-Curie Grant Agreement No. 722619 (FOIE GRAS), Grant Agreement No. 734719 (mtFOIE GRAS), Grant Regione Puglia, CUP H99C20000340002 (*Fever Apulia*), and Grant EUROSEEDS Uniba—S56—By-Products Sustainable Recovery 4 Health (BSR-4H): University of Bari Aldo Moro, 2022.

Institutional Review Board Statement: Not applicable.

Informed Consent Statement: Not applicable.

Acknowledgments: The authors are indebted to Rosa De Venuto, Paola De Benedictis, Domenica Di Palo, Ilaria Farella, and Elisa Lanza for their skillful technical assistance and to Danute Razuka-Ebela for their English-language revision.

Conflicts of Interest: The authors declare no conflict of interest.

Abbreviations

ATP	Adenosine triphosphate
ATP III	Adult Treatment Panel III
AMPK	AMP-activated protein kinase
APG	Apigenin
ApoJ	Apolipoprotein J
BMI	Body mass index
Bcl2	B-cell lymphoma 2
Bax	BCl2-associated X
CV	Cardiovascular
CoQ	Coenzyme Q
CoA	Coenzyme A
CPT-1	Carnitine palmitoyltransferase-1
CYP2E1	Mitochondrial Cytochrome P450 2E1
CVD	Cardiovascular disease
CRP	C-reactive protein
CGA	Chlorogenic acid
Cox	Cytochrome C oxidase
Cycs	Cytochrome C
CO_2	Carbon dioxide
CPT1b	Carnitine palmitoyltransferase 1B
DNA	Deoxyribonucleic acid
DHAs	Docosahexaenoic acids
Drp1	Dynamin-related protein 1
DHA	Doxosahexaenoic acid
ETC	Electron transport chain
ERR	Estrogen-related receptors
EPAs	Eicosapentaenoic acids
EPC	Endothelial progenitor cells
EVOO	Extra virgin olive oil
ER stress	Endoplasmic reticulum stress
EPA	Eicosapenteanoic acid
ECM	Extracellular matrix
EGCG	Epigallocatechin-3-gallate
ENDOG	Endonuclease G
FAD	Flavin adenine dinucleotide
FFA	Free fatty cid
FOXO3a	Forkhead box O3 (transcription factors)
FBA	*N*-(1-carbamoyl-2-phenyl-ethyl) butyramide
Fis1	Mitochondrial fission protein1

FA	Ferulic acid
FABPL	Fatty acid binding protein—liver type
FASTKD2	FAST kinase domain-containing protein 2
H2O2	Hydrogen peroxide
HFD	High-fat diet
HUVECs	Human endothelial cells
HT	Hydroxytyrosol
HepG2	Human liver cancer cell line
HDL	High-density lipoprotein
IDF	International Diabetes Federation
IMM	Inner mitochondrial membrane
IL-1β	Interleukin 1 beta
IL-6	Interleukin 6
LYC	Lycopene
LFD	Low-fat diet
MetS	Metabolic Syndrome
MD	Mediterranean diet
MHO	Metabolically healthy obesity
MAO	Metabolically altered obesity
MAFLD	Metabolic dysfunction associated fatty liver disease
mtROS	Mitochondrial reactive oxygen species
mtDNA	Mitochondrial Deoxyribonucleic Acid
Mfn1	Mitofusin-1
Mfn2	Mitofusin-2
MWCNTs	Multi-walled carbon nanotubes
NAFLD	Non-alcoholic fatty liver disease
NAD	Nicotinamide adenine dinucleotide
NRF	Nuclear respiratory factors
NRF-1	Nuclear respiratory factor 1
NASH	Non-alcoholic steatohepatitis
NADPH	Nicotinamide adenine dinucleotide phosphate oxidase
NF-κB	Nuclear factor kappa-light-chain-enhancer of activated B cells
NO2-OA	Nitro-fatty acids
OMM	Outer mitochondrial membrane
OXPHOS	Oxidative phosphorylation
ox-LDL	Oxidized low-density lipoprotein
PGC-1α	Peroxisome proliferator–activated receptor gamma coactivator-1 α
PUFA	Polyunsaturated fatty acid
PPAR-α	Peroxisome proliferator–activated receptor-α
PolG	DNA polymerase subunit gamma
PC-12	Pheochromocytoma
PBMC	Peripheral blood mononuclear cell
PRDX6	Peroxiredoxin 6

PGRMC1	Progesterone receptor membrane component 1 protein
PEBP1	Phosphatidylethanolamine-binding protein 1
ROS	Reactive oxidative atress
RBEE	Rice bran enzymatic extract
SO	Sarcopenic obesity
SIRT1	Sirtuin 1
SH-SY5Y	Human neuroblastoma cells
SDH	Succinate dehydrogenase
SCFAs	Short-chain fatty acids
T2DM	Type 2 diabetes mellitus
TOFI	Thin-outside-fat-inside
t-RNA	Transfer ribonucleic acid
TFAM	Mitochondrial transcription factor A
TFB2M	Mitochondrial transcription factor B2
TNF-α	Tumor necrosis factor
Tfb2m	Transcription factor B2, mitochondria
UCP3	Uncoupling protein 3
UQCRC1	Ubiquinol cytochrome c reductase
VEGF	Vascular endothelial growth factor
WHO	World Health Organization

References

1. Di Ciaula, A.; Krawczyk, M.; Filipiak, K.J.; Geier, A.; Bonfrate, L.; Portincasa, P. Noncommunicable diseases, climate change and iniquities: What COVID-19 has taught us about syndemic. *Eur. J. Clin. Investig.* **2021**, *51*, e13682. [CrossRef]
2. Chaudhuri, R.; Thompson, M.A.; Pabelick, C.; Agrawal, A.; Prakash, Y.S. Obesity, mitochondrial dysfunction, and obstructive lung disease. In *Mechanisms and Manifestations of Obesity in Lung Disease*; Johnston, R.A., Suratt, B.T., Eds.; Academic Press: Cambridge, MA, USA, 2019; pp. 143–167.
3. Mitchell, T.; Darley-Usmar, V. Metabolic syndrome and mitochondrial dysfunction: Insights from preclinical studies with a mitochondrially targeted antioxidant. *Free Radic. Biol. Med.* **2012**, *52*, 838–840. [CrossRef]
4. Isomaa, B.; Almgren, P.; Tuomi, T.; Forsen, B.; Lahti, K.; Nissen, M.; Taskinen, M.R.; Groop, L. Cardiovascular morbidity and mortality associated with the metabolic syndrome. *Diabetes Care* **2001**, *24*, 683–689. [CrossRef]
5. Reaven, G.M. Banting lecture 1988. Role of insulin resistance in human disease. *Diabetes* **1988**, *37*, 1595–1607. [CrossRef]
6. Rogge, M.M. The role of impaired mitochondrial lipid oxidation in obesity. *Biol. Res. Nurs.* **2009**, *10*, 356–373. [CrossRef] [PubMed]
7. Osellame, L.D.; Blacker, T.S.; Duchen, M.R. Cellular and molecular mechanisms of mitochondrial function. *Best Pract. Res. Clin. Endocrinol. Metab.* **2012**, *26*, 711–723. [CrossRef]
8. Kusminski, C.M.; Scherer, P.E. Mitochondrial dysfunction in white adipose tissue. *Trends Endocrinol. Metab. TEM* **2012**, *23*, 435–443. [CrossRef]
9. Bournat, J.C.; Brown, C.W. Mitochondrial dysfunction in obesity. *Curr. Opin. Endocrinol. Diabetes Obes.* **2010**, *17*, 446–452. [CrossRef]
10. Brand, M.D.; Nicholls, D.G. Assessing mitochondrial dysfunction in cells. *Biochem. J.* **2011**, *435*, 297–312. [CrossRef]
11. Liesa, M.; Shirihai, O.S. Mitochondrial dynamics in the regulation of nutrient utilization and energy expenditure. *Cell Metab.* **2013**, *17*, 491–506. [CrossRef]
12. Zorzano, A.; Liesa, M.; Palacin, M. Role of mitochondrial dynamics proteins in the pathophysiology of obesity and type 2 diabetes. *Int. J. Biochem. Cell Biol.* **2009**, *41*, 1846–1854. [CrossRef] [PubMed]
13. Yin, X.; Lanza, I.R.; Swain, J.M.; Sarr, M.G.; Nair, K.S.; Jensen, M.D. Adipocyte mitochondrial function is reduced in human obesity independent of fat cell size. *J. Clin. Endocrinol. Metab.* **2014**, *99*, E209–E216. [CrossRef]
14. Putti, R.; Sica, R.; Migliaccio, V.; Lionetti, L. Diet impact on mitochondrial bioenergetics and dynamics. *Front. Physiol.* **2015**, *6*, 109. [CrossRef]
15. Heinonen, S.; Buzkova, J.; Muniandy, M.; Kaksonen, R.; Ollikainen, M.; Ismail, K.; Hakkarainen, A.; Lundbom, J.; Lundbom, N.; Vuolteenaho, K.; et al. Impaired Mitochondrial Biogenesis in Adipose Tissue in Acquired Obesity. *Diabetes* **2015**, *64*, 313–3145. [CrossRef] [PubMed]

16. Simopoulos, A.P. The importance of the omega-6/omega-3 fatty acid ratio in cardiovascular disease and other chronic diseases. *Exp. Biol. Med.* **2008**, *233*, 674–688. [CrossRef]

17. Marventano, S.; Kolacz, P.; Castellano, S.; Galvano, F.; Buscemi, S.; Mistretta, A.; Grosso, G. A review of recent evidence in human studies of n-3 and n-6 PUFA intake on cardiovascular disease, cancer, and depressive disorders: Does the ratio really matter? *Int. J. Food Sci. Nutr.* **2015**, *66*, 611–622. [CrossRef]

18. Huo, R.; Du, T.; Xu, Y.; Xu, W.; Chen, X.; Sun, K.; Yu, X. Effects of Mediterranean-style diet on glycemic control, weight loss and cardiovascular risk factors among type 2 diabetes individuals: A meta-analysis. *Eur. J. Clin. Nutr.* **2015**, *69*, 1200–1208. [CrossRef]

19. Kastorini, C.M.; Milionis, H.J.; Esposito, K.; Giugliano, D.; Goudevenos, J.A.; Panagiotakos, D.B. The effect of Mediterranean diet on metabolic syndrome and its components: A meta-analysis of 50 studies and 534,906 individuals. *J. Am. Coll. Cardiol.* **2011**, *57*, 1299–1313. [CrossRef] [PubMed]

20. Rees, K.; Hartley, L.; Clarke, A.; Thorogood, M.; Stranges, S. 'Mediterranean' dietary pattern for the primary prevention of cardiovascular disease. *Cochrane Database Syst. Rev.* **2012**, *8*, CD009825. [CrossRef]

21. Rosato, V.; Temple, N.J.; La Vecchia, C.; Castellan, G.; Tavani, A.; Guercio, V. Mediterranean diet and cardiovascular disease: A systematic review and meta-analysis of observational studies. *Eur. J. Nutr.* **2019**, *58*, 173–191. [CrossRef]

22. Estruch, R.; Ros, E.; Salas-Salvado, J.; Covas, M.I.; Corella, D.; Aros, F.; Gomez-Gracia, E.; Ruiz-Gutierrez, V.; Fiol, M.; Lapetra, J.; et al. Primary Prevention of Cardiovascular Disease with a Mediterranean Diet Supplemented with Extra-Virgin Olive Oil or Nuts. *N. Engl. J. Med.* **2018**, *378*, e34. [CrossRef] [PubMed]

23. Dontas, A.S.; Zerefos, N.S.; Panagiotakos, D.B.; Vlachou, C.; Valis, D.A. Mediterranean diet and prevention of coronary heart disease in the elderly. *Clin. Interv. Aging* **2007**, *2*, 109–115. [CrossRef] [PubMed]

24. Laubinger, W.; Dimroth, P. Characterization of the ATP synthase of *Propionigenium modestum* as a primary sodium pump. *Biochemistry* **1988**, *27*, 7531–7537. [CrossRef]

25. Medina-Remon, A.; Casas, R.; Tresserra-Rimbau, A.; Ros, E.; Martinez-Gonzalez, M.A.; Fito, M.; Corella, D.; Salas-Salvado, J.; Lamuela-Raventos, R.M.; Estruch, R.; et al. Polyphenol intake from a Mediterranean diet decreases inflammatory biomarkers related to atherosclerosis: A substudy of the PREDIMED trial. *Br. J. Clin. Pharm.* **2017**, *83*, 114–128. [CrossRef]

26. World Health Organization. Fact Sheet: Obesity and Overweight. Available online: http://apps.who.int/bmi/index.jsp?introPage=intro_3.html (accessed on 15 June 2022).

27. Schwartz, M.W.; Seeley, R.J.; Zeltser, L.M.; Drewnowski, A.; Ravussin, E.; Redman, L.M.; Leibel, R.L. Obesity Pathogenesis: An Endocrine Society Scientific Statement. *Endocr. Rev.* **2017**, *38*, 267–296. [CrossRef]

28. Upadhyay, J.; Farr, O.; Perakakis, N.; Ghaly, W.; Mantzoros, C. Obesity as a Disease. *Med. Clin. North Am.* **2018**, *102*, 13–33. [CrossRef]

29. Nam, S.Y. Obesity-related digestive diseases and their pathophysiology. *Gut Liver* **2017**, *11*, 323. [CrossRef]

30. Catalan, V.; Aviles-Olmos, I.; Rodriguez, A.; Becerril, S.; Fernandez-Formoso, J.A.; Kiortsis, D.; Portincasa, P.; Gomez-Ambrosi, J.; Fruhbeck, G. Time to Consider the "Exposome Hypothesis" in the Development of the Obesity Pandemic. *Nutrients* **2022**, *14*, 1597. [CrossRef]

31. Matsuzawa, Y. Therapy Insight: Adipocytokines in metabolic syndrome and related cardiovascular disease. *Nat. Clin. Pract. Cardiovasc. Med.* **2006**, *3*, 35–42. [CrossRef] [PubMed]

32. Tilg, H.; Moschen, A.R. Adipocytokines: Mediators linking adipose tissue, inflammation and immunity. *Nat. Rev. Immunol.* **2006**, *6*, 772–783. [CrossRef]

33. Despres, J.P.; Lemieux, I.; Bergeron, J.; Pibarot, P.; Mathieu, P.; Larose, E.; Rodes-Cabau, J.; Bertrand, O.F.; Poirier, P. Abdominal obesity and the metabolic syndrome: Contribution to global cardiometabolic risk. *Arterioscler. Thromb. Vasc. Biol.* **2008**, *28*, 1039–1049. [CrossRef]

34. Deschênes, D.; Couture, P.; Dupont, P.; Tchernof, A. Subdivision of the Subcutaneous Adipose Tissue Compartment and Lipid-Lipoprotein Levels in Women. *Obesity* **2003**, *11*, 469–476. [CrossRef]

35. Abate, N.; Garg, A.; Peshock, R.M.; Stray-Gundersen, J.; Grundy, S.M. Relationships of generalized and regional adiposity to insulin sensitivity in men. *J. Clin. Investig.* **1995**, *96*, 88. [CrossRef]

36. Vecchie, A.; Dallegri, F.; Carbone, F.; Bonaventura, A.; Liberale, L.; Portincasa, P.; Fruhbeck, G.; Montecucco, F. Obesity phenotypes and their paradoxical association with cardiovascular diseases. *Eur. J. Intern. Med.* **2018**, *48*, 6–17. [CrossRef]

37. Gonzalez-Muniesa, P.; Martinez-Gonzalez, M.A.; Hu, F.B.; Despres, J.P.; Matsuzawa, Y.; Loos, R.J.F.; Moreno, L.A.; Bray, G.A.; Martinez, J.A. Obesity. *Nat. Rev. Dis. Primers* **2017**, *3*, 17034. [CrossRef]

38. Aung, K.; Lorenzo, C.; Hinojosa, M.A.; Haffner, S.M. Risk of developing diabetes and cardiovascular disease in metabolically unhealthy normal-weight and metabolically healthy obese individuals. *J. Clin. Endocrinol. Metab.* **2014**, *99*, 462–468. [CrossRef]

39. Wei, Y.; Rector, R.S.; Thyfault, J.P.; Ibdah, J.A. Nonalcoholic fatty liver disease and mitochondrial dysfunction. *World J. Gastroenterol. WJG* **2008**, *14*, 193–199. [CrossRef]

40. Sharma, A.M.; Kushner, R.F. A proposed clinical staging system for obesity. *Int. J. Obes.* **2009**, *33*, 289–295. [CrossRef]

41. Blundell, J.E.; Dulloo, A.G.; Salvador, J.; Fruhbeck, G.; BMI, E.S.W.G.o. Beyond BMI—Phenotyping the obesities. *Obes. Facts* **2014**, *7*, 322–328. [CrossRef] [PubMed]

42. Fruhbeck, G.; Busetto, L.; Dicker, D.; Yumuk, V.; Goossens, G.H.; Hebebrand, J.; Halford, J.G.C.; Farpour-Lambert, N.J.; Blaak, E.E.; Woodward, E.; et al. The ABCD of Obesity: An EASO Position Statement on a Diagnostic Term with Clinical and Scientific Implications. *Obes. Facts* **2019**, *12*, 131–136. [CrossRef] [PubMed]

43. Mechanick, J.I.; Farkouh, M.E.; Newman, J.D.; Garvey, W.T. Cardiometabolic-Based Chronic Disease, Addressing Knowledge and Clinical Practice Gaps: JACC State-of-the-Art Review. *J. Am. Coll. Cardiol.* **2020**, *75*, 539–555. [CrossRef]

44. Donini, L.M.; Busetto, L.; Bauer, J.M.; Bischoff, S.; Boirie, Y.; Cederholm, T.; Cruz-Jentoft, A.J.; Dicker, D.; Fruhbeck, G.; Giustina, A.; et al. Critical appraisal of definitions and diagnostic criteria for sarcopenic obesity based on a systematic review. *Clin. Nutr.* **2020**, *39*, 2368–2388. [CrossRef]

45. Landecho, M.F.; Tuero, C.; Valentí, V.; Bilbao, I.; de la Higuera, M.; Frühbeck, G. Relevance of leptin and other adipokines in obesity-associated cardiovascular risk. *Nutrients* **2019**, *11*, 2664. [CrossRef] [PubMed]

46. Cuthbertson, D.J.; Wilding, J.P.H. Metabolically healthy obesity: Time for a change of heart? *Nat. Rev. Endocrinol.* **2021**, *17*, 519–520. [CrossRef] [PubMed]

47. Guglielmetti, S.; Bernardi, S.; Del Bo, C.; Cherubini, A.; Porrini, M.; Gargari, G.; Hidalgo-Liberona, N.; Gonzalez-Dominguez, R.; Peron, G.; Zamora-Ros, R.; et al. Effect of a polyphenol-rich dietary pattern on intestinal permeability and gut and blood microbiomics in older subjects: Study protocol of the MaPLE randomised controlled trial. *BMC Geriatr.* **2020**, *20*, 77. [CrossRef] [PubMed]

48. Primeau, V.; Coderre, L.; Karelis, A.D.; Brochu, M.; Lavoie, M.E.; Messier, V.; Sladek, R.; Rabasa-Lhoret, R. Characterizing the profile of obese patients who are metabolically healthy. *Int. J. Obes.* **2011**, *35*, 971–981. [CrossRef]

49. Chang, Y.; Ryu, S.; Suh, B.S.; Yun, K.E.; Kim, C.W.; Cho, S.I. Impact of BMI on the incidence of metabolic abnormalities in metabolically healthy men. *Int. J. Obes.* **2012**, *36*, 1187–1194. [CrossRef]

50. Fortuno, A.; Rodriguez, A.; Gomez-Ambrosi, J.; Muniz, P.; Salvador, J.; Diez, J.; Fruhbeck, G. Leptin inhibits angiotensin II-induced intracellular calcium increase and vasoconstriction in the rat aorta. *Endocrinology* **2002**, *143*, 3555–3560. [CrossRef]

51. Alvarez-Sola, G.; Uriarte, I.; Latasa, M.U.; Urtasun, R.; Barcena-Varela, M.; Elizalde, M.; Jimenez, M.; Rodriguez-Ortigosa, C.M.; Corrales, F.J.; Fernandez-Barrena, M.G.; et al. Fibroblast Growth Factor 15/19 in Hepatocarcinogenesis. *Dig. Dis.* **2017**, *35*, 158–165. [CrossRef]

52. Gomez-Ambrosi, J.; Gallego-Escuredo, J.M.; Catalan, V.; Rodriguez, A.; Domingo, P.; Moncada, R.; Valenti, V.; Salvador, J.; Giralt, M.; Villarroya, F.; et al. FGF19 and FGF21 serum concentrations in human obesity and type 2 diabetes behave differently after diet- or surgically-induced weight loss. *Clin. Nutr.* **2017**, *36*, 861–868. [CrossRef]

53. Bell, J.A.; Sabia, S.; Singh-Manoux, A.; Hamer, M.; Kivimaki, M. Healthy obesity and risk of accelerated functional decline and disability. *Int. J. Obes.* **2017**, *41*, 866–872. [CrossRef] [PubMed]

54. Rossi, A.P.; Fantin, F.; Caliari, C.; Zoico, E.; Mazzali, G.; Zanardo, M.; Bertassello, P.; Zanandrea, V.; Micciolo, R.; Zamboni, M. Dynapenic abdominal obesity as predictor of mortality and disability worsening in older adults: A 10-year prospective study. *Clin. Nutr.* **2016**, *35*, 199–204. [CrossRef]

55. Donini, L.M.; Busetto, L.; Bischoff, S.; Cederholm, T.; Ballesteros-Pomar, M.D.; Batsis, J.A.; Bauer, J.M.; Boirie, Y.; Cruz-Jentoft, A.J.; Dicker, D.; et al. Definition and diagnostic criteria for sarcopenic obesity: ESPEN and EASO consensus statement. *Clin. Nutr. ESPEN* **2022**, *41*, 990–1000. [CrossRef]

56. Cauley, J.A. An Overview of Sarcopenic Obesity. *J. Clin. Densitom.* **2015**, *18*, 499–505. [CrossRef]

57. Goisser, S.; Kemmler, W.; Porzel, S.; Volkert, D.; Sieber, C.C.; Bollheimer, L.C.; Freiberger, E. Sarcopenic obesity and complex interventions with nutrition and exercise in community-dwelling older persons—A narrative review. *Clin. Interv. Aging* **2015**, *10*, 1267–1282. [CrossRef] [PubMed]

58. Kim, T.N.; Choi, K.M. The implications of sarcopenia and sarcopenic obesity on cardiometabolic disease. *J. Cell. Biochem.* **2015**, *116*, 1171–1178. [CrossRef] [PubMed]

59. Tian, S.; Xu, Y. Association of sarcopenic obesity with the risk of all-cause mortality: A meta-analysis of prospective cohort studies. *Geriatr. Gerontol. Int.* **2016**, *16*, 155–166. [CrossRef] [PubMed]

60. De Meyts, P.; Delzenne, N. The Brain—Gut—Microbiome Network in Metabolic Regulation and Dysregulation. *Front. Endocrinol.* **2021**, *12*, 760558. [CrossRef] [PubMed]

61. Ladabaum, U.; Mannalithara, A.; Myer, P.A.; Singh, G. Obesity, Abdominal Obesity, Physical Activity, and Caloric Intake in US Adults: 1988 to 2010. *Am. J. Med.* **2014**, *127*, 717–727.e712. [CrossRef]

62. Collaborators, G.B.D.O.; Afshin, A.; Forouzanfar, M.H.; Reitsma, M.B.; Sur, P.; Estep, K.; Lee, A.; Marczak, L.; Mokdad, A.H.; Moradi-Lakeh, M.; et al. Health Effects of Overweight and Obesity in 195 Countries over 25 Years. *N. Engl. J. Med.* **2017**, *377*, 13–27. [CrossRef] [PubMed]

63. Hales, C.M.; Carroll, M.D.; Fryar, C.D.; Ogden, C.L. Prevalence of Obesity and Severe Obesity Among Adults: United States, 2017–2018. *NCHS Data Brief* **2020**, *360*, 1–8.

64. Centers for Disease Control and Prevention. Overweight and obesity: Adult obesity facts. 2021. Available online: https://www.cdc.gov/obesity/data/adult.html (accessed on 28 August 2021).

65. Alberti, K.G.; Eckel, R.H.; Grundy, S.M.; Zimmet, P.Z.; Cleeman, J.I.; Donato, K.A.; Fruchart, J.C.; James, W.P.; Loria, C.M.; Smith, S.C., Jr.; et al. Harmonizing the metabolic syndrome: A joint interim statement of the International Diabetes Federation Task Force on Epidemiology and Prevention; National Heart, Lung, and Blood Institute; American Heart Association; World Heart Federation; International Atherosclerosis Society; and International Association for the Study of Obesity. *Circulation* **2009**, *120*, 1640–1645. [CrossRef] [PubMed]

66. Eckel, R.H.; Alberti, K.G.; Grundy, S.M.; Zimmet, P.Z. The metabolic syndrome. *Lancet* **2010**, *375*, 181–183. [CrossRef]

67. Poirier, P.; Giles, T.D.; Bray, G.A.; Hong, Y.; Stern, J.S.; Pi-Sunyer, F.X.; Eckel, R.H. Obesity and cardiovascular disease: Pathophysiology, evaluation, and effect of weight loss: An update of the 1997 American Heart Association Scientific Statement on Obesity and Heart Disease from the Obesity Committee of the Council on Nutrition, Physical Activity, and Metabolism. *Circulation* **2006**, *113*, 898–918. [CrossRef] [PubMed]

68. Grattagliano, I.; Di Ciaula, A.; Baj, J.; Molina-Molina, E.; Shanmugam, H.; Garruti, G.; Wang, D.Q.; Portincasa, P. Protocols for Mitochondria as the Target of Pharmacological Therapy in the Context of Nonalcoholic Fatty Liver Disease (NAFLD). *Methods Mol. Biol.* **2021**, *2310*, 201–246. [CrossRef] [PubMed]

69. Di Ciaula, A.; Calamita, G.; Shanmugam, H.; Khalil, M.; Bonfrate, L.; Wang, D.Q.; Baffy, G.; Portincasa, P. Mitochondria Matter: Systemic Aspects of Nonalcoholic Fatty Liver Disease (NAFLD) and Diagnostic Assessment of Liver Function by Stable Isotope Dynamic Breath Tests. *Int. J. Mol. Sci.* **2021**, *22*, 7702. [CrossRef]

70. Grattagliano, I.; Montezinho, L.P.; Oliveira, P.J.; Fruhbeck, G.; Gomez-Ambrosi, J.; Montecucco, F.; Carbone, F.; Wieckowski, M.R.; Wang, D.Q.; Portincasa, P. Targeting mitochondria to oppose the progression of nonalcoholic fatty liver disease. *Biochem. Pharmacol.* **2019**, *160*, 34–45. [CrossRef]

71. Giorgi, C.; Marchi, S.; Simoes, I.C.M.; Ren, Z.; Morciano, G.; Perrone, M.; Patalas-Krawczyk, P.; Borchard, S.; Jedrak, P.; Pierzynowska, K.; et al. Mitochondria and Reactive Oxygen Species in Aging and Age-Related Diseases. *Int. Rev. Cell Mol. Biol.* **2018**, *340*, 209–344. [CrossRef]

72. Lu, Y.; Hajifathalian, K.; Ezzati, M.; Woodward, M.; Rimm, E.B.; Danaei, G. Metabolic mediators of the effects of body-mass index, overweight, and obesity on coronary heart disease and stroke: A pooled analysis of 97 prospective cohorts with 1.8 million participants. *Lancet* **2014**, *383*, 970–983. [CrossRef]

73. Faienza, M.F.; Chiarito, M.; Molina-Molina, E.; Shanmugam, H.; Lammert, F.; Krawczyk, M.; D'Amato, G.; Portincasa, P. Childhood obesity, cardiovascular and liver health: A growing epidemic with age. *World J. Pediatrics WJP* **2020**, *16*, 438–445. [CrossRef]

74. Faienza, M.F.; Wang, D.Q.H.; Frühbeck, G.; Garruti, G.; Portincasa, P. The dangerous link between childhood and adulthood predictors of obesity and metabolic syndrome. *Intern. Emerg. Med.* **2016**, *11*, 175–182. [CrossRef]

75. Akil, L.; Ahmad, H.A. Relationships between obesity and cardiovascular diseases in four southern states and Colorado. *J. Health Care Poor Underserved* **2011**, *22*, 61. [CrossRef]

76. Bhaskaran, K.; Douglas, I.; Forbes, H.; dos-Santos-Silva, I.; Leon, D.A.; Smeeth, L. Body-mass index and risk of 22 specific cancers: A population-based cohort study of 5.24 million UK adults. *Lancet* **2014**, *384*, 755–765. [CrossRef]

77. American Heart, A.; National Heart, L.; Blood, I.; Grundy, S.M.; Cleeman, J.I.; Daniels, S.R.; Donato, K.A.; Eckel, R.H.; Franklin, B.A.; Gordon, D.J.; et al. Diagnosis and management of the metabolic syndrome. An American Heart Association/National Heart, Lung, and Blood Institute Scientific Statement. Executive summary. *Cardiol. Rev.* **2005**, *13*, 322–327.

78. International Diabetes Federation. The IDF Consensus Worldwide Definition of the Metabolic Syndrome. Available online: https://www.idf.org/e-library/consensus-statements/60-idfconsensus-worldwide-definitionof-the-metabolic-syndrome.html (accessed on 23 June 2022).

79. Ford, E.S. Prevalence of the metabolic syndrome defined by the International Diabetes Federation among adults in the U.S. *Diabetes Care* **2005**, *28*, 2745–2749. [CrossRef]

80. Beltran-Sanchez, H.; Harhay, M.O.; Harhay, M.M.; McElligott, S. Prevalence and trends of metabolic syndrome in the adult U.S. population, 1999–2010. *J. Am. Coll. Cardiol.* **2013**, *62*, 697–703. [CrossRef]

81. Ravikiran, M.; Bhansali, A.; Ravikumar, P.; Bhansali, S.; Dutta, P.; Thakur, J.S.; Sachdeva, N.; Bhadada, S.; Walia, R. Prevalence and risk factors of metabolic syndrome among Asian Indians: A community survey. *Diabetes Res. Clin. Pract.* **2010**, *89*, 181–188. [CrossRef]

82. Zuo, H.; Shi, Z.; Hu, X.; Wu, M.; Guo, Z.; Hussain, A. Prevalence of metabolic syndrome and factors associated with its components in Chinese adults. *Metab. Clin. Exp.* **2009**, *58*, 1102–1108. [CrossRef]

83. Ford, E.S.; Giles, W.H.; Dietz, W.H. Prevalence of the metabolic syndrome among US adults: Findings from the third National Health and Nutrition Examination Survey. *JAMA* **2002**, *287*, 356–359. [CrossRef] [PubMed]

84. Wang, H.H.; Lee, D.K.; Liu, M.; Portincasa, P.; Wang, D.Q. Novel Insights into the Pathogenesis and Management of the Metabolic Syndrome. *Pediatric Gastroenterol. Hepatol. Nutr.* **2020**, *23*, 189–230. [CrossRef]

85. Alberti, K.G.M.M.; Zimmet, P.; Shaw, J.; IDF Epidemiology Task Force Consensus Group. The metabolic syndrome—A new worldwide definition. *Lancet* **2005**, *366*, 1059–1062. [CrossRef]

86. Gastaldi, G.; Giacobino, J.P.; Ruiz, J. Metabolic syndrome, a mitochondrial disease? *Rev. Med. Suisse* **2008**, *4*, 1387–1388. [PubMed]

87. Beuther, D.A. Recent insight into obesity and asthma. *Curr. Opin. Pulm. Med.* **2010**, *16*, 64–70. [CrossRef] [PubMed]

88. Strazzullo, P.; Barbato, A.; Siani, A.; Cappuccio, F.P.; Versiero, M.; Schiattarella, P.; Russo, O.; Avallone, S.; della Valle, E.; Farinaro, E. Diagnostic criteria for metabolic syndrome: A comparative analysis in an unselected sample of adult male population. *Metab. Clin. Exp.* **2008**, *57*, 355–361. [CrossRef] [PubMed]

89. Nisoli, E.; Clementi, E.; Carruba, M.O.; Moncada, S. Defective mitochondrial biogenesis: A hallmark of the high cardiovascular risk in the metabolic syndrome? *Circ. Res.* **2007**, *100*, 795–806. [CrossRef]

90. Huang, P.L. A comprehensive definition for metabolic syndrome. *Dis. Model Mech.* **2009**, *2*, 231–237. [CrossRef]

91. Krawczyk, M.; Wang, D.Q.; Portincasa, P.; Lammert, F. Dissecting the genetic heterogeneity of gallbladder stone formation. *Semin. Liver Dis.* **2011**, *31*, 157–172. [CrossRef] [PubMed]

92. Grundy, S.M. Cholesterol gallstones: A fellow traveler with metabolic syndrome? *Am. J. Clin. Nutr.* **2004**, *80*, 1–2. [CrossRef]
93. Portincasa, P.; Moschetta, A.; Palasciano, G. *Cholesterol gallstone* disease. *Lancet* **2006**, *368*, 230–239. [CrossRef]
94. Wang, D.Q.H.; Portincasa, P.; Wang, H.H. Bile Formation and Pathophysiology of Gallstones. In *Encyclopedia of Gastroenterology*, 2nd ed.; Kujpers, E.J., Ed.; Elsevier: Amsterdam, The Netherlands; Academic Press: Oxford, UK, 2020; pp. 287–306.
95. Portincasa, P.; Molina-Molina, E.; Garruti, G.; Wang, D.Q. Critical Care Aspects of Gallstone Disease. *J. Crit. Care Med.* **2019**, *5*, 6–18. [CrossRef]
96. Portincasa, P.; van Erpecum, K.J.; Di Ciaula, A.; Wang, D.Q. The physical presence of gallstone modulates ex vivo cholesterol crystallization pathways of human bile. *Gastroenterol. Rep.* **2019**, *7*, 32–41. [CrossRef] [PubMed]
97. Di Ciaula, A.; Wang, D.Q.; Portincasa, P. Cholesterol cholelithiasis: Part of a systemic metabolic disease, prone to primary prevention. *Expert Rev. Gastroenterol. Hepatol.* **2019**, *13*, 157–171. [CrossRef]
98. Mendez-Sanchez, N.; Bugianesi, E.; Gish, R.G.; Lammert, F.; Tilg, H.; Nguyen, M.H.; Sarin, S.K.; Fabrellas, N.; Zelber-Sagi, S.; Fan, J.G.; et al. Global multi-stakeholder endorsement of the MAFLD definition. *Lancet. Gastroenterol. Hepatol.* **2022**, *7*, 388–390. [CrossRef]
99. Powell, E.E.; Wong, V.W.; Rinella, M. Non-alcoholic fatty liver disease. *Lancet* **2021**, *397*, 2212–2224. [CrossRef]
100. Portincasa, P.; Bonfrate, L.; Khalil, M.; Angelis, M.; Calabrese, F.M.; D'Amato, M.; Wang, D.Q.; Di Ciaula, A. Intestinal Barrier and Permeability in Health, Obesity and NAFLD. *Biomedicines* **2021**, *10*, 83. [CrossRef] [PubMed]
101. Di Ciaula, A.; Bonfrate, L.; Portincasa, P. The role of microbiota in nonalcoholic fatty liver disease. *Eur. J. Clin. Investig.* **2022**, *52*, e13768. [CrossRef] [PubMed]
102. Di Ciaula, A.; Passarella, S.; Shanmugam, H.; Noviello, M.; Bonfrate, L.; Wang, D.Q.-H.; Portincasa, P. Nonalcoholic Fatty Liver Disease (NAFLD). Mitochondria as Players and Targets of Therapies? *Int. J. Mol. Sci.* **2021**, *22*, 5375. [CrossRef]
103. Di Ciaula, A.; Baj, J.; Garruti, G.; Celano, G.; De Angelis, M.; Wang, H.H.; Di Palo, D.M.; Bonfrate, L.; Wang, D.Q.; Portincasa, P. Liver Steatosis, Gut-Liver Axis, Microbiome and Environmental Factors. A Never-Ending Bidirectional Cross-Talk. *J. Clin. Med.* **2020**, *9*, 2648. [CrossRef]
104. Molina-Molina, E.; Shanmugam, H.; Di Palo, D.; Grattagliano, I.; Portincasa, P. Exploring Liver Mitochondrial Function by (13)C-Stable Isotope Breath Tests: Implications in Clinical Biochemistry. *Methods Mol. Biol.* **2021**, *2310*, 179–199. [CrossRef] [PubMed]
105. Baldini, F.; Fabbri, R.; Eberhagen, C.; Voci, A.; Portincasa, P.; Zischka, H.; Vergani, L. Adipocyte hypertrophy parallels alterations of mitochondrial status in a cell model for adipose tissue dysfunction in obesity. *Life Sci.* **2021**, *265*, 118812. [CrossRef]
106. Bugger, H.; Abel, E.D. Molecular mechanisms for myocardial mitochondrial dysfunction in the metabolic syndrome. *Clin. Sci.* **2008**, *114*, 195–210. [CrossRef] [PubMed]
107. Spinelli, J.B.; Haigis, M.C. The multifaceted contributions of mitochondria to cellular metabolism. *Nat. Cell Biol.* **2018**, *20*, 745–754. [CrossRef] [PubMed]
108. Passarella, S.; Schurr, A.; Portincasa, P. Mitochondrial Transport in Glycolysis and Gluconeogenesis: Achievements and Perspectives. *Int. J. Mol. Sci.* **2021**, *22*, 2620. [CrossRef] [PubMed]
109. Walsh, C.T.; Tu, B.P.; Tang, Y. Eight Kinetically Stable but Thermodynamically Activated Molecules that Power Cell Metabolism. *Chem. Rev.* **2018**, *118*, 1460–1494. [CrossRef] [PubMed]
110. Sun, F.; Zhou, Q.; Pang, X.; Xu, Y.; Rao, Z. Revealing various coupling of electron transfer and proton pumping in mitochondrial respiratory chain. *Curr. Opin. Struct. Biol.* **2013**, *23*, 526–538. [CrossRef]
111. Sazanov, L.A. A giant molecular proton pump: Structure and mechanism of respiratory complex I. *Nat. Rev. Mol. Cell Biol.* **2015**, *16*, 375–388. [CrossRef] [PubMed]
112. Alcazar-Fabra, M.; Navas, P.; Brea-Calvo, G. Coenzyme Q biosynthesis and its role in the respiratory chain structure. *Biochim. Biophys. Acta* **2016**, *1857*, 1073–1078. [CrossRef] [PubMed]
113. Kuhlbrandt, W. Structure and function of mitochondrial membrane protein complexes. *BMC Biol.* **2015**, *13*, 89. [CrossRef] [PubMed]
114. Watt, I.N.; Montgomery, M.G.; Runswick, M.J.; Leslie, A.G.; Walker, J.E. Bioenergetic cost of making an adenosine triphosphate molecule in animal mitochondria. *Proc. Natl. Acad. Sci. USA* **2010**, *107*, 16823–16827. [CrossRef]
115. Schatz, G. Mitochondrial oxidative phosphorylation. *Angew Chem. Int. Ed. Engl.* **1967**, *6*, 1035–1046. [CrossRef]
116. Brookes, P.S. Mitochondrial H(+) leak and ROS generation: An odd couple. *Free Radic. Biol. Med.* **2005**, *38*, 12–23. [CrossRef]
117. Venditti, P.; Di Stefano, L.; Di Meo, S. Mitochondrial metabolism of reactive oxygen species. *Mitochondrion* **2013**, *13*, 71–82. [CrossRef]
118. Turrens, J.F. Mitochondrial formation of reactive oxygen species. *J. Physiol.* **2003**, *552*, 335–344. [CrossRef] [PubMed]
119. Pastor, N.; Weinstein, H.; Jamison, E.; Brenowitz, M. A detailed interpretation of OH radical footprints in a TBP-DNA complex reveals the role of dynamics in the mechanism of sequence-specific binding. *J. Mol. Biol.* **2000**, *304*, 55–68. [CrossRef] [PubMed]
120. Chen, Y.R.; Zweier, J.L. Cardiac mitochondria and reactive oxygen species generation. *Circ. Res.* **2014**, *114*, 524–537. [CrossRef] [PubMed]
121. Lipinski, B. Hydroxyl radical and its scavengers in health and disease. *Oxid. Med. Cell Longev.* **2011**, *2011*, 809696. [CrossRef]
122. Karnati, S.; Lüers, G.; Pfreimer, S.; Baumgart-Vogt, E. Mammalian SOD2 is exclusively located in mitochondria and not present in peroxisomes. *Histochem. Cell Biol.* **2013**, *140*, 105–117. [CrossRef]
123. Murphy, M.P. How mitochondria produce reactive oxygen species. *Biochem. J.* **2009**, *417*, 1–13. [CrossRef]

124. Wang, B.; Van Veldhoven, P.P.; Brees, C.; Rubio, N.; Nordgren, M.; Apanasets, O.; Kunze, M.; Baes, M.; Agostinis, P.; Fransen, M. Mitochondria are targets for peroxisome-derived oxidative stress in cultured mammalian cells. *Free Radic. Biol. Med.* **2013**, *65*, 882–894. [CrossRef]

125. Stowe, D.F.; Bienengraeber, M.; Camara, A.K.S. Mitochondrial Approaches to Protect Against Cardiac Ischemia and Reperfusion Injury. *Front. Physiol.* **2011**, *2*, 13. [CrossRef]

126. Stowe, D.F.; Camara, A.K.S. Mitochondrial Reactive Oxygen Species Production in Excitable Cells: Modulators of Mitochondrial and Cell Function. *Antioxid. Redox Signal.* **2009**, *11*, 1373–1414. [CrossRef]

127. Camara, A.K.S.; Lesnefsky, E.J.; Stowe, D.F. Potential Therapeutic Benefits of Strategies Directed to Mitochondria. *Antioxid. Redox Signal.* **2010**, *13*, 279–347. [CrossRef]

128. Storz, G.; Imlay, J.A. Oxidative stress. *Curr. Opin. Microbiol.* **1999**, *2*, 188–194. [CrossRef]

129. Bhatti, J.S.; Bhatti, G.K.; Reddy, P.H. Mitochondrial dysfunction and oxidative stress in metabolic disorders—A step towards mitochondria based therapeutic strategies. *Biochim. Biophys. Acta Mol. Basis Dis.* **2017**, *1863*, 1066–1077. [CrossRef] [PubMed]

130. James, A.M.; Collins, Y.; Logan, A.; Murphy, M.P. Mitochondrial oxidative stress and the metabolic syndrome. *Trends Endocrinol. Metab.* **2012**, *23*, 429–434. [CrossRef]

131. Wallace, D.C. Mitochondrial genetic medicine. *Nat. Genet.* **2018**, *50*, 1642–1649. [CrossRef] [PubMed]

132. Austin, S.; St-Pierre, J. PGC1alpha and mitochondrial metabolism—Emerging concepts and relevance in ageing and neurodegenerative disorders. *J. Cell Sci.* **2012**, *125*, 4963–4971. [CrossRef]

133. Di, W.; Lv, J.; Jiang, S.; Lu, C.; Yang, Z.; Ma, Z.; Hu, W.; Yang, Y.; Xu, B. PGC-1: The Energetic Regulator in Cardiac Metabolism. *Curr. Issues Mol. Biol.* **2018**, *28*, 29–46. [CrossRef]

134. Islam, H.; Edgett, B.A.; Gurd, B.J. Coordination of mitochondrial biogenesis by PGC-1alpha in human skeletal muscle: A re-evaluation. *Metab. Clin. Exp.* **2018**, *79*, 42–51. [CrossRef]

135. Jornayvaz, F.R.; Shulman, G.I. Regulation of mitochondrial biogenesis. *Essays Biochem.* **2010**, *47*, 69–84. [CrossRef] [PubMed]

136. Scarpulla, R.C.; Vega, R.B.; Kelly, D.P. Transcriptional integration of mitochondrial biogenesis. *Trends Endocrinol. Metab.* **2012**, *23*, 459–466. [CrossRef]

137. Barshad, G.; Marom, S.; Cohen, T.; Mishmar, D. Mitochondrial DNA Transcription and Its Regulation: An Evolutionary Perspective. *Trends Genet.* **2018**, *34*, 682–692. [CrossRef]

138. Canto, C.; Auwerx, J. PGC-1alpha, SIRT1 and AMPK, an energy sensing network that controls energy expenditure. *Curr. Opin. Lipidol.* **2009**, *20*, 98–105. [CrossRef] [PubMed]

139. Hardie, D.G. Minireview: The AMP-activated protein kinase cascade: The key sensor of cellular energy status. *Endocrinology* **2003**, *144*, 5179–5183. [CrossRef]

140. Kahn, B.B.; Alquier, T.; Carling, D.; Hardie, D.G. AMP-activated protein kinase: Ancient energy gauge provides clues to modern understanding of metabolism. *Cell Metab.* **2005**, *1*, 15–25. [CrossRef]

141. Dominy, J.E., Jr.; Lee, Y.; Gerhart-Hines, Z.; Puigserver, P. Nutrient-dependent regulation of PGC-1alpha's acetylation state and metabolic function through the enzymatic activities of Sirt1/GCN5. *Biochim. Biophys. Acta* **2010**, *1804*, 1676–1683. [CrossRef]

142. Benton, C.R.; Wright, D.C.; Bonen, A. PGC-1alpha-mediated regulation of gene expression and metabolism: Implications for nutrition and exercise prescriptions. *Appl. Physiol. Nutr. Metab.* **2008**, *33*, 843–862. [CrossRef]

143. Bonen, A. PGC-1alpha-induced improvements in skeletal muscle metabolism and insulin sensitivity. *Appl. Physiol. Nutr. Metab.* **2009**, *34*, 307–314. [CrossRef] [PubMed]

144. Kang, C.; Li Ji, L. Role of PGC-1alpha signaling in skeletal muscle health and disease. *Ann. N. Y. Acad. Sci.* **2012**, *1271*, 110–117. [CrossRef]

145. Wu, Z.; Boss, O. Targeting PGC-1 alpha to control energy homeostasis. *Expert Opin. Targets* **2007**, *11*, 1329–1338. [CrossRef]

146. Handschin, C.; Spiegelman, B.M. Peroxisome proliferator-activated receptor gamma coactivator 1 coactivators, energy homeostasis, and metabolism. *Endocr. Rev.* **2006**, *27*, 728–735. [CrossRef] [PubMed]

147. Komen, J.C.; Thorburn, D.R. Turn up the power—Pharmacological activation of mitochondrial biogenesis in mouse models. *Br. J. Pharmacol.* **2014**, *171*, 1818–1836. [CrossRef] [PubMed]

148. Valero, T. Mitochondrial biogenesis: Pharmacological approaches. *Curr. Pharm. Des.* **2014**, *20*, 5507–5509. [CrossRef]

149. Radika, M.K.; Anuradha, C.V. Activation of insulin signaling and energy sensing network by AICAR, an AMPK activator in insulin resistant rat tissues. *J. Basic Clin. Physiol. Pharm.* **2015**, *26*, 563–574. [CrossRef]

150. Yarnoz-Esquiroz, P.; Olazaran, L.; Aguas-Ayesa, M.; Perdomo, C.M.; Garcia-Goni, M.; Silva, C.; Fernandez-Formoso, J.A.; Escalada, J.; Montecucco, F.; Portincasa, P.; et al. 'Obesities': Position statement on a complex disease entity with multifaceted drivers. *Eur. J. Clin. Investig.* **2022**, *52*, e13811. [CrossRef] [PubMed]

151. de Mello, A.H.; Costa, A.B.; Engel, J.D.G.; Rezin, G.T. Mitochondrial dysfunction in obesity. *Life Sci.* **2018**, *192*, 26–32. [CrossRef] [PubMed]

152. Anderson, E.J.; Lustig, M.E.; Boyle, K.E.; Woodlief, T.L.; Kane, D.A.; Lin, C.T.; Price, J.W., 3rd; Kang, L.; Rabinovitch, P.S.; Szeto, H.H.; et al. Mitochondrial H_2O_2 emission and cellular redox state link excess fat intake to insulin resistance in both rodents and humans. *J. Clin. Investig.* **2009**, *119*, 573–581. [CrossRef] [PubMed]

153. Rong, J.X.; Qiu, Y.; Hansen, M.K.; Zhu, L.; Zhang, V.; Xie, M.; Okamoto, Y.; Mattie, M.D.; Higashiyama, H.; Asano, S.; et al. Adipose Mitochondrial Biogenesis Is Suppressed in db/db and High-Fat Diet–Fed Mice and Improved by Rosiglitazone. *Diabetes* **2007**, *56*, 1751–1760. [CrossRef]

154. Chen, C.C.; Lee, T.Y.; Kwok, C.F.; Hsu, Y.P.; Shih, K.C.; Lin, Y.J.; Ho, L.T. Cannabinoid receptor type 1 mediates high-fat diet-induced insulin resistance by increasing forkhead box O1 activity in a mouse model of obesity. *Int. J. Mol. Med.* **2016**, *37*, 743–754. [CrossRef]

155. Konopka, A.R.; Asante, A.; Lanza, I.R.; Robinson, M.M.; Johnson, M.L.; Dalla Man, C.; Cobelli, C.; Amols, M.H.; Irving, B.A.; Nair, K.S. Defects in mitochondrial efficiency and H_2O_2 emissions in obese women are restored to a lean phenotype with aerobic exercise training. *Diabetes* **2015**, *64*, 2104–2115. [CrossRef]

156. Semple, R.K.; Crowley, V.C.; Sewter, C.P.; Laudes, M.; Christodoulides, C.; Considine, R.V.; Vidal-Puig, A.; O'Rahilly, S. Expression of the thermogenic nuclear hormone receptor coactivator PGC-1alpha is reduced in the adipose tissue of morbidly obese subjects. *Int. J. Obes. Relat. Metab. Disord. J. Int. Assoc. Study Obes.* **2004**, *28*, 176–179. [CrossRef]

157. Diogo, C.V.; Grattagliano, I.; Oliveira, P.J.; Bonfrate, L.; Portincasa, P. Re-wiring the circuit: Mitochondria as a pharmacological target in liver disease. *Curr. Med. Chem.* **2011**, *18*, 5448–5465. [CrossRef] [PubMed]

158. Grattagliano, I.; Russmann, S.; Diogo, C.; Bonfrate, L.; Oliveira, P.J.; Wang, D.Q.; Portincasa, P. Mitochondria in chronic liver disease. *Curr. Drug Targets* **2011**, *12*, 879–893. [CrossRef] [PubMed]

159. Caldwell, S.H.; Swerdlow, R.H.; Khan, E.M.; Iezzoni, J.C.; Hespenheide, E.E.; Parks, J.K.; Parker, W.D., Jr. Mitochondrial abnormalities in non-alcoholic steatohepatitis. *J. Hepatol.* **1999**, *31*, 430–434. [CrossRef]

160. Szczepaniak, L.S.; Nurenberg, P.; Leonard, D.; Browning, J.D.; Reingold, J.S.; Grundy, S.; Hobbs, H.H.; Dobbins, R.L. Magnetic resonance spectroscopy to measure hepatic triglyceride content: Prevalence of hepatic steatosis in the general population. *Am. J. Physiol. Endocrinol. Metab.* **2005**, *288*, E462–E468. [CrossRef]

161. Simoes, I.C.M.; Karkucinska-Wieckowska, A.; Janikiewicz, J.; Szymanska, S.; Pronicki, M.; Dobrzyn, P.; Dabrowski, M.; Dobrzyn, A.; Oliveira, P.J.; Zischka, H.; et al. Western Diet Causes Obesity-Induced Nonalcoholic Fatty Liver Disease Development by Differentially Compromising the Autophagic Response. *Antioxidants* **2020**, *9*, 995. [CrossRef]

162. Oyewole, A.O.; Birch-Machin, M.A. Mitochondria-targeted antioxidants. *FASEB J.* **2015**, *29*, 4766–4771. [CrossRef]

163. Longo, M.; Meroni, M.; Paolini, E.; Macchi, C.; Dongiovanni, P. Mitochondrial dynamics and nonalcoholic fatty liver disease (NAFLD): New perspectives for a fairy-tale ending? *Metab. Clin. Exp.* **2021**, *117*, 154708. [CrossRef]

164. Ajaz, S.; McPhail, M.J.; Gnudi, L.; Trovato, F.M.; Mujib, S.; Napoli, S.; Carey, I.; Agarwal, K. Mitochondrial dysfunction as a mechanistic biomarker in patients with non-alcoholic fatty liver disease (NAFLD). *Mitochondrion* **2021**, *57*, 119–130. [CrossRef] [PubMed]

165. Shannon, C.E.; Ragavan, M.; Palavicini, J.P.; Fourcaudot, M.; Bakewell, T.M.; Valdez, I.A.; Ayala, I.; Jin, E.S.; Madesh, M.; Han, X.; et al. Insulin resistance is mechanistically linked to hepatic mitochondrial remodeling in non-alcoholic fatty liver disease. *Mol. Metab.* **2021**, *45*, 101154. [CrossRef]

166. Li, Y.; Wu, J.; Yang, M.; Wei, L.; Wu, H.; Wang, Q.; Shi, H. Physiological evidence of mitochondrial permeability transition pore opening caused by lipid deposition leading to hepatic steatosis in db/db mice. *Free Radic. Biol. Med.* **2021**, *162*, 523–532. [CrossRef] [PubMed]

167. Garcia-Martinez, I.; Santoro, N.; Chen, Y.; Hoque, R.; Ouyang, X.; Caprio, S.; Shlomchik, M.J.; Coffman, R.L.; Candia, A.; Mehal, W.Z. Hepatocyte mitochondrial DNA drives nonalcoholic steatohepatitis by activation of TLR9. *J. Clin. Investig.* **2016**, *126*, 859–864. [CrossRef]

168. Pan, J.; Ou, Z.; Cai, C.; Li, P.; Gong, J.; Ruan, X.Z.; He, K. Fatty acid activates NLRP3 inflammasomes in mouse Kupffer cells through mitochondrial DNA release. *Cell Immunol.* **2018**, *332*, 111–120. [CrossRef]

169. Pirola, C.J.; Garaycoechea, M.; Flichman, D.; Castano, G.O.; Sookoian, S. Liver mitochondrial DNA damage and genetic variability of Cytochrome b—A key component of the respirasome—Drive the severity of fatty liver disease. *J. Intern. Med.* **2021**, *289*, 84–96. [CrossRef] [PubMed]

170. Malik, A.N.; Czajka, A. Is mitochondrial DNA content a potential biomarker of mitochondrial dysfunction? *Mitochondrion* **2013**, *13*, 481–492. [CrossRef] [PubMed]

171. Zhao, R.Z.; Jiang, S.; Zhang, L.; Yu, Z.B. Mitochondrial electron transport chain, ROS generation and uncoupling. *Int. J. Mol. Med.* **2019**, *44*, 3–15. [CrossRef] [PubMed]

172. Aubert, J.; Begriche, K.; Knockaert, L.; Robin, M.A.; Fromenty, B. Increased expression of cytochrome P450 2E1 in nonalcoholic fatty liver disease: Mechanisms and pathophysiological role. *Clin. Res. Hepatol. Gastroenterol.* **2011**, *35*, 630–637. [CrossRef]

173. Weltman, M.D.; Farrell, G.C.; Liddle, C. Increased hepatocyte CYP2E1 expression in a rat nutritional model of hepatic steatosis with inflammation. *Gastroenterology* **1996**, *111*, 1645–1653. [CrossRef]

174. Chalasani, N.; Gorski, J.C.; Asghar, M.S.; Asghar, A.; Foresman, B.; Hall, S.D.; Crabb, D.W. Hepatic cytochrome P450 2E1 activity in nondiabetic patients with nonalcoholic steatohepatitis. *Hepatology* **2003**, *37*, 544–550. [CrossRef]

175. Piccinin, E.; Villani, G.; Moschetta, A. Metabolic aspects in NAFLD, NASH and hepatocellular carcinoma: The role of PGC1 coactivators. *Nat. Rev. Gastroenterol. Hepatol.* **2019**, *16*, 160–174. [CrossRef]

176. Kelley, D.E.; He, J.; Menshikova, E.V.; Ritov, V.B. Dysfunction of mitochondria in human skeletal muscle in type 2 diabetes. *Diabetes* **2002**, *51*, 2944–2950. [CrossRef] [PubMed]

177. Jezek, P.; Hlavata, L. Mitochondria in homeostasis of reactive oxygen species in cell, tissues, and organism. *Int. J. Biochem. Cell Biol.* **2005**, *37*, 2478–2503. [CrossRef] [PubMed]

178. Antonetti, D.A.; Reynet, C.; Kahn, C.R. Increased expression of mitochondrial-encoded genes in skeletal muscle of humans with diabetes mellitus. *J. Clin. Invest.* **1995**, *95*, 1383–1388. [CrossRef]

179. Prasun, P. Mitochondrial dysfunction in metabolic syndrome. *Biochim. Biophys. Acta Mol. Basis Dis.* **2020**, *1866*, 165838. [CrossRef]
180. Lahera, V.; de Las Heras, N.; López-Farré, A.; Manucha, W.; Ferder, L. Role of Mitochondrial Dysfunction in Hypertension and Obesity. *Curr. Hypertens. Rep.* **2017**, *19*, 11. [CrossRef] [PubMed]
181. Furukawa, S.; Fujita, T.; Shimabukuro, M.; Iwaki, M.; Yamada, Y.; Nakajima, Y.; Nakayama, O.; Makishima, M.; Matsuda, M.; Shimomura, I. Increased oxidative stress in obesity and its impact on metabolic syndrome. *J. Clin. Invest.* **2004**, *114*, 1752–1761. [CrossRef]
182. Dikalov, S. Cross talk between mitochondria and NADPH oxidases. *Free Radic. Biol. Med.* **2011**, *51*, 1289–1301. [CrossRef]
183. Choo, H.J.; Kim, J.H.; Kwon, O.B.; Lee, C.S.; Mun, J.Y.; Han, S.S.; Yoon, Y.S.; Yoon, G.; Choi, K.M.; Ko, Y.G. Mitochondria are impaired in the adipocytes of type 2 diabetic mice. *Diabetologia* **2006**, *49*, 784–791. [CrossRef]
184. Petersen, K.F.; Befroy, D.; Dufour, S.; Dziura, J.; Ariyan, C.; Rothman, D.L.; DiPietro, L.; Cline, G.W.; Shulman, G.I. Mitochondrial dysfunction in the elderly: Possible role in insulin resistance. *Science* **2003**, *300*, 1140–1142. [CrossRef]
185. Lee, H.Y.; Choi, C.S.; Birkenfeld, A.L.; Alves, T.C.; Jornayvaz, F.R.; Jurczak, M.J.; Zhang, D.; Woo, D.K.; Shadel, G.S.; Ladiges, W.; et al. Targeted expression of catalase to mitochondria prevents age-associated reductions in mitochondrial function and insulin resistance. *Cell Metab.* **2010**, *12*, 668–674. [CrossRef]
186. Steckhan, N.; Hohmann, C.D.; Kessler, C.; Dobos, G.; Michalsen, A.; Cramer, H. Effects of different dietary approaches on inflammatory markers in patients with metabolic syndrome: A systematic review and meta-analysis. *Nutrition* **2016**, *32*, 338–348. [CrossRef]
187. Keane, D.; Kelly, S.; Healy, N.P.; McArdle, M.A.; Holohan, K.; Roche, H.M. Diet and metabolic syndrome: An overview. *Curr. Vasc. Pharmacol.* **2013**, *11*, 842–857. [CrossRef] [PubMed]
188. Drake, I.; Sonestedt, E.; Ericson, U.; Wallström, P.; Orho-Melander, M. A Western dietary pattern is prospectively associated with cardio-metabolic traits and incidence of the metabolic syndrome. *Br. J. Nutr.* **2018**, *119*, 1168–1176. [CrossRef] [PubMed]
189. Marrone, G.; Guerriero, C.; Palazzetti, D.; Lido, P.; Marolla, A.; Di Daniele, F.; Noce, A. Vegan Diet Health Benefits in Metabolic Syndrome. *Nutrients* **2021**, *13*, 817. [CrossRef]
190. Martinez, K.B.; Leone, V.; Chang, E.B. Western diets, gut dysbiosis, and metabolic diseases: Are they linked? *Gut Microbes* **2017**, *8*, 130–142. [CrossRef] [PubMed]
191. Mirmiran, P.; Bahadoran, Z.; Vakili, A.Z.; Azizi, F. Western dietary pattern increases risk of cardiovascular disease in Iranian adults: A prospective population-based study. *Appl. Physiol. Nutr. Metab.* **2017**, *42*, 326–332. [CrossRef] [PubMed]
192. Turner-McGrievy, G.; Harris, M. Key elements of plant-based diets associated with reduced risk of metabolic syndrome. *Curr. Diabetes Rep.* **2014**, *14*, 524. [CrossRef] [PubMed]
193. Shah, B.; Newman, J.D.; Woolf, K.; Ganguzza, L.; Guo, Y.; Allen, N.; Zhong, J.; Fisher, E.A.; Slater, J. Anti-Inflammatory Effects of a Vegan Diet Versus the American Heart Association-Recommended Diet in Coronary Artery Disease Trial. *J. Am. Heart Assoc.* **2018**, *7*, e011367. [CrossRef]
194. Magriplis, E.; Chourdakis, M. Special Issue "Mediterranean Diet and Metabolic Diseases". *Nutrients* **2021**, *13*, 2680. [CrossRef]
195. D'Innocenzo, S.; Biagi, C.; Lanari, M. Obesity and the Mediterranean Diet: A Review of Evidence of the Role and Sustainability of the Mediterranean Diet. *Nutrients* **2019**, *11*, 1306. [CrossRef]
196. Carbone, F.; Ciaula, A.D.; Pagano, S.; Minetti, S.; Ansaldo, A.M.; Ferrara, D.; Belfiore, A.; Elia, E.; Pugliese, S.; Ostilio Palmieri, V.; et al. Anti-ApoA-1 IgGs predict resistance to waist circumference reduction after Mediterranean diet. *Eur. J. Clin. Invest.* **2021**, *51*, e13410. [CrossRef]
197. Finicelli, M.; Squillaro, T.; Di Cristo, F.; Di Salle, A.; Melone, M.A.B.; Galderisi, U.; Peluso, G. Metabolic syndrome, Mediterranean diet, and polyphenols: Evidence and perspectives. *J. Cell. Physiol.* **2019**, *234*, 5807–5826. [CrossRef] [PubMed]
198. Velázquez-López, L.; Santiago-Díaz, G.; Nava-Hernández, J.; Muñoz-Torres, A.V.; Medina-Bravo, P.; Torres-Tamayo, M. Mediterranean-style diet reduces metabolic syndrome components in obese children and adolescents with obesity. *BMC Pediatrics* **2014**, *14*, 175. [CrossRef]
199. Sayon-Orea, C.; Razquin, C.; Bullo, M.; Corella, D.; Fito, M.; Romaguera, D.; Vioque, J.; Alonso-Gomez, A.M.; Warnberg, J.; Martinez, J.A.; et al. Effect of a Nutritional and Behavioral Intervention on Energy-Reduced Mediterranean Diet Adherence Among Patients with Metabolic Syndrome: Interim Analysis of the PREDIMED-Plus Randomized Clinical Trial. *JAMA* **2019**, *322*, 1486–1499. [CrossRef]
200. Bach-Faig, A.; Berry, E.M.; Lairon, D.; Reguant, J.; Trichopoulou, A.; Dernini, S.; Medina, F.X.; Battino, M.; Belahsen, R.; Miranda, G.; et al. Mediterranean diet pyramid today. Science and cultural updates. *Public Health Nutr.* **2011**, *14*, 2274–2284. [CrossRef] [PubMed]
201. Willett, W.C.; Sacks, F.; Trichopoulou, A.; Drescher, G.; Ferro-Luzzi, A.; Helsing, E.; Trichopoulos, D. Mediterranean diet pyramid: A cultural model for healthy eating. *Am. J. Clin. Nutr.* **1995**, *61*, 1402S–1406S. [CrossRef]
202. Evert, A.B.; Boucher, J.L.; Cypress, M.; Dunbar, S.A.; Franz, M.J.; Mayer-Davis, E.J.; Neumiller, J.J.; Nwankwo, R.; Verdi, C.L.; Urbanski, P.; et al. Nutrition therapy recommendations for the management of adults with diabetes. *Diabetes Care* **2014**, *37* (Suppl. S1), S120–S143. [CrossRef]
203. Portincasa, P.; Wang, D.Q.H. Il valore di una dieta "prudente" nell'individuo adulto. Considerazioni per vivere meglio e forse piu' a lungo. In *Atti e Relazioni Accademia Pugliese delle Scienze*; Adda Editrice: Bari, Italy, 2022; Volume LVII, pp. 85–108.
204. Fung, T.T.; Rexrode, K.M.; Mantzoros, C.S.; Manson, J.E.; Willett, W.C.; Hu, F.B. Mediterranean diet and incidence of and mortality from coronary heart disease and stroke in women. *Circulation* **2009**, *119*, 1093–1100. [CrossRef]

205. Gao, X.; Chen, H.; Fung, T.T.; Logroscino, G.; Schwarzschild, M.A.; Hu, F.B.; Ascherio, A. Prospective study of dietary pattern and risk of Parkinson disease. *Am. J. Clin. Nutr.* **2007**, *86*, 1486–1494. [CrossRef]
206. Scarmeas, N.; Luchsinger, J.A.; Schupf, N.; Brickman, A.M.; Cosentino, S.; Tang, M.X.; Stern, Y. Physical activity, diet, and risk of Alzheimer disease. *JAMA* **2009**, *302*, 627–637. [CrossRef]
207. Buckland, G.; Agudo, A.; Lujan, L.; Jakszyn, P.; Bueno-de-Mesquita, H.B.; Palli, D.; Boeing, H.; Carneiro, F.; Krogh, V.; Sacerdote, C.; et al. Adherence to a Mediterranean diet and risk of gastric adenocarcinoma within the European Prospective Investigation into Cancer and Nutrition (EPIC) cohort study. *Am. J. Clin. Nutr.* **2010**, *91*, 381–390. [CrossRef] [PubMed]
208. Buckland, G.; Travier, N.; Cottet, V.; Gonzalez, C.A.; Lujan-Barroso, L.; Agudo, A.; Trichopoulou, A.; Lagiou, P.; Trichopoulos, D.; Peeters, P.H.; et al. Adherence to the mediterranean diet and risk of breast cancer in the European prospective investigation into cancer and nutrition cohort study. *Int. J. Cancer* **2013**, *132*, 2918–2927. [CrossRef]
209. La Vecchia, C. Association between Mediterranean dietary patterns and cancer risk. *Nutr. Rev.* **2009**, *67* (Suppl. S1), S126–S129. [CrossRef]
210. Panagiotakos, D.B.; Polystipioti, A.; Papairakleous, N.; Polychronopoulos, E. Long-term adoption of a Mediterranean diet is associated with a better health status in elderly people; a cross-sectional survey in Cyprus. *Asia Pac. J. Clin. Nutr.* **2007**, *16*, 331–337.
211. Nunez-Cordoba, J.M.; Valencia-Serrano, F.; Toledo, E.; Alonso, A.; Martinez-Gonzalez, M.A. The Mediterranean diet and incidence of hypertension: The Seguimiento Universidad de Navarra (SUN) Study. *Am. J. Epidemiol.* **2009**, *169*, 339–346. [CrossRef]
212. Martinez-Gonzalez, M.A.; de la Fuente-Arrillaga, C.; Nunez-Cordoba, J.M.; Basterra-Gortari, F.J.; Beunza, J.J.; Vazquez, Z.; Benito, S.; Tortosa, A.; Bes-Rastrollo, M. Adherence to Mediterranean diet and risk of developing diabetes: Prospective cohort study. *BMJ* **2008**, *336*, 1348–1351. [CrossRef]
213. Schroder, H.; Marrugat, J.; Vila, J.; Covas, M.I.; Elosua, R. Adherence to the traditional mediterranean diet is inversely associated with body mass index and obesity in a spanish population. *J. Nutr.* **2004**, *134*, 3355–3361. [CrossRef] [PubMed]
214. Tektonidis, T.G.; Akesson, A.; Gigante, B.; Wolk, A.; Larsson, S.C. A Mediterranean diet and risk of myocardial infarction, heart failure and stroke: A population-based cohort study. *Atherosclerosis* **2015**, *243*, 93–98. [CrossRef] [PubMed]
215. Lopez-Garcia, E.; Rodriguez-Artalejo, F.; Li, T.Y.; Fung, T.T.; Li, S.; Willett, W.C.; Rimm, E.B.; Hu, F.B. The Mediterranean-style dietary pattern and mortality among men and women with cardiovascular disease. *Am. J. Clin. Nutr.* **2014**, *99*, 172–180. [CrossRef]
216. Fox, C.S.; Golden, S.H.; Anderson, C.; Bray, G.A.; Burke, L.E.; de Boer, I.H.; Deedwania, P.; Eckel, R.H.; Ershow, A.G.; Fradkin, J.; et al. Update on Prevention of Cardiovascular Disease in Adults with Type 2 Diabetes Mellitus in Light of Recent Evidence: A Scientific Statement from the American Heart Association and the American Diabetes Association. *Diabetes Care* **2015**, *38*, 1777–1803. [CrossRef]
217. Zito, F.P.; Polese, B.; Vozzella, L.; Gala, A.; Genovese, D.; Verlezza, V.; Medugno, F.; Santini, A.; Barrea, L.; Cargiolli, M.; et al. Good adherence to mediterranean diet can prevent gastrointestinal symptoms: A survey from Southern Italy. *World J. Gastrointest. Pharm.* **2016**, *7*, 564–571. [CrossRef]
218. Khalil, M.; Hayek, S.; Khalil, N.; Serale, N.; Vergani, L.; Calasso, M.; De Angelis, M.; Portincasa, P. Role of Sumac (*Rhus coriaria* L.) in the management of metabolic syndrome and related disorders: Focus on NAFLD-atherosclerosis interplay. *J. Funct. Foods* **2021**, *87*, 104811. [CrossRef]
219. Khalil, M.; Rita Caponio, G.; Diab, F.; Shanmugam, H.; Di Ciaula, A.; Khalifeh, H.; Vergani, L.; Calasso, M.; De Angelis, M.; Portincasa, P. Unraveling the beneficial effects of herbal Lebanese mixture "Za'atar". History, studies, and properties of a potential healthy food ingredient. *J. Funct. Foods* **2022**, *90*, 104993. [CrossRef]
220. Khalil, M.; Khalifeh, H.; Baldini, F.; Serale, N.; Parodi, A.; Voci, A.; Vergani, L.; Daher, A. Antitumor Activity of Ethanolic Extract from Thymbra Spicata L. aerial Parts: Effects on Cell Viability and Proliferation, Apoptosis Induction, STAT3, and NF-kB Signaling. *Nutr. Cancer* **2021**, *73*, 1193–1206. [CrossRef]
221. Khalil, M.; Bazzi, A.; Zeineddine, D.; Jomaa, W.; Daher, A.; Awada, R. Repressive effect of *Rhus coriaria* L. fruit extracts on microglial cells-mediated inflammatory and oxidative stress responses. *J. Ethnopharmacol.* **2021**, *269*, 113748. [CrossRef]
222. Farooqui, A.A.; Farooqui, T.; Panza, F.; Frisardi, V. Metabolic syndrome as a risk factor for neurological disorders. *Cell. Mol. Life Sci. CMLS* **2012**, *69*, 741–762. [CrossRef]
223. Fernandez-Sanchez, A.; Madrigal-Santillan, E.; Bautista, M.; Esquivel-Soto, J.; Morales-Gonzalez, A.; Esquivel-Chirino, C.; Durante-Montiel, I.; Sanchez-Rivera, G.; Valadez-Vega, C.; Morales-Gonzalez, J.A. Inflammation, oxidative stress, and obesity. *Int. J. Mol. Sci.* **2011**, *12*, 3117–3132. [CrossRef]
224. Freeman, B.A.; Baker, P.R.; Schopfer, F.J.; Woodcock, S.R.; Napolitano, A.; d'Ischia, M. Nitro-fatty acid formation and signaling. *J. Biol. Chem.* **2008**, *283*, 15515–15519. [CrossRef]
225. Farooqui, A.A. *Phytochemicals, Signal Transduction, and Neurological Disorders*; Springer: New York, NY, USA, 2012.
226. Calder, P.C. Polyunsaturated fatty acids and inflammatory processes: New twists in an old tale. *Biochimie* **2009**, *91*, 791–795. [CrossRef]
227. Lagouge, M.; Argmann, C.; Gerhart-Hines, Z.; Meziane, H.; Lerin, C.; Daussin, F.; Messadeq, N.; Milne, J.; Lambert, P.; Elliott, P.; et al. Resveratrol improves mitochondrial function and protects against metabolic disease by activating SIRT1 and PGC-1alpha. *Cell* **2006**, *127*, 1109–1122. [CrossRef]
228. Hardie, D.G.; Ross, F.A.; Hawley, S.A. AMPK: A nutrient and energy sensor that maintains energy homeostasis. *Nat. Rev. Mol. Cell Biol.* **2012**, *13*, 251–262. [CrossRef]

229. Scarmeas, N.; Stern, Y.; Mayeux, R.; Luchsinger, J.A. Mediterranean diet, Alzheimer disease, and vascular mediation. *Arch. Neurol.* **2006**, *63*, 1709–1717. [CrossRef] [PubMed]

230. Castro-Quezada, I.; Roman-Vinas, B.; Serra-Majem, L. The Mediterranean diet and nutritional adequacy: A review. *Nutrients* **2014**, *6*, 231–248. [CrossRef] [PubMed]

231. Wang, J.; Persuitte, G.; Olendzki, B.C.; Wedick, N.M.; Zhang, Z.; Merriam, P.A.; Fang, H.; Carmody, J.; Olendzki, G.F.; Ma, Y. Dietary magnesium intake improves insulin resistance among non-diabetic individuals with metabolic syndrome participating in a dietary trial. *Nutrients* **2013**, *5*, 3910–3919. [CrossRef]

232. Lu, J.; Gu, Y.; Guo, M.; Chen, P.; Wang, H.; Yu, X. Serum Magnesium Concentration Is Inversely Associated with Albuminuria and Retinopathy among Patients with Diabetes. *J. Diabetes Res.* **2016**, *2016*, 1260141. [CrossRef]

233. Paletas, K.; Athanasiadou, E.; Sarigianni, M.; Paschos, P.; Kalogirou, A.; Hassapidou, M.; Tsapas, A. The protective role of the Mediterranean diet on the prevalence of metabolic syndrome in a population of Greek obese subjects. *J. Am. Coll. Nutr.* **2010**, *29*, 41–45. [CrossRef]

234. Dominguez, L.J.; Di Bella, G.; Veronese, N.; Barbagallo, M. Impact of Mediterranean Diet on Chronic Non-Communicable Diseases and Longevity. *Nutrients* **2021**, *13*, 2028. [CrossRef]

235. Farooqui, A.A. Essential Fatty Acid Metabolism in Metabolic Syndrome and Neurological Disorders. In *Metabolic Syndrome*; Springer: New York, NY, USA, 2013; pp. 67–101.

236. Chrysohoou, C.; Panagiotakos, D.B.; Pitsavos, C.; Das, U.N.; Stefanadis, C. Adherence to the Mediterranean diet attenuates inflammation and coagulation process in healthy adults: The ATTICA Study. *J. Am. Coll. Cardiol.* **2004**, *44*, 152–158. [CrossRef]

237. Richard, C.; Couture, P.; Desroches, S.; Lamarche, B. Effect of the Mediterranean diet with and without weight loss on markers of inflammation in men with metabolic syndrome. *Obesity* **2013**, *21*, 51–57. [CrossRef] [PubMed]

238. Georgousopoulou, E.N.; Kastorini, C.M.; Milionis, H.J.; Ntziou, E.; Kostapanos, M.S.; Nikolaou, V.; Vemmos, K.N.; Goudevenos, J.A.; Panagiotakos, D.B. Association between mediterranean diet and non-fatal cardiovascular events, in the context of anxiety and depression disorders: A case/case-control study. *Hell. J. Cardiol.* **2014**, *55*, 24–31.

239. Couto, E.; Boffetta, P.; Lagiou, P.; Ferrari, P.; Buckland, G.; Overvad, K.; Dahm, C.C.; Tjonneland, A.; Olsen, A.; Clavel-Chapelon, F.; et al. Mediterranean dietary pattern and cancer risk in the EPIC cohort. *Br. J. Cancer* **2011**, *104*, 1493–1499. [CrossRef]

240. Salas-Salvado, J.; Bullo, M.; Estruch, R.; Ros, E.; Covas, M.I.; Ibarrola-Jurado, N.; Corella, D.; Aros, F.; Gomez-Gracia, E.; Ruiz-Gutierrez, V.; et al. Prevention of diabetes with Mediterranean diets: A subgroup analysis of a randomized trial. *Ann. Intern. Med.* **2014**, *160*, 1–10. [CrossRef] [PubMed]

241. Salamone, F.; Li Volti, G.; Titta, L.; Puzzo, L.; Barbagallo, I.; La Delia, F.; Zelber-Sagi, S.; Malaguarnera, M.; Pelicci, P.G.; Giorgio, M.; et al. Moro orange juice prevents fatty liver in mice. *World J. Gastroenterol.* **2012**, *18*, 3862–3868. [CrossRef]

242. Garcia-Marcos, L.; Castro-Rodriguez, J.A.; Weinmayr, G.; Panagiotakos, D.B.; Priftis, K.N.; Nagel, G. Influence of Mediterranean diet on asthma in children: A systematic review and meta-analysis. *Pediatr. Allergy Immunol.* **2013**, *24*, 330–338. [CrossRef]

243. Salas-Salvado, J.; Bullo, M.; Babio, N.; Martinez-Gonzalez, M.A.; Ibarrola-Jurado, N.; Basora, J.; Estruch, R.; Covas, M.I.; Corella, D.; Aros, F.; et al. Reduction in the incidence of type 2 diabetes with the Mediterranean diet: Results of the PREDIMED-Reus nutrition intervention randomized trial. *Diabetes Care* **2011**, *34*, 14–19. [CrossRef] [PubMed]

244. Wong, C.Y.; Yiu, K.H.; Li, S.W.; Lee, S.; Tam, S.; Lau, C.P.; Tse, H.F. Fish-oil supplement has neutral effects on vascular and metabolic function but improves renal function in patients with Type 2 diabetes mellitus. *Diabet Med.* **2010**, *27*, 54–60. [CrossRef]

245. Seidl, S.E.; Santiago, J.A.; Bilyk, H.; Potashkin, J.A. The emerging role of nutrition in Parkinson's disease. *Front Aging Neurosci* **2014**, *6*, 36. [CrossRef] [PubMed]

246. Martinez-Gonzalez, M.A.; Bes-Rastrollo, M.; Serra-Majem, L.; Lairon, D.; Estruch, R.; Trichopoulou, A. Mediterranean food pattern and the primary prevention of chronic disease: Recent developments. *Nutr. Rev.* **2009**, *67* (Suppl. S1), S111–S116. [CrossRef]

247. Sofi, F.; Cesari, F.; Abbate, R.; Gensini, G.F.; Casini, A. Adherence to Mediterranean diet and health status: Meta-analysis. *BMJ* **2008**, *337*, a1344. [CrossRef]

248. Tsai, K.L.; Hung, C.H.; Chan, S.H.; Hsieh, P.L.; Ou, H.C.; Cheng, Y.H.; Chu, P.M. Chlorogenic Acid Protects Against oxLDL-Induced Oxidative Damage and Mitochondrial Dysfunction by Modulating SIRT1 in Endothelial Cells. *Mol. Nutr. Food Res.* **2018**, *62*, e1700928. [CrossRef] [PubMed]

249. Duluc, L.; Jacques, C.; Soleti, R.; Andriantsitohaina, R.; Simard, G. Delphinidin inhibits VEGF induced-mitochondrial biogenesis and Akt activation in endothelial cells. *Int. J. Biochem. Cell Biol.* **2014**, *53*, 9–14. [CrossRef]

250. Wang, J.; Zou, Q.; Suo, Y.; Tan, X.; Yuan, T.; Liu, Z.; Liu, X. Lycopene ameliorates systemic inflammation-induced synaptic dysfunction via improving insulin resistance and mitochondrial dysfunction in the liver-brain axis. *Food Funct.* **2019**, *10*, 2125–2137. [CrossRef]

251. Feng, C.; Luo, T.; Zhang, S.; Liu, K.; Zhang, Y.; Luo, Y.; Ge, P. Lycopene protects human SHSY5Y neuroblastoma cells against hydrogen peroxideinduced death via inhibition of oxidative stress and mitochondriaassociated apoptotic pathways. *Mol. Med. Rep.* **2016**, *13*, 4205–4214. [CrossRef]

252. Liu, J.; Wang, Y.; Hao, Y.; Wang, Z.; Yang, Z.; Wang, Z.; Wang, J. 5-Heptadecylresorcinol attenuates oxidative damage and mitochondria-mediated apoptosis through activation of the SIRT3/FOXO3a signaling pathway in neurocytes. *Food Funct.* **2020**, *11*, 2535–2542. [CrossRef] [PubMed]

253. Gueguen, N.; Desquiret-Dumas, V.; Leman, G.; Chupin, S.; Baron, S.; Nivet-Antoine, V.; Vessieres, E.; Ayer, A.; Henrion, D.; Lenaers, G.; et al. Resveratrol Directly Binds to Mitochondrial Complex I and Increases Oxidative Stress in Brain Mitochondria of Aged Mice. *PLoS ONE* **2015**, *10*, e0144290. [CrossRef]

254. Baur, J.A.; Pearson, K.J.; Price, N.L.; Jamieson, H.A.; Lerin, C.; Kalra, A.; Prabhu, V.V.; Allard, J.S.; Lopez-Lluch, G.; Lewis, K.; et al. Resveratrol improves health and survival of mice on a high-calorie diet. *Nature* **2006**, *444*, 337–342. [CrossRef]

255. Gao, Z.; Yin, J.; Zhang, J.; Ward, R.E.; Martin, R.J.; Lefevre, M.; Cefalu, W.T.; Ye, J. Butyrate improves insulin sensitivity and increases energy expenditure in mice. *Diabetes* **2009**, *58*, 1509–1517. [CrossRef]

256. Mollica, M.P.; Mattace Raso, G.; Cavaliere, G.; Trinchese, G.; De Filippo, C.; Aceto, S.; Prisco, M.; Pirozzi, C.; Di Guida, F.; Lama, A.; et al. Butyrate Regulates Liver Mitochondrial Function, Efficiency, and Dynamics in Insulin-Resistant Obese Mice. *Diabetes* **2017**, *66*, 1405–1418. [CrossRef]

257. Deng, X.; Zhang, S.; Wu, J.; Sun, X.; Shen, Z.; Dong, J.; Huang, J. Promotion of Mitochondrial Biogenesis via Activation of AMPK-PGC1a Signaling Pathway by Ginger (*Zingiber officinale* Roscoe) Extract, and Its Major Active Component 6-Gingerol. *J. Food Sci.* **2019**, *84*, 2101–2111. [CrossRef] [PubMed]

258. Perez-Ternero, C.; Werner, C.M.; Nickel, A.G.; Herrera, M.D.; Motilva, M.J.; Bohm, M.; Alvarez de Sotomayor, M.; Laufs, U. Ferulic acid, a bioactive component of rice bran, improves oxidative stress and mitochondrial biogenesis and dynamics in mice and in human mononuclear cells. *J. Nutr. Biochem.* **2017**, *48*, 51–61. [CrossRef] [PubMed]

259. Liu, R.; Chen, L.; Wang, Y.; Zhang, G.; Cheng, Y.; Feng, Z.; Bai, X.; Liu, J. High ratio of omega-3/omega-6 polyunsaturated fatty acids targets mTORC1 to prevent high-fat diet-induced metabolic syndrome and mitochondrial dysfunction in mice. *J. Nutr. Biochem.* **2020**, *79*, 108330. [CrossRef]

260. Sanchez-Calvo, B.; Cassina, A.; Mastrogiovanni, M.; Santos, M.; Trias, E.; Kelley, E.E.; Rubbo, H.; Trostchansky, A. Olive oil-derived nitro-fatty acids: Protection of mitochondrial function in non-alcoholic fatty liver disease. *J. Nutr. Biochem.* **2021**, *94*, 108646. [CrossRef]

261. Cao, K.; Xu, J.; Zou, X.; Li, Y.; Chen, C.; Zheng, A.; Li, H.; Li, H.; Szeto, I.M.; Shi, Y.; et al. Hydroxytyrosol prevents diet-induced metabolic syndrome and attenuates mitochondrial abnormalities in obese mice. *Free Radic. Biol. Med.* **2014**, *67*, 396–407. [CrossRef]

262. Dong, Y.Z.; Li, L.; Espe, M.; Lu, K.L.; Rahimnejad, S. Hydroxytyrosol Attenuates Hepatic Fat Accumulation via Activating Mitochondrial Biogenesis and Autophagy through the AMPK Pathway. *J. Agric. Food Chem.* **2020**, *68*, 9377–9386. [CrossRef]

263. Keshtzar, E.; Khodayar, M.J.; Javadipour, M.; Ghaffari, M.A.; Bolduc, D.L.; Rezaei, M. Ellagic acid protects against arsenic toxicity in isolated rat mitochondria possibly through the maintaining of complex II. *Hum. Exp. Toxicol.* **2016**, *35*, 1060–1072. [CrossRef]

264. Zamani, F.; Samiei, F.; Mousavi, Z.; Azari, M.R.; Seydi, E.; Pourahmad, J. Apigenin ameliorates oxidative stress and mitochondrial damage induced by multiwall carbon nanotubes in rat kidney mitochondria. *J. Biochem. Mol. Toxicol.* **2021**, *35*, 1–7. [CrossRef]

265. Wang, D.; Yang, Y.; Zou, X.; Zhang, J.; Zheng, Z.; Wang, Z. Antioxidant Apigenin Relieves Age-Related Muscle Atrophy by Inhibiting Oxidative Stress and Hyperactive Mitophagy and Apoptosis in Skeletal Muscle of Mice. *J. Gerontol. A Biol. Sci. Med. Sci.* **2020**, *75*, 2081–2088. [CrossRef]

266. Daussin, F.N.; Cuillerier, A.; Touron, J.; Bensaid, S.; Melo, B.; Al Rewashdy, A.; Vasam, G.; Menzies, K.J.; Harper, M.E.; Heyman, E.; et al. Dietary Cocoa Flavanols Enhance Mitochondrial Function in Skeletal Muscle and Modify Whole-Body Metabolism in Healthy Mice. *Nutrients* **2021**, *13*, 3466. [CrossRef]

267. Okada, L.; Oliveira, C.P.; Stefano, J.T.; Nogueira, M.A.; Silva, I.; Cordeiro, F.B.; Alves, V.A.F.; Torrinhas, R.S.; Carrilho, F.J.; Puri, P.; et al. Omega-3 PUFA modulate lipogenesis, ER stress, and mitochondrial dysfunction markers in NASH—Proteomic and lipidomic insight. *Clin. Nutr.* **2018**, *37*, 1474–1484. [CrossRef]

268. Anderson, E.J.; Thayne, K.A.; Harris, M.; Shaikh, S.R.; Darden, T.M.; Lark, D.S.; Williams, J.M.; Chitwood, W.R.; Kypson, A.P.; Rodriguez, E. Do fish oil omega-3 fatty acids enhance antioxidant capacity and mitochondrial fatty acid oxidation in human atrial myocardium via PPARgamma activation? *Antioxid. Redox Signal* **2014**, *21*, 1156–1163. [CrossRef] [PubMed]

269. Yoshino, J.; Smith, G.I.; Kelly, S.C.; Julliand, S.; Reeds, D.N.; Mittendorfer, B. Effect of dietary n-3 PUFA supplementation on the muscle transcriptome in older adults. *Physiol. Rep.* **2016**, *4*, e12785. [CrossRef] [PubMed]

270. Capo, X.; Martorell, M.; Sureda, A.; Llompart, I.; Tur, J.A.; Pons, A. Diet supplementation with DHA-enriched food in football players during training season enhances the mitochondrial antioxidant capabilities in blood mononuclear cells. *Eur. J. Nutr.* **2015**, *54*, 35–49. [CrossRef] [PubMed]

271. de Ligt, M.; Bruls, Y.M.H.; Hansen, J.; Habets, M.F.; Havekes, B.; Nascimento, E.B.M.; Moonen-Kornips, E.; Schaart, G.; Schrauwen-Hinderling, V.B.; van Marken Lichtenbelt, W.; et al. Resveratrol improves ex vivo mitochondrial function but does not affect insulin sensitivity or brown adipose tissue in first degree relatives of patients with type 2 diabetes. *Mol. Metab.* **2018**, *12*, 39–47. [CrossRef]

272. Pollack, R.M.; Barzilai, N.; Anghel, V.; Kulkarni, A.S.; Golden, A.; O'Broin, P.; Sinclair, D.A.; Bonkowski, M.S.; Coleville, A.J.; Powell, D.; et al. Resveratrol Improves Vascular Function and Mitochondrial Number but Not Glucose Metabolism in Older Adults. *J. Gerontol. A Biol. Sci. Med. Sci.* **2017**, *72*, 1703–1709. [CrossRef] [PubMed]

273. Most, J.; Timmers, S.; Warnke, I.; Jocken, J.W.; van Boekschoten, M.; de Groot, P.; Bendik, I.; Schrauwen, P.; Goossens, G.H.; Blaak, E.E. Combined epigallocatechin-3-gallate and resveratrol supplementation for 12 wk increases mitochondrial capacity and fat oxidation, but not insulin sensitivity, in obese humans: A randomized controlled trial. *Am. J. Clin. Nutr.* **2016**, *104*, 215–227. [CrossRef] [PubMed]

274. Joy, J.M.; Vogel, R.M.; Moon, J.R.; Falcone, P.H.; Mosman, M.M.; Pietrzkowski, Z.; Reyes, T.; Kim, M.P. Ancient peat and apple extracts supplementation may improve strength and power adaptations in resistance trained men. *BMC Complement Altern. Med.* **2016**, *16*, 224. [CrossRef]
275. Srinivasan, S.; Guha, M.; Kashina, A.; Avadhani, N.G. Mitochondrial dysfunction and mitochondrial dynamics-The cancer connection. *Biochim. Biophys. Acta Bioenerg.* **2017**, *1858*, 602–614. [CrossRef] [PubMed]
276. Portincasa, P.; Bonfrate, L.; Vacca, M.; De Angelis, M.; Farella, I.; Lanza, E.; Khalil, M.; Wang, D.Q.; Sperandio, M.; Di Ciaula, A. Gut Microbiota and Short Chain Fatty Acids: Implications in Glucose Homeostasis. *Int. J. Mol. Sci.* **2022**, *23*, 1105. [CrossRef] [PubMed]
277. Nadtochiy, S.M.; Redman, E.K. Mediterranean diet and cardioprotection: The role of nitrite, polyunsaturated fatty acids, and polyphenols. *Nutrition* **2011**, *27*, 733–744. [CrossRef] [PubMed]
278. Stanley, W.C.; Khairallah, R.J.; Dabkowski, E.R. Update on lipids and mitochondrial function: Impact of dietary n-3 polyunsaturated fatty acids. *Curr. Opin. Clin. Nutr. Metab. Care* **2012**, *15*, 122–126. [CrossRef] [PubMed]

Review

The Role of Diet Quality in Mediating the Association between Ultra-Processed Food Intake, Obesity and Health-Related Outcomes: A Review of Prospective Cohort Studies

Samuel J. Dicken [1] and Rachel L. Batterham [1,2,3,*]

[1]   Centre for Obesity Research, Department of Medicine, University College London (UCL), London WC1E 6JF, UK; samuel.dicken.20@ucl.ac.uk
[2]   Bariatric Centre for Weight Management and Metabolic Surgery, University College London Hospital (UCLH), London NW1 2BU, UK
[3]   National Institute for Health Research, Biomedical Research Centre, University College London Hospital (UCLH), London W1T 7DN, UK
*   Correspondence: r.batterham@ucl.ac.uk

Abstract: Prospective cohort studies show that higher intakes of ultra-processed food (UPF) increase the risk of obesity and obesity-related outcomes, including cardiovascular disease, cancer and type 2 diabetes. Whether ultra-processing itself is detrimental, or whether UPFs just have a lower nutritional quality, is debated. Higher UPF intakes are inversely associated with fruit, vegetables, legumes and seafood consumption. Therefore, the association between UPFs and poor health could simply be from excess nutrient intake or from a less healthful dietary pattern. If so, adjustment for dietary quality or pattern should explain or greatly reduce the size of the significant associations between UPFs and health-related outcomes. Here, we provide an overview of the literature and by using a novel approach, review the relative impact of adjusting for diet quality/patterns on the reported associations between UPF intake and health-related outcomes in prospective cohort studies. We find that the majority of the associations between UPFs, obesity and health-related outcomes remain significant and unchanged in magnitude after adjustment for diet quality or pattern. Our findings suggest that the adverse consequences of UPFs are independent of dietary quality or pattern, questioning the utility of reformulation to mitigate against the obesity pandemic and wider negative health outcomes of UPFs.

Keywords: obesity; diet; ultra-processed food; NOVA classification; diet quality; dietary pattern; non-communicable disease

Citation: Dicken, S.J.; Batterham, R.L. The Role of Diet Quality in Mediating the Association between Ultra-Processed Food Intake, Obesity and Health-Related Outcomes: A Review of Prospective Cohort Studies. *Nutrients* **2022**, *14*, 23. https://doi.org/10.3390/nu14010023

Academic Editor: Javier Gómez-Ambrosi

Received: 30 November 2021
Accepted: 16 December 2021
Published: 22 December 2021

Publisher's Note: MDPI stays neutral with regard to jurisdictional claims in published maps and institutional affiliations.

1. Introduction

Obesity (defined as an excess accumulation of fat that may result in adverse health [1]) is a leading cause of poor health, increasing the risk of non-communicable disease (NCD), all-cause mortality and negatively impacting on quality of life [2–4]. Management strategies for obesity prevention and treatment are therefore important.

Diet has long been a cornerstone of weight management, with dietary policies being a core feature of government and health organisation strategies to reduce obesity worldwide. Indeed, poor diets are a leading cause of preventable obesity-related death and NCD, including cancer, cardiovascular disease (CVD) and type 2 diabetes (T2DM), accounting for 11 million deaths annually [5,6]. As such, dietary improvements could prevent one in every five deaths [5]. There is converging evidence that a healthy diet consists predominantly of whole, plant-based foods, including fruit, vegetables, pulses, nuts, whole grains and oily fish [7–14]. Such diets, as exemplified by the Mediterranean diet and Dietary Approaches to Stop Hypertension (DASH), are high in fibre and limit saturated fat, sodium and added sugar intake. In contrast, Western diets high in refined grains, red and processed meat, sweets and sugar-sweetened beverages are rich in saturated fat, sodium and added sugar and associated with an increased risk of disease [13–17].

Despite the importance of specific nutrients and food groups within overall dietary patterns for health, it is becoming increasingly clear that other dimensions of diets are important [18]. In recent decades, a nutrition transition has resulted in a global shift away from consuming minimally processed foods, and towards ultra-processed alternatives [19,20], away from home-prepared dishes, and towards ready-to-eat meals and snacks [21]. This same period has seen a rapid rise in the global prevalence of obesity in children and adults [22]. Besides their nutrient content, healthy dietary patterns such as the Mediterranean diet tend to be minimally processed [15], and unhealthy dietary patterns such as the Western diet tend to be ultra-processed [11,16,23].

Whether ultra-processed diets are detrimental to health simply because they are of a poor nutritional quality, or whether the nature and extent of processing itself has health consequences is an ongoing debate [24]. Several recent systematic reviews, meta-analyses and reviews have discussed the prevalence of UPF consumption and its impact on health-related outcomes. However, no reviews to date have considered how dietary adjustment in prospective cohort studies may alter the significance and magnitude of effect estimates. This review provides a brief overview of the current state of the literature as well as the current key discussion points regarding mechanisms of action, before reviewing in detail the prospective analyses adjusting for dietary quality, which provides important insights into the relative role of nutrient content compared with ultra-processing on obesity risk and adverse health-related outcomes.

2. NOVA Classification

Several classification systems have been developed to categorise food and drink based on levels of processing, including the International Food Information Council, International Agency for Research on Cancer and NOVA classifications [25]. The most commonly used is the NOVA classification, which considers the nature, extent and purpose of processing, not the act of processing itself, to be important [26]. The NOVA food classification consists of four groups: minimally processed foods (MPF), processed culinary ingredients (PCI), processed foods (PF) and ultra-processed foods (UPF) (Table 1) [27]. UPFs are industrial formulations, typically with five or more ingredients including additives, flavourings and colours that no longer resemble their original constituent ingredients [28]. Nutritional quality, such as nutrients to limit content, is not a core aspect of the NOVA classification.

In recent decades, the contribution of UPFs to diets worldwide has been increasing year on year [29]. In the US and UK, over 55% of the average daily energy intake now comes from UPFs, and those in the highest quintiles consume over 75% of their daily energy intake from UPFs [30]. Additionally, UPFs are becoming increasingly more prevalent in the diets of infants, children and adolescents [31,32].

Table 1. Definition of NOVA classifications, from Monteiro et al., 2019 [28].

Group	Definition	Examples
1. Unprocessed and minimally processed foods	Unprocessed foods altered by processes such as the removal of inedible or unwanted parts, drying, crushing, grinding, fractioning, roasting, boiling, pasteurisation, refrigeration, freezing, placement in containers, vacuum packaging or non-alcoholic fermentation. Salt, sugar, oils or fats, or other food substances are not added. The primary aim is to extend the life of the food, enabling storage for longer use, and to make preparation easier or more diverse.	Fresh, squeezed, chilled, frozen, or dried fruit, leafy and root vegetables, brown rice, white rice, corn cob, beans, lentils, chickpeas, potatoes, sweet potatoes, mushrooms, meat, poultry, fish, seafood, meat cuts, eggs, fresh or pasteurised milk or plain yoghurt, fresh or pasteurised fruit or vegetable juices (with no added sugar, sweeteners or flavours), grits, flakes or flour made from corn, wheat, oats, or cassava, nuts and other oily seeds (with no added salt or sugar), herbs and spices used in culinary preparations, such as thyme, oregano and pepper, tea, coffee, and water.
2. Processed culinary ingredients	Substances derived from unprocessed and minimally processed foods, or from nature. They are created by industrial processes including pressing, centrifuging, refining, extracting or mining, and used in the preparation, seasoning and cooking of group 1 foods.	Oils and fats, sugar and salt.
3. Processed foods	Industrial products made by adding processed culinary ingredients found in group 2 to group 1 foods, using preservation methods such as canning and bottling. For breads and cheeses, non-alcoholic fermentation is used. Food processing in group 3 aims to increase the durability of group 1 foods and make them more enjoyable, by modifying or enhancing their sensory qualities.	Canned or bottled vegetables and legumes in brine, salted or sugared nuts and seeds, salted, dried, cured, or smoked meats and fish, canned fish (with or without added preservatives), fruits in syrup (with or without added antioxidants), freshly made unpackaged breads and cheeses.
4. Ultra-processed foods	Formulations of ingredients, mostly of exclusive industrial use, resulting from a series of industrial processes, many requiring sophisticated equipment and technology. Processes enabling the manufacture of ultra-processed foods include the fractioning of whole foods into substances, chemical modifications of these substances, assembly of unmodified and modified food substances using industrial techniques such as extrusion, moulding and pre-frying, frequent application of additives whose function is to make the final product palatable or hyper-palatable ('cosmetic additives'), and sophisticated packaging, usually with synthetic materials.	Carbonated soft drinks, sweet or savoury packaged snacks, chocolate, confectionery, ice cream, mass-produced packaged breads and buns, margarines, biscuits, pastries, cakes, breakfast 'cereals', pre-prepared pies and pasta and pizza dishes, poultry or fish nuggets, sausages, burgers, hot dogs and other reconstituted meat products, powdered and packaged 'instant' soups, noodles and desserts.

3. UPFs, Obesity Risk and Health-Related Outcomes

Systematic reviews and meta-analyses of prospective cohort studies and cross-sectional studies show that UPF consumption is associated with an increased risk of weight gain, overweight and obesity [33–38], as well as other obesity-related health outcomes [33,34], including hypertension, type 2 diabetes (T2DM) [38,39], cancer [33], cardiovascular disease (CVD) [33,34], depression and all-cause mortality [33,35–37,40]. In Europe, a 1% increase in the national household availability of UPFs is associated with a 0.25% increase in the national prevalence of obesity, after adjusting for income, physical inactivity and smoking [41]. Additionally, increases in ultra-processed food and drink volume sales per capita are associated with population-level BMI trajectories [42]. The rising contribution of UPFs to diets worldwide poses a significant threat to addressing the obesity epidemic.

4. Mechanisms and Current Debates around Ultra-Processing: Correlation or Causation?

There is increasing evidence showing that UPFs are linked with obesity and other adverse health-related outcomes. However, the potential mechanisms that lead to these adverse health outcomes are diverse, and still largely debated (readers are directed to other comprehensive overviews for further reading on potential mechanisms [43–45]). These mechanisms can be broadly considered as being as a result of nutrient content, or as a result of ultra-processing [43].

From a nutrient perspective, UPFs have on average a higher energy density (2.3 vs. 1.1 kcal/g) and lower nutrient density than minimally processed foods [44,46]. UPFs tend to be high in saturated fat, added sugar and sodium [47], with meta-analyses demonstrating that diets higher in UPFs tend to contain greater intakes of total energy, free sugars and total and saturated fat, and lower intakes of fibre, protein and some micronutrients [30,48]. The high palatability of UPFs has the potential to promote a faster eating rate and energy overconsumption [44], with daily energy intake increasing as the proportion of daily energy intake from UPFs increases (3.47 kcal increase per 1% increase in daily energy intake from UPFs) [30].

However, aspects of ultra-processing may also increase the risk of obesity and other adverse health-related outcomes. Textural and structural changes to the food matrix as a result of ultra-processing can also allow for UPFs to be consumed more quickly [49–51]. Reducing the oro-sensory exposure (OSE) time of a food can delay the onset of satiation [52], and UPFs have been shown to be less satiating than minimally processed foods [53,54]. The delayed satiation from faster eating rates can promote increased energy intake [55]. Food matrix changes can also alter nutrient bioavailability, and the harm from UPFs may come from the fact that they tend to be more hyperglycaemic than MPFs [53,54]. Besides the nutritional quality of UPFs and degradation of the food matrix, the additive content and excessive heat treatment of UPFs have also been proposed to lead to changes in gut microbiota and promote inflammation [56,57].

Beyond nutrients and ultra-processing, behavioural aspects of UPFs and the local, environmental and systemic drivers influencing food choice are also important [58]. The heavy marketing [59,60], low cost [46], high availability [61] and large portion sizes of UPFs [43,62] can make them preferable choices over minimally processed options.

5. UPF Removal or UPF Reformulation: The Case for 'Healthy' UPFs?

Actions to reduce the risks associated with UPF intake have largely been either reformulation to improve the nutrient profile of UPFs, or avoidance of UPFs altogether. Whether experts support UPF reformulation or UPF avoidance is dependent on the views regarding which mechanisms link UPFs with poor health.

Both those in favour of limiting UPFs [63], and those against the NOVA classification [64–66], acknowledge that the nutritional quality of UPFs is an important factor. Even some proponents of NOVA and reducing UPF intake have suggested that the saturated fat, added sugar and sodium content of UPFs is important, despite this not being a core aspect of the UPF definition [28]. For example, authors have focussed on the impact of reducing UPF intake on changes in saturated fat, added sugar and sodium intake and dietary quality, and the subsequent benefit of these changes on disease risk [67–70]. Critics of NOVA/supporters of reformulation argue that any link between UPFs and adverse health is solely due to their nutrient content; that some UPFs are just high in saturated fat, added sugar and sodium and that some UPFs are not nutritionally inferior, with some studies showing no difference in saturated fat, added sugar and sodium intakes across extremes of UPF intake [64,71,72].

Indeed, many UPFs are nutritionally poor and energy dense, but not all are. Studies demonstrate that UPFs tend to relate with existing nutrient profile indices, based on saturated fat, added sugar and sodium content [73]. In comparison with the Nutri-Score (ranking foods from class A to E, where A is high and E is low nutritional quality) used across several European countries, the majority of UPFs are class C, D or E, whereas the majority of unprocessed or minimally processed foods are class A or B [74]. However, 26% of class A foods are UPFs, largely being UPF ready meals or dairy products. Studies comparing UPFs with other nutrient indices (such as the Nutrient Rich Foods index, based on the protein, fibre, vitamins, minerals, saturated fat, added sugar and sodium content of food) show similar findings; most UPFs are low in nutritional quality, but some are high, and most MPFs are high in nutritional quality, but some are still low in quality [46]. Indeed, a range of UPFs have been identified as being 'healthy', based on nutrient profiling [65]. 'Healthy' UPFs are often reformulations and plant-based alternatives [65,74], which carry nutritional claims such as 'fat free', 'reduced salt', 'low sugars or 'added fibre' according to European Food Safety Authority guidelines [75]. Other 'healthy' UPFs such as fortified bread have been suggested to be important sources of vitamins and minerals [64,65,76], and avoidance of such UPFs may result in micronutrient deficiencies [77]. Therefore, two foods can be defined as having a high level of nutritional quality, but with very different levels of processing [78].

Given that particular UPFs, such as reformulations, can be considered to be of a similar or greater nutritional quality than some MPFs, it has been suggested that these UPFs are therefore healthy and nutritious [65,76]. Experts proposing that reformulations are sufficient to address all issues surrounding UPFs are making the assumption that the association between UPF intake and adverse health is mediated solely by their content of specific nutrients [71]. Experts proposing avoidance of all UPFs and arguing that reformulations are insufficient to significantly improve health are making the assumption that no UPFs can be considered to be healthy [79]. Such experts argue that reformulation does not address aspects of ultra-processing [80–82]; reformulated UPFs still have a degraded food matrix [83,84], and components of the raw constituent foods are still lost [85].

In summary, there is agreement that energy dense foods high in saturated fat, sodium and added sugar are harmful to health and should be limited. Such foods also tend to be ultra-processed, but not all are [73,74]. Despite the mounting evidence showing the adverse impacts of UPFs, the argument between nutrients and ultra-processing, and therefore between reformulation or avoidance of UPFs, is ongoing [24,81,82]. Further research understanding the relative impact of nutrients vs. ultra-processing is therefore warranted. However, largely overlooked to date, is the fact that many published prospective cohort studies have already considered the overlap between nutrition and ultra-processing, performing dietary adjustments of models to delineate the association between UPF intake, obesity and adverse health-related outcomes.

6. Review of Prospective Studies Adjusting for Dietary Quality

One of the main criticisms against the NOVA classification is that UPFs simply capture nutrient poor foods high in saturated fat, sodium and added sugar [71,73]. Furthermore, it is well established that the overall dietary pattern is important for health [10]. Higher UPF intakes are inversely associated with MPF intake, including fruit, vegetables, cereals, beans, legumes and seafood intake [30]. Therefore, the association between high UPF intake and poor health could simply be from excess nutrient intake, or from a less healthful dietary pattern. If this were the case, adjustment for participants' dietary saturated fat, sugar and sodium intake, or adjustment for their overall dietary pattern should explain the significant associations found between higher intakes of UPF and adverse health-related outcomes in prospective cohort studies, either rendering the association to be non-significant, or greatly reducing the size of the association.

Many prospective studies in adults have performed dietary adjustments, with only a small proportion not adjusting for aspects of dietary quality [86–91]. A greater proportion of prospective studies during gestation [92,93], or in children [94–101], have not performed dietary adjustments. These dietary adjustments can be broadly classed as adjustment for fat (typically saturated fat), carbohydrate (typically sugar) and sodium, adjustment for dietary patterns (including Mediterranean diet, Healthy Eating Index (HEI), Alternate Healthy Eating Index (AHEI), Dietary Guidelines for Americans Adherence Index (DGAI), healthy and Western dietary patterns and Food Standards Agency Nutrient Profiling System Dietary Index (FSA-NPS-DI)), or other dietary adjustments (typically for specific food groups such as fruit and vegetables).

Table 2 presents the 37 longitudinal, observational studies that report some form of adjustment for diet quality/pattern in their analyses investigating the association between UPF intake as defined by NOVA, and health-related outcomes (the search process and criteria for the review is detailed in the Supplementary Materials). Table 2 also presents the association between UPF intake and health-related outcomes from adjusted models preceding the dietary adjustment, or where not reported, the adjusted model including the dietary adjustment.

Across 37 studies, 87 health-related outcomes were assessed using 91 models. Of the 66 models that demonstrate a significant association between UPF and a health-related outcome, 64 remained significant following adjustment for diet quality or diet pattern. In total, 136/142 dietary adjustments did not explain the significance of the association between UPF intake and a health-related outcome. Across four studies, all four models demonstrated higher UPF intakes were significantly associated with all-cause mortality [102–105]. No dietary adjustments (15/15) altered the significance of UPF intake with all-cause mortality. Across 13 models within five studies, 11 were significantly associated with a CVD outcome [104,105,117–119]. 29/31 dietary adjustments did not alter the significance of UPF intake with CVD outcomes. Across two studies, UPF intake was significantly associated with cancer outcomes in 2/5 models [105,126]. 8/8 dietary adjustments did not alter the significance of the association between UPF intake and the two cancer outcomes. In two models significantly associated with T2DM, 7/7 dietary adjustments did not alter the significance [123,124]. Across nine studies, 23/26 models demonstrated a significant association between UPF intake and adult and child anthropometrics (weight/body mass index (BMI)/fat mass index (FMI) gain, other measures of adiposity and risk of overweight/obesity) [106–113,115]. 40/43 dietary adjustments did not alter the significance of these associations.

6.1. Adjustment for Saturated Fat, Sugar and Sodium, and for Dietary Pattern

Table 3 presents the adjustments for saturated fat, sodium and added sugar. Table 4 presents the adjustments for dietary pattern. All but one study retained the significant association between UPF and the health-related outcome after adjustments for saturated fat, sodium and added sugar. All but two studies retained the significant association between UPF and the health-related outcome after adjustment for dietary pattern.

Table 2. Characteristics of prospective studies assessing the association between UPF intake and health-related outcomes whilst also reporting dietary adjustments.

Author, Year	Cohort	Sample	Country	Sample Size	Outcome	Method of Analysis	Effect Estimate (95%CI)
Schnabel 2019 [102]	Nutri-Net Santé	Adults ≥ 45	France	44,551	All-cause mortality	HR per 10% increase in UPF	1.15 (1.04, 1.27) [1]
Rico-Campa 2019 [103]	SUN	University graduates	Spain	19,899	All-cause mortality	HR 1st vs. 4th quartile	1.62 (1.13, 2.33) [2]
Kim 2019 [104]	NHANES III	Adults ≥ 20	US	11,898	All-cause mortality	HR 1st vs. 4th quartile	1.31 (1.09, 1.58) [3]
				11,898	CVD mortality	HR 1st vs. 4th quartile	1.10 (0.74, 1.67) [3]
Bonaccio 2021 [105]	Moli-sani	Adults	Italy	22,475	All-cause mortality	HR 1st vs. 4th quartile	1.32 (1.15, 1.53) [4]
				22,475	Other cause mortality (exc. CVD and cancer)	HR 1st vs. 4th quartile	1.36 (1.01, 1.83) [4]
				22,475	Cancer mortality	HR 1st vs. 4th quartile	1.00 (0.80, 1.26) [4]
				22,475	CVD mortality	HR 1st vs. 4th quartile	1.65 (1.29, 2.11) [4]
				22,475	IHD/cerebrovascular mortality	HR 1st vs. 4th quartile	1.63 (1.19, 2.25) [4]
Beslay 2020 [106]	Nutri-Net Santé	Adults ≥ 18	France	110,260	BMI change (kg/m^2)	Beta per 10% increase in UPF	0.02 (0.01, 0.02) [5]
				55,307	Overweight	HR per 10% increase in UPF	1.11 (1.08, 1.14) [5]
				71,871	Obesity	HR per 10% increase in UPF	1.09 (1.05, 1.13) [5]
Mendonca 2016 [107]	SUN	Middle-aged University graduates	Spain	8451	Overweight/obesity	HR 1st vs. 4th quartile	1.26 (1.10, 1.35) [6]
Li 2021 [108]	CHNS	Adults > 20	China	12,451	Overweight/obesity	OR none vs. ≥ 50g/day	1.85 (1.58, 2.17) [7]
				12,451	Central obesity	OR none vs. ≥ 50g/day	2.04 (1.79, 2.33) [7]
Koniecnzna 2021 [109]	PREDIMED-Plus	Adults aged 55–75 with overweight/obesity and metabolic syndrome	Spain	1485	Total fat mass (z-score)	Beta per 10% increase in UPF	0.09 (0.06, 0.13) [8]
				1485	Visceral fat mass (z-score)	Beta per 10% increase in UPF	0.09 (0.05, 0.13) [8]
				1485	Android:gynoid fat ratio (z-score)	Beta per 10% increase in UPF	0.05 (0.00, 0.09) ($p = 0.031$) [8]
Sandoval-Insausti 2020 [110]	Seniors-ENRICA-1	Older adults	Spain	652	Abdominal obesity	OR 1st vs. 3rd tertile	1.62 (1.04, 2.54) [9]
Cordova 2021 [111]	EPIC	Adults aged 25–70	Multi-national (nine countries)	348,748	Weight gain (kg)	Beta per 1SD increase in UPF/day	0.12 (0.09, 0.15) [10]
				191,255	Overweight/obesity	RR per 1SD increase in UPF/day	1·05 (1·04, 1.06) [10]
				103,259	Obesity	RR per 1SD increase in UPF/day	1·05 (1.03, 1.07) [10]
Canhada 2020 [112]	ELSA-Brazil	Civil servants aged 35–74	Brazil	11,827	Large weight gain (≥90th percentile: ≥1.68 kg/year)	RR 1st vs. 4th quartile	1.27 (1.07, 1.50) [11]
				11,827	Large WC gain (≥90th percentile: ≥2.42 cm/year)	RR 1st vs. 4th quartile	1.33 (1.12, 1.58) [11]
				4527	Incident overweight/obesity	RR 1st vs. 4th quartile	1.20 (1.03, 1.40) [11]
				4771	Incident obesity	RR 1st vs. 4th quartile	1.02 (0.85, 1.21) [11]

Table 2. *Cont.*

Author, Year	Cohort	Sample	Country	Sample Size	Outcome	Method of Analysis	Effect Estimate (95%CI)
Rohatgi 2017 [113]	Women's Health Center and Obstetrics & Gynecology Clinic		MO, US	45	Gestational weight gain (kg)	Beta per 1% increase in UPF intake	1.3 (0.3, 2.4) [12]
				45	Neonate thigh skinfold thickness (mm)	Beta per 1% increase in UPF intake	0.20 (0.005, 0.40) [12]
				45	Neonate subscapular skinfold thickness (mm)	Beta per 1% increase in UPF intake	0.10 (0.02, 0.30) [12]
				45	Neonate body fat percentage (%)	Beta per 1% increase in UPF intake	0.60 (0.04, 1.20) [12]
Leone 2021 [114]	SUN	Females Females < 30 Females $\geq$ 30	Spain	3730 2538 1192	Gestational diabetes Gestational diabetes Gestational diabetes	OR 1st vs. 3rd tertile OR 1st vs. 3rd tertile OR 1st vs. 3rd tertile	1.10 (0.74, 1.64) [13] 0.89 (0.54, 1.46) [13] 2.05 (1.03, 4.07) [13]
Chang 2021 [115]	ALSPAC	Children	Britain	9020	BMI (kg/m^2)/year	Beta 1st vs. 5th quintile	0.06 (0.04, 0.08) [14]
				8078	Fat mass index (kg/m^2)/year	Beta 1st vs. 5th quintile	0.03 (0.01, 0.05) [14]
				8078	Lean mass index (kg/m^2)/year	Beta 1st vs. 5th quintile	0.004 (−0.007, 0.01) [14]
				8078	Body fat percentage (%)/year	Beta 1st vs. 5th quintile	0.004 (−0.05, 0.06) [14]
Costa 2021 [116]	Pelotas-Brazil 2004 Birth Cohort	6–11-year-olds	Brazil	4231	Fat mass index (kg/m^2)	Beta/100 g increase in UPF intake	0.09 (0.07, 0.10) [15]
Srour 2019 [117]	Nutri-Net Santé	Adults $\geq$ 18	France	105,159 105,159 105,159	All CVD Coronary heart disease Cerebrovascular disease	HR per 10% increase in UPF HR per 10% increase in UPF HR per 10% increase in UPF	1.12 (1.05, 1.20) [16] 1.13 (1.02, 1.24) [16] 1.11 (1.01, 1.21) [16]
Juul 2021 [118]	Framingham Offspring Cohort	Adults	US	3003	Overall CVD	HR per serving UPF/day	1.05 (1.02, 1.08) [17]
				3003	CVD mortality	HR per serving UPF/day	1.09 (1.02, 1.16) [17]
				3003	Incident hard CVD	HR per serving UPF/day	1.07 (1.03, 1.12) [17]
				3003	Hard coronary heart disease	HR per serving UPF/day	1.09 (1.04, 1.15) [17]
Zhong 2021 [119]	Prostate, Lung, Colorectal, and Ovarian Cancer Screening Trial	Adults aged 55–74 at baseline	US	91,891	CVD mortality	HR 1st vs. 5th quintile	1.50 (1.36, 1.64) [18]
				91,891	Heart disease mortality	HR 1st vs. 5th quintile	1.68 (1.50, 1.87) [18]
				91,891	Cerebrovascular disease mortality	HR 1st vs. 5th quintile	0.94 (0.76, 1.17) [18]
Scaranni 2021 [120]	ELSA-Brasil	Civil servants aged 35–74 at baseline	Brazil	8754	Incident hypertension	OR 1st vs. 3rd tertile	1.20 (1.04, 1.40) [19]
				8754	Change in SBP	Beta 1st vs. 3rd tertile	−0.37 (−1.05, 0.30) [19]
				8754	Change in DBP	Beta 1st vs. 3rd tertile	0.19 (−0.28, 0.66) [19]

Table 2. *Cont.*

Author, Year	Cohort	Sample	Country	Sample Size	Outcome	Method of Analysis	Effect Estimate (95%CI)
Monge 2021 [121]	Mexican Teachers' Cohort	Females aged ≥ 25 at baseline	Mexico	64,934	Incident hypertension	≤20% vs. >45% of energy from any UPF	0.99 (0.85, 1.15) [20]
				64,934	Incident hypertension	≤20% vs. >45% of energy from liquid UPF	1.33 (1.09, 1.63) [20]
				64,934	Incident hypertension	≤20% vs. >45% of energy from solid UPF	0.91 (0.82, 1.01) [20]
Mendonca 2017 [122]	SUN	Middle-aged University graduates	Spain	14,790	Hypertension	HR 1st vs. 3rd tertile	1.23 (1.09, 1.38) [21]
Llavero-Valero 2021 [123]	SUN	Middle-aged University graduates	Spain	20,060	T2DM	HR 1st vs. 3rd tertile	1.53 (1.06, 2.22) [22]
Srour 2020 [124]	Nutri-Net Santé	Adults ≥ 18	France	104,707	T2DM	HR per 10% increase in UPF	1.15 (1.06, 1.25) [23]
Zhang 2021 [125]	TCLSIH	Adults aged 18–90	China	16,168	NAFLD	HR 1st vs. 4th quartile	1.17 (1.07, 1.29) [24]
Fiolet 2018 [126]	Nutri-Net Santé	Adults ≥ 18	France	104,980	All cancers	HR per 10% increase in UPF	1.12 (1.06, 1.18) [25]
				104,980	Breast cancer	HR per 10% increase in UPF	1.11 (1.02, 1.22) [25]
				104,980	Prostate cancer	HR per 10% increase in UPF	0.98 (0.83, 1.16) [25]
				104,980	Colorectal cancer	HR per 10% increase in UPF	1.13 (0.92, 1.38) [25]
Vasseur 2021 [127]	Nutri-Net Santé	Adults ≥ 18	France	105,832	IBD	RR 1st vs. 3rd tertile	1.32 (0.75, 2.34) [26]
Narula 2021 [128]	PURE	Adults aged 35–70	21 low, middle, and high income countries	116,087	IBD	HR <1 vs. ≥5 servings UPF/day	1.82 (1.22, 2.72) [27]
				116,087	Crohn's disease	HR <1 vs. ≥5 servings UPF/day	4.50 (1.67, 12.13) [27]
				116,087	Ulcerative Colitis	HR <1 vs. ≥5 servings UPF/day	1.46 (0.93, 2.28) [27]
Schnabel 2018 [129]	Nutri-Net Santé	Adults	France	33,343	Irritable bowel syndrome	OR 1st vs. 4th quartile	1.24 (1.12, 1.38) [28]
				33,343	Functional Constipation	OR 1st vs. 4th quartile	1.00 (0.87, 1.15) [28]
				33,343	Functional diarrhoea	OR 1st vs. 4th quartile	0.94 (0.71, 1.26) [28]
				33,343	Functional dyspepsia	OR 1st vs. 4th quartile	1.26 (1.07, 1.48) [28]
Lo 2021 [130]	NHS, NHS II, HPFS	Adult health professionals	US	245,112	Crohn's disease	HR 1st vs. 4th quartile	1.75 (1.29, 2.35) [29]
				245,112	Ulcerative Colitis	HR 1st vs. 4th quartile	1.25 (0.97, 1.62) [29]
Adjibade 2019 [131]	Nutri-Net Santé	Adults aged 18–86	France	26,730	Depressive symptoms	HR per 10% increase in UPF	1.21 (1.15, 1.27) [30]
Gómez-Donoso 2020 [132]	SUN	Middle-aged University graduates	Spain	14,907	Incident depression	HR 1st vs. 4th quartile	1.41 (1.15, 1.73) [31]
Rey-Garcia 2021 [133]	Seniors-ENRICA-1	Adult ≥ 60	Spain	1312	Renal function	OR 1st vs. 3rd tertile	1.75 (1.16, 2.64) [32]
Zhang 2021 [134]	TCLSIH	Adults ≥ 18	China	18,444	Hyperuricemia	HR 1st vs. 4th quartile	1.21 (1.10, 1.33) [33]

90

Table 2. *Cont.*

Author, Year	Cohort	Sample	Country	Sample Size	Outcome	Method of Analysis	Effect Estimate (95%CI)
Leffa 2020 [135]	Impact of the "Ten Steps for Healthy Feeding of Children Younger Than Two Years" in Health Centers	Children	Porto Alegre, Brazil	308	Total cholesterol at age 6	Beta per 10% increase in UPF intake at age 3	0.07 (0.00, 0.15) $p = 0.044$ [34]
				308	LDL-cholesterol at age 6	Beta per 10% increase in UPF intake at age 3	0.03 (−0.02, 0.09) [34]
				308	HDL-cholesterol at age 6	Beta per 10% increase in UPF intake at age 3	0.01 (−0.01, 0.06) [34]
				308	TAG at age 6	Beta per 10% increase in UPF intake at age 3	0.03 (0.00, 0.07) $p = 0.034$ [34]
Donat-Vargas 2021 [136]	ENRICA	Adults > 60	Spain	895	Incident hypertriglyceridemia ($\geq$150 mg/dL)	OR 1st vs. 3rd tertile	2.00 (1.04, 3.85) [35]
				878	Low HDL-cholesterol (<40 in men or <50 mg/dL in women)	OR 1st vs. 3rd tertile	2.04 (1.22, 3.41) [35]
				472	High LDL-cholesterol (>129 mg/dL)	OR 1st vs. 3rd tertile	0.95 (0.46, 1.97) [35]
				895	Δtriglycerides (mg/dL)	Beta 1st vs. 3rd tertile	6.11 (1.30, 10.91) [35]
				878	ΔHDL cholesterol (mg/dL)	Beta 1st vs. 3rd tertile	0.03 (−1.38, 1.44) [35]
				472	ΔLDL cholesterol (mg/dL)	Beta 1st vs. 3rd tertile	−4.52 (−9.40, 0.36) [35]
Borge 2021 [137]	Norwegian Mother, Father and Child Cohort Study	Mother and child	Norway	46,976	ADHD diagnosis at 8 years	RR per 1 SD increase in UPF	1.00 (0.93, 1.08) [36]
				31,152	ADHD symptoms (absolute) at 8 years	Beta per 1 SD increase in UPF	0.38 (0.27, 0.49) [36] *
				31,152	ADHD symptoms (relative %) at 8 years	Beta per 1 SD increase in UPF	4.5 (3.3, 4.9) [36] *
Zhang 2021 [138]	TCLSIH	Adults $\geq$ 40	China	5409	Annual change in grip strength (kg per year)	Beta per 10% increase in UPF	−0.2955 (−0.4992, −0.0919) [37]
				5409	Annual change in weight-adjusted grip strength (kg/kg per year)	Beta per 10% increase in UPF	−0.0043 (−0.0073, −0.0014) [37]

Effect estimates are presented from adjusted models preceding the dietary adjustment, or where not performed, the adjusted model including dietary adjustment. OR, odds ratio; HR, hazard ratio; RR, relative risk; SD, standard deviation; ADHD, attention deficit hyperactivity disorder; CVD, cardiovascular disease; IHD, ischaemic heart disease; BMI, body mass index; T2DM, type 2 diabetes mellitus; UPF, ultra-processed food; SBP, systolic blood pressure; DBP, diastolic blood pressure; LDL, low-density lipoprotein; HDL, high-density lipoprotein; TAG, triacylglycerol; SUN, Seguimiento University of Navarra; NHANES III, Third National Health and Nutrition Examination Survey; CHNS, China Health and Nutrition Survey; Seniors-ENRICA 1, Seniors Study on Nutrition and Cardiovascular Risk in Spain; EPIC, European Prospective Investigation into Cancer and Nutrition; ELSA-Brazil, Brazilian Longitudinal Study of Adult Health; ALSPAC, Avon Longitudinal Study of Parents and Children; TCLSIH, Tianjin Chronic Low-grade Systemic Inflammation and Health; PURE, Prospective Urban Rural Epidemiology; NHS, Nurses' Health Study 1986–2014); NHS II, Nurses' Health Study II (1991–2017); HPFS, Health Professionals Follow-up Study; [1] Adjusted for sex, age, income level, education level, marital status, residence, BMI, physical activity level, smoking status, energy intake, alcohol intake, season of food records, first-degree family history of cancer or cardiovascular diseases and number of food records. [2] Adjusted for age, sex, marital status, physical activity, smoking status, snacking, special diet at baseline, BMI, total energy intake, alcohol consumption, family history of CVD, diabetes or hypertension at baseline, self-reported hypercholesterolaemia, baseline CVD, cancer or depression, education level and lifelong smoking stratified by recruitment period, deciles of age, sedentary index (sum of hours each day spent watching television, using a computer and driving) and television viewing. [3] Adjusted for age, sex, race/ethnicity, total energy intake, poverty level, education level, smoking status, physical activity and alcohol intake. [4] Adjusted for sex, age, energy intake, educational level, housing tenure, smoking, BMI, leisure-time physical activity, history of cancer, CVD, diabetes, hypertension and hyperlipidaemia and residence. [5] Adjusted for age, sex, marital status, educational level, physical activity, smoking status, alcohol consumption, energy intake and number of dietary records (overweight and obesity outcomes further adjusted for baseline BMI). [6] Adjusted for age, sex, marital status, educational status, physical activity, television watching, siesta sleep, smoking status, snacking between meals, following a special diet at baseline and baseline BMI. [7] Adjusted for age, sex and energy intake. [8] Adjusted for age, sex, study arm, follow-up time, educational level, marital status, smoking habits, type 2 diabetes prevalence, height and repeated measures of physical activity and sedentary behaviour. [9] Adjusted for age, sex, educational level, marital status, smoking, ex-drinker status, physical activity in the household and at leisure time, number of medications consumed per day and number of chronic diseases diagnosed by a doctor (chronic obstructive pulmonary disease/asthma, coronary heart disease, stroke, heart failure, osteoarthritis or depression). [10] Adjusted for age, sex, baseline BMI, educational level, physical activity, baseline alcohol intake, baseline smoking status and plausibility of dietary energy reporting. Overweight and obesity outcomes further adjusted for country/centre, follow-up time in years, smoking status at follow-up (instead of baseline smoking status) and for the modified relative Mediterranean diet score. [11] Adjusted for age, sex, colour/race, centre, income, school achievement, smoking and physical activity. Additionally for incident overweight/obesity and weight gain: baseline BMI. Additionally for waist gain: waist circumference at baseline. [12] Adjusted for maternal age, race, socioeconomic status, weight status, average daily energy intake, time spent in moderate physical activity and fat intake. [13] Adjusted for age, BMI, education, smoking status, physical activity, family history of diabetes, recruitment year, time between recruitment and the first pregnancy or gestational diabetes, number of pregnancies during follow-up, parity, multiple pregnancies, time spent watching TV, hypertension, following a nutritional therapy and energy intake. [14] Adjusted for age, baseline UPF, age*baseline UPF interaction term, child sex, race, birth weight, physical activity, quintiles of Index of Multiple Deprivation, the mother's prepregnancy BMI, marital status, highest educational attainment, socioeconomic status and the child's baseline total energy intake. [15] Adjusted for skin colour, maternal age and schooling, birthweight and sex, screen time and energy intake/expenditure ratio. [16] Adjusted for age, sex, energy intake, number of 24-h dietary records, smoking status, educational level, physical activity, BMI, alcohol intake and family history of CVD. [17] Adjusted for age, sex, education, smoking status, alcohol intake and physical activity. [18] Adjusted for age, sex, race, educational level, marital status, study centre, aspirin use, history of hypertension or diabetes, smoking status, alcohol consumption, BMI, physical activity and energy intake. [19] Adjusted for age, sex, colour or race, education, time since baseline and SBP/DBP/hypertension. [20] Adjusted for age, indigenous, internet access, insurance, family history of hypertension, menopausal status, smoking status and physical activity. [21] Adjusted for age, sex, physical activity, hours of TV watching, baseline BMI, smoking status, use of analgesics, following a special diet at baseline, family history of hypertension, hypercholesterolaemia and alcohol consumption. [22] Adjusted for age, sex, BMI, educational status, family history of diabetes, smoking status, snacking between meals, active and sedentary lifestyle score and following a special diet at baseline. [23] Adjusted for age, sex, educational level, baseline BMI, physical activity, smoking status, alcohol intake, number of 24-h dietary records, energy intake, Food Standards Agency nutrient profiling system dietary index score and family history of T2DM. [24] Adjusted for age, sex and BMI. [25] Adjusted for age, sex, energy intake without alcohol, number of 24-h dietary records, smoking status, educational level, physical activity, height, BMI, alcohol intake and family history of cancers (and for breast cancer outcome, additionally adjusted for menopausal status, hormonal treatment for menopause, oral contraception, and number of children). [26] Adjusted for age and sex. [27] Adjusted for age, sex, geographical region, education, alcohol intake, smoking status, BMI, total energy intake and location. [28] Adjusted for age, sex, income level, education level, marital status, residence, BMI, physical activity, smoking status, energy intake, season of food records and time between food record and functional gastrointestinal disorders questionnaire. [29] Adjusted for age, cohort and calendar year. [30] Adjusted for age, sex, BMI, marital status, educational level, occupational categories, household income per consumption unit, residential area, number of 24-h dietary records, inclusion month, energy intake without alcohol, alcohol intake, smoking status and physical activity. [31] Adjusted for sex, age and year of entrance to the cohort. [32] Adjusted for sex, age and total energy intake. [33] Adjusted for age, sex, BMI, smoking status, alcohol consumption status, education levels, employment status, household income, physical activity, depression symptoms, family history of disease (including cardiovascular disease, hypertension, hyperlipidaemia and diabetes), hypertension, hyperlipidaemia, diabetes and metabolic syndrome. [34] Adjusted for sex, group status in the early phase (intervention and control), family income, pre-pregnancy BMI, child birth weight and BMI z-scores at 3 years. [35] Adjusted for age and sex. [36] Crude model. [37] Adjusted for baseline age, sex and BMI. * 95% credible intervals.

Table 3. Prospective studies adjusting for fat, added sugar/carbohydrate and sodium content.

Author, Year	Outcome	Method of Analysis	Diet Adjustment	Effect Estimate (95%CI)
Rico-Campa 2019 [103]	All-cause mortality	HR 1st vs. 4th quartile	SFA, sodium, added sugar and TFA	1.69 (1.12, 2.56)
Bonaccio 2021 [105]	All-cause mortality	HR 1st vs. 4th quartile	SFA, sodium, sugar, cholesterol and energy intake	1.28 (1.09, 1.49)
	CVD mortality	HR 1st vs. 4th quartile	SFA, sodium, sugar, cholesterol and energy intake	1.56 (1.19, 2.03)
	IHD/cerebrovascular mortality	HR 1st vs. 4th quartile	SFA, sodium, sugar, cholesterol and energy intake	1.33 (0.94, 1.90)
Beslay 2020 [106]	BMI change (kg/m^2)	Beta per 10% increase in UPF	SFA, sodium, sugar and fibre	0.02 (0.01, 0.02)
	Overweight	HR per 10% increase in UPF	SFA, sodium, sugar and fibre	1.10 (1.08, 1.13)
	Obesity	HR per 10% increase in UPF	SFA, sodium, sugar and fibre	1.10 (1.06, 1.14)
Koniecnzna 2021 [109]	Total fat mass (z-score)	Beta per 10% increase in UPF	SFA, sodium, glycaemic index, TFA, alcohol and fibre	0.06 (0.03, 0.09)
	Visceral fat mass (z-score)	Beta per 10% increase in UPF	SFA, sodium, glycaemic index, TFA, alcohol and fibre	0.06 (0.01, 0.10)
	Android:gynoid fat ratio (z-score)	Beta per 10% increase in UPF	SFA, sodium, glycaemic index, TFA, alcohol and fibre	0.02 (−0.02, 0.07)
Srour 2019 [117]	All CVD	HR per 10% increase in UPF	SFA, sodium and sugar	1.13 (1.05, 1.20)
	Coronary heart disease	HR per 10% increase in UPF	SFA, sodium and sugar	1.14 (1.03, 1.26)
	Cerebrovascular disease	HR per 10% increase in UPF	SFA, sodium and sugar	1.12 (1.02, 1.22)
Zhong 2021 [119]	CVD mortality	HR 1st vs. 5th quintile	SFA, sodium and added sugar	1.48 (1.34, 1.63)
	Heart disease mortality	HR 1st vs. 5th quintile	SFA, sodium and added sugar	1.65 (1.47, 1.85)
	Cerebrovascular disease mortality	HR 1st vs. 5th quintile	SFA, sodium and added sugar	0.93 (0.74, 1.17)
Srour 2020 [124]	T2DM	HR per 10% increase in UPF	SFA, sodium, sugar and fibre	1.19 (1.09, 1.30)
Fiolet 2018 [126]	All cancers	HR per 10% increase in UPF	Lipids, sodium and carbohydrates	1.12 (1.07, 1.18)
	Breast cancer	HR per 10% increase in UPF	Lipids, sodium and carbohydrates	1.11 (1.01, 1.21)
	Prostate cancer	HR per 10% increase in UPF	Lipids, sodium and carbohydrates	0.98 (0.83, 1.16)
	Colorectal cancer	HR per 10% increase in UPF	Lipids, sodium and carbohydrates	1.16 (0.95, 1.42)
Chang 2021 [115]	BMI (kg/m^2)/year	Beta 1st vs. 5th quintile	SFA, sodium, sugar and fibre	0.07 (0.04, 0.08)
	Fat mass index (kg/m^2)/year	Beta 1st vs. 5th quintile	SFA, sodium, sugar and fibre	0.03 (0.01, 0.05)
	Lean mass index (kg/m^2)/year	Beta 1st vs. 5th quintile	SFA, sodium, sugar and fibre	0.005 (−0.007, 0.010)
	Body fat percentage (%)/year	Beta 1st vs. 5th quintile	SFA, sodium, sugar and fibre	0.002 (−0.05, 0.05)

SFA, saturated fatty acids; TFA, trans fatty acids; HR, hazard ratio; CVD, cardiovascular disease; IHD, ischemic heart disease; BMI, body mass index; T2DM, type 2 diabetes mellitus.

Table 4. Prospective studies adjusting for dietary pattern.

Author, Year	Outcome	Method of Analysis	Diet Adjustment	Effect Estimate (95%CI)
Schnabel 2019 [102]	All-cause mortality	HR per 10% increase in UPF	French dietary guidelines	1.14 (1.04, 1.27)
	All-cause mortality	HR per 10% increase in UPF	French dietary guidelines and Western dietary pattern	1.19 (1.05, 1.35)
Rico-Campa 2019 [103]	All-cause mortality	HR 1st vs. 4th quartile	Mediterranean dietary pattern	1.58 (1.10, 2.28)
Kim 2019 [104]	All-cause mortality	P-trend	Dietary quality score	p-trend only 0.001 [1]
	CVD mortality	P-trend	Dietary quality score	p-trend only 0.540 [1]
Bonaccio 2021 [105]	All-cause mortality	HR 1st vs. 4th quartile	Mediterranean dietary pattern	1.26 (1.09, 1.46)
	Other cause mortality (exc. CVD and cancer)	HR 1st vs. 4th quartile	Mediterranean dietary pattern	1.26 (0.94, 1.69)
	CVD mortality	HR 1st vs. 4th quartile	Mediterranean dietary pattern	1.58 (1.23, 2.03)
	IHD/cerebrovascular mortality	HR 1st vs. 4th quartile	Mediterranean dietary pattern	1.52 (1.10, 2.09)
	Cancer mortality	HR 1st vs. 4th quartile	Mediterranean dietary pattern	0.97 (0.77, 1.22)
Beslay 2020 [106]	BMI change (kg/m^2)	Beta per 10% increase in UPF	Healthy and Western dietary patterns	0.02 (0.01, 0.02)
	Overweight	HR per 10% increase in UPF	Healthy and Western dietary patterns	1.10 (1.07, 1.13)
	Obesity	HR per 10% increase in UPF	Healthy and Western dietary patterns	1.11 (1.07, 1.15)
Li 2021 [108]	Overweight/obesity	OR none vs. $\geq$50 g/day	Traditional and modern dietary patterns	1.45 (1.21, 1.74) [2]
	Central obesity	OR none vs. $\geq$50 g/day	Traditional and modern dietary patterns	1.50 (1.29, 1.74) [2]
Koniecnzna 2021 [109]	Total fat mass (z-score)	Beta per 10% increase in UPF	Mediterranean dietary pattern adherence	0.06 (0.02, 0.09)
	Visceral fat mass (z-score)	Beta per 10% increase in UPF	Mediterranean dietary pattern adherence	0.06 (0.01, 0.10)
	Android:gynoid fat ratio (z-score)	Beta per 10% increase in UPF	Mediterranean dietary pattern adherence	0.02 (−0.02, 0.06)
Sandoval-Insausti 2020 [110]	Abdominal obesity	OR 1st vs. 3rd tertile	Mediterranean dietary pattern, fibre and very long chain omega-3 fatty acid intake	1.61 (1.01, 2.56)
Cordova 2021 [111]	Weight gain (kg)	Beta per 1SD increase in UPF/day	Mediterranean dietary pattern	0.118 (0.085, 0.151)
	Overweight/obesity	RR per 1SD increase in UPF/day	Mediterranean dietary pattern	1.05 (1.04, 1.06)
	Obesity	RR per 1SD increase in UPF/day	Mediterranean dietary pattern	1.05 (1.03, 1.07)
Leone 2021 [114]	Gestational diabetes pooled	OR 1st vs. 3rd tertile	Mediterranean dietary pattern	1.10 [0.74, 1.65]
	Gestational diabetes < 30	OR 1st vs. 3rd tertile	Mediterranean dietary pattern	0.89 [0.53, 1.47]
	Gestational diabetes $\geq$ 30	OR 1st vs. 3rd tertile	Mediterranean dietary pattern	2.06 (1.05, 4.06)

Table 4. *Cont.*

Author, Year	Outcome	Method of Analysis	Diet Adjustment	Effect Estimate (95%CI)
Costa 2021 [116]	Fat mass index (kg/m^2)	Beta/100g daily increase in UPF intake	Unprocessed or minimally processed foods, processed culinary ingredients and processed foods intake	0.14 (0.13, 0.15)
Srour 2019 [117]	All CVD Coronary heart disease Cerebrovascular disease	HR per 10% increase in UPF HR per 10% increase in UPF HR per 10% increase in UPF	Healthy dietary pattern Healthy dietary pattern Healthy dietary pattern	1.11 (1.03, 1.19) 1.11 (1.00, 1.23) $p = 0.04$ 1.10 (1.00, 1.20) $p = 0.04$
Juul 2021 [118]	Overall CVD CVD mortality Incident hard CVD Hard coronary heart disease	HR per serving UPF/day HR per serving UPF/day HR per serving UPF/day HR per serving UPF/day	Dietary Guidelines Adherence Index (DGAI) 2010 Dietary Guidelines Adherence Index (DGAI) 2010 Dietary Guidelines Adherence Index (DGAI) 2010 Dietary Guidelines Adherence Index (DGAI) 2010	1.04 (1.01, 1.07) 1.09 (1.02, 1.16) 1.06 (1.02, 1.11) 1.09 (1.03, 1.15)
Zhong 2021 [119]	CVD mortality Heart disease mortality Cerebrovascular disease mortality	HR 1st vs. 5th quintile HR 1st vs. 5th quintile HR 1st vs. 5th quintile	Healthy Eating Index (HEI) 2005 Healthy Eating Index (HEI) 2005 Healthy Eating Index (HEI) 2005	1.48 (1.35, 1.63) 1.67 (1.49, 1.86) 0.94 (0.75, 1.16)
Llavero-Valero 2021 [123]	T2DM	HR 1st vs. 3rd tertile	Mediterranean dietary pattern	1.50 (1.02, 2.21)
Srour 2020 [124]	T2DM	HR per 10% increase in UPF	Healthy and Western dietary patterns	1.13 (1.04, 1.24)
Zhang 2021 [125]	NAFLD	HR 1st vs. 4th quartile	Healthy diet score	1.19 (1.08, 1.31) [3]
Fiolet 2018 [126]	All cancers Breast cancer Prostate cancer Colorectal cancer	HR per 10% increase in UPF HR per 10% increase in UPF HR per 10% increase in UPF HR per 10% increase in UPF	Western dietary pattern Western dietary pattern Western dietary pattern Western dietary pattern	1.12 (1.06, 1.18) 1.11 (1.02, 1.22) 0.98 (0.83, 1.15) 1.13 (0.92, 1.38)
Vasseur 2021 [127]	IBD	RR 1st vs. 3rd tertile	Healthy dietary pattern	1.44 (0.70, 2.94) [4]
Narula 2021 [128]	IBD Crohn's disease Ulcerative Colitis	HR <1 vs. ≥5 servings UPF/day HR <1 vs. ≥5 servings UPF/day HR <1 vs. ≥5 servings UPF/day	Alternate Healthy Eating Index (AHEI) 2010 Alternate Healthy Eating Index (AHEI) 2010 Alternate Healthy Eating Index (AHEI) 2010	1.92 (1.28, 2.90) 4.90 (1.78, 13.45) 1.52 (0.96, 2.41)
Lo 2021 [130]	Crohn's disease Ulcerative Colitis	HR 1st vs. 4th quartile HR 1st vs. 4th quartile	Alternate Healthy Eating Index (AHEI) 2010 Alternate Healthy Eating Index (AHEI) 2010	1.70 (1.23, 2.35) [5] 1.20 (0.91, 1.58) [5]
Schnabel 2018 [129]	Irritable bowel syndrome Functional Constipation Functional diarrhoea Functional dyspepsia	OR 1st vs. 4th quartile OR 1st vs. 4th quartile OR 1st vs. 4th quartile OR 1st vs. 4th quartile	French dietary guidelines French dietary guidelines French dietary guidelines French dietary guidelines	1.25 (1.12, 1.39) 0.98 (0.85, 1.12) 0.92 (0.69, 1.24) 1.25 (1.05, 1.47)

Table 4. *Cont.*

Author, Year	Outcome	Method of Analysis	Diet Adjustment	Effect Estimate (95%CI)
Gómez-Donoso 2020 [132]	Incident depression	HR 1st vs. 4th quartile	Mediterranean dietary pattern	1.33 (1.07, 1.64) [6]
Zhang 2021 [134]	Hyperuricemia	HR 1st vs. 4th quartile	Sweet, animal and healthy dietary patterns	1.17 (1.06, 1.30) [7]
Donat-Vargas 2021 [136]	Incident hypertriglyceridemia ($\geq$150 mg/dL)	OR 1st vs. 3rd tertile	Unprocessed or minimally processed food intake	2.66 (1.20, 5.90) [8]
	Low HDL-cholesterol (<40 in men or <50 mg/dL in women)	OR 1st vs. 3rd tertile	Unprocessed or minimally processed food intake	2.23 (1.22, 4.05) [8]
	High LDL-cholesterol (>129 mg/dL)	OR 1st vs. 3rd tertile	Unprocessed or minimally processed food intake	1.03 (0.43, 2.47) [8]
	Δtriglycerides (mg/dL)	Beta 1st vs. 3rd tertile	Unprocessed or minimally processed food intake	6.87 (1.48, 12.27) [8]
	ΔHDL cholesterol (mg/dL)	Beta 1st vs. 3rd tertile	Unprocessed or minimally processed food intake	0.13 ($-$1.46, 1.71) [8]
	ΔLDL cholesterol (mg/dL)	Beta 1st vs. 3rd tertile	Unprocessed or minimally processed food intake	$-$2.03 ($-$7.86, 3.80) [8]
Borge 2021 [137]	ADHD diagnosis at 8 years	RR per 1 SD increase in UPF	Child diet quality score at 3 years	1.07 (0.99, 1.18) [9]
	ADHD symptoms (absolute) at 8 years	Beta per 1 SD increase in UPF	Child diet quality score at 3 years	0.25 (0.13, 0.38) [9,*]
	ADHD symptoms (relative %) at 8 years	Beta per 1 SD increase in UPF	Child diet quality score at 3 years	3.0 (1.5, 4.5) [9,*]
Zhang 2021 [138]	Change in grip strength (kg/year)	Beta per 10% increase in UPF	Healthy diet score	$-$0.3207 ($-$0.5281, $-$0.1133) [10]
	Change in weight-adjusted grip strength (kg/kg/year)	Beta per 10% increase in UPF	Healthy diet score	$-$0.0046 ($-$0.0076, $-$0.0016) [10]

OR, odds ratio; HR, hazard ratio; RR, relative risk; ADHD, attention deficit hyperactivity disorder; CVD, cardiovascular disease; IHD, ischemic heart disease; BMI, body mass index; T2DM, type 2 diabetes mellitus; IBD, inflammatory bowel disease; NAFLD, non-alcoholic fatty liver disease. [1] Further adjusted for body mass index, hypertension status, total cholesterol, and estimated glomerular filtration rate. [2] Further adjusted for fat intake, income, education, urbanisation, alcohol, smoking and physical activity. [3] Further adjusted for total energy intake, smoking status, alcohol drinking status, educational level, occupation, monthly household income, physical activity, family history of disease (including cardiovascular disease, hypertension, hyperlipidaemia and diabetes) and depressive symptoms. [4] Further adjusted for income level, education level, marital status, residence, BMI, physical activity, smoking status, hormonal contraception, number of 24-h dietary records and energy intake. [5] Further adjusted for race, family history of IBD, smoking, BMI, physical activity, total energy intake, regular NSAIDs use, oral contraceptives use, and menopausal hormone therapy. [6] Further adjusted for baseline BMI, total energy intake, physical activity, smoking status, marital status, living alone, employment status, working hours per week, health-related career, years of education and baseline self-perception of competitiveness, anxiety and dependence levels. [7] Further adjusted for energy intake. [8] Further adjusted for fibre intake, total energy intake, educational level, marital status, smoking status, BMI, physical activity, alcohol consumption, number of medications and number of chronic conditions. [9] Further adjusted for maternal pre-pregnancy BMI, maternal education, smoking and alcohol intake during pregnancy, maternal symptoms of depression and ADHD, maternal age, parity, child sex and childbirth quarter. [10] Further adjusted for smoking status, alcohol drinking status, education level, employment, monthly household income, physical activity, family history of disease (including CVD, hypertension, hyperlipidaemia and diabetes), depressive symptoms, hypertension, hyperlipidaemia, diabetes, total energy intake, dietary supplement use, total protein intake and milk intake. * 95% credible intervals.

Within the NutriNet-Santé cohort, several studies have performed dietary adjustments for the associations between UPF intake and all-cause mortality, CVD, overweight/obesity incidence, T2DM, cancer and functional gastrointestinal disorders [102,106,117,124,126,127,129]. Schnabel et al. found a 15% (95% confidence interval: 1.04, 1.27) increased risk of all-cause mortality per 10% increase in UPF intake in the diet [102]. Adjusting for French dietary guideline adherence or for both French dietary guideline adherence and for Western dietary pattern still resulted in each 10% increment in UPF intake being associated with a 14% (1.04, 1.27) or 19% (1.05, 1.35) increased risk, respectively, of all-cause mortality [102].

Srour et al. reported a 12% (1.05, 1.20), 13% (1.02, 1.24) and 11% (1.01, 1.21) increased risk of CVD, coronary heart disease (CHD) and cerebrovascular disease, respectively, per 10% increase in UPF in the diet [117]. Multiple dietary adjustments did not alter these risk estimates. First, adjusting for saturated fat, sodium and added sugar intake resulted in a 13% (1.05, 1.20), 14% (1.03, 1.26) and 12% (1.02, 1.22) increased risk of CVD, CHD and cerebrovascular disease, respectively [117]. Second, adjusting instead for a healthy dietary pattern still resulted in an 11% (1.03, 1.19), 11% (1.00, 1.23, p = 0.04) and 10% (1.00, 1.20, p = 0.04) increased risk of CVD, CHD and cerebrovascular disease, respectively [117]. Third, adjusting for intakes of sugary products, red and processed meat, salty snacks, beverages, fats and sauces also still resulted in a 12% (1.04, 1.20), 12% (1.01, 1.24) and 11% (1.01, 1.22) increased risk of CVD, CHD and cerebrovascular disease, respectively, per 10% increase in UPF in the diet [117].

In a separate study, Srour et al. reported a 15% (1.06, 1.25) increased risk of T2DM with each 10% increase in UPF in the diet, which included adjustment for dietary quality using the FSA-NPS-DI [124]. Again, subsequent dietary adjustments did not alter the increased risk of T2DM. A 10% increase in UPF in the diet was still associated with a 19% (1.09, 1.30) increased risk when further adjusting for saturated fat, sodium, sugar and dietary fibre intake, a 13% (1.04, 1.24) increased risk after adjusting for healthy and Western dietary patterns, and a 14% (1.04, 1.25) increased risk after adjusting for intakes of red and processed meat, sugary drinks, fruits and vegetables, whole grains, nuts, and yogurt in place of the FSA-NPS-DI adjustment [124]. Srour et al. also adjusted for absolute amounts of unprocessed or minimally processed food intake, which few studies have performed to date. This adjustment also did not alter the increased risk of T2DM (hazard ratio (HR) per 100g/day increase in UPF intake: 1.05 (1.02, 1.08) [124].

Fiolet et al. reported a 12% (1.06, 1.18) and 11% (1.02, 1.22) increased risk of all cancer and breast cancer, respectively, per 10% increase in UPF in the diet [126]. Adjustment for lipids (including fat), sodium and carbohydrate intake had no impact on the risk of all cancer (HR: 1.12 (1.07, 1.18)) or breast cancer (HR: 1.11 (1.01, 1.21)) per 10% increase in UPF in the diet, respectively [126]. Adjustment instead for Western dietary pattern also did not change the 12% (1.06, 1.18) and 11% (1.02, 1.22) increased risks [126].

Beslay et al. reported a greater BMI gain (β: 0.02 kg/m^2 (0.01, 0.02)) and increased risk of overweight (HR: 1.11 (1.08, 1.14)) or obesity (HR: 1.09 (1.05, 1.13)), per 10% increase in UPF in the diet [106]. Adjusting for healthy and Western dietary patterns did not alter the greater BMI gain (β: 0.02 kg/m^2 (0.01, 0.02)), or increased risk of overweight (HR: 1.10 (1.07, 1.13)) or obesity (HR: 1.11 (1.07, 1.15)), and neither did adjustment for saturated fat, sugar, sodium and dietary fibre intake, which also resulted in a greater BMI gain (β: 0.02 kg/m^2 (0.01, 0.02)), and increased risk of overweight (HR: 1.10 (1.08, 1.13)) or obesity (HR: 1.10 (1.06, 1.14), per 10% increase in UPF intake [106].

Schnabel et al. identified an increased risk of irritable bowel syndrome (IBS) (odds ratio (OR): 1.24 (1.12, 1.38)) and functional dyspepsia (OR: 1.26 (1.07, 1.48)) when comparing the highest vs. lowest quartiles of UPF intake [129]. Adjustment for adherence to French dietary guidelines did not alter the increased risk of IBS (OR: 1.25 (1.12, 1.39)) or functional dyspepsia (OR: 1.25 (1.05, 1.47)) across extreme quartiles of UPF intake [129].

Four studies within the Seguimiento Universidad de Navarra (SUN) cohort have adjusted for fat, added sugar and sodium intake, or for dietary pattern. Rico-Campà et al. demonstrated that the highest vs. lowest quartile of UPF intake was associated with a

62% (1.13, 2.33) increased risk of all-cause mortality [103]. Adjustment for saturated and trans fats, added sugar and sodium intake still resulted in a 69% (1.12, 2.56) increased risk of all-cause mortality. A 58% (1.10, 2.28) increased risk still remained after adjusting for Mediterranean diet pattern adherence instead [103].

Llavero-Valero et al. reported that the highest vs. lowest tertile of UPF intake was associated with a 53% (1.06, 2.22) increased risk of T2DM, which was unaltered (HR: 1.50 (1.02, 2.21)) after adjusting for Mediterranean diet pattern adherence [123].

Gómez-Donoso et al. found a 41% (1.15, 1.73) increased risk of incident depression in the highest vs. lowest quartile of UPF intake, which was still associated with a 33% (1.07, 1.64) higher risk of incident depression after further adjustment for other covariates, including Mediterranean diet pattern adherence [132].

Leone et al. identified an increased risk of gestational diabetes in females aged 30 and over (OR 1st vs. 3rd tertile: 2.05 (1.03, 4.07)), which was unaltered after adjustment for Mediterranean diet pattern adherence (OR 1st vs. 3rd tertile: 2.06 (1.05, 4.06)) [114].

In the US Third National Health and Nutrition Examination Survey (NHANES III) cohort, there was a 31% (1.09, 1.58) increased risk of all-cause mortality in the highest vs. lowest quartile of UPF intake, which remained significant after further adjustment for dietary quality score using the HEI-2000 (p-trend = 0.001) [104]. However, diet-adjusted risk estimates were not provided.

In the Italian Moli-sani cohort, the highest vs. lowest quartile of UPF intake was associated with a 32% (1.15, 1.53) higher risk of all-cause mortality, 65% (1.29, 2.11) higher risk of CVD mortality, and a 63% (1.19, 2.25) higher risk of ischemic heart disease (IHD)/cerebrovascular mortality [105]. Adjusting for saturated fat, sugar, sodium and dietary cholesterol intake resulted in a 28% (1.09, 1.49), 56% (1.19, 2.03) and 33% (0.94, 1.90) increased risk of all-cause, CVD and IHD/cerebrovascular mortality, respectively, in the highest vs. lowest quartile of UPF intake. Bonaccio et al. also individually adjusted for saturated fat, sugar, sodium and dietary cholesterol in turn, with UPF intake remaining significantly associated with all-cause, CVD and IHD/cerebrovascular mortality in all adjustments, except for sugar intake and IHD/cerebrovascular mortality (HR: 1.37 (0.98, 1.90)). Adjusting instead for Mediterranean diet pattern adherence resulted in a 26% (1.09, 1.46), 58% (1.23, 2.03) and 52% (1.10, 2.09) increased risk of all-cause, CVD and IHD/cerebrovascular mortality [105].

In the Framingham Offspring cohort, each additional serving of UPF per day was associated with a 5% (1.02, 1.08), 9% (1.02, 1.16), 7% (1.03, 1.12) and 9% (1.04, 1.15) increased risk of overall CVD, CVD mortality, hard CVD and hard coronary heart disease, respectively [118]. Further adjustment for diet quality using the DGAI-2010 still resulted in a 4% (1.01, 1.07), 9% (1.02, 1.16), 6% (1.02, 1.11) and 9% (1.03, 1.15) increased risk of overall CVD, CVD mortality, hard CVD and hard coronary heart disease [118].

In the US Prostate, Lung, Colorectal, and Ovarian Cancer Screening Trial cohort, the highest vs. lowest quintile of UPF intake was associated with a 50% (1.36, 1.64) increased risk of CVD mortality, and a 68% (1.50, 1.87) increased risk of heart disease mortality [119]. Multiple dietary adjustments did not alter this risk; adjustment for saturated fat, added sugar and sodium resulted in a 48% (1.34, 1.63) and 65% (1.47, 1.85) increased risk of CVD mortality and heart disease mortality, adjustment for diet quality using HEI-2005 resulted in a 48% (1.35, 1.63) and 67% (1.49, 1.86) increased risk of CVD mortality and heart disease mortality, and adjustment instead for red meat, processed meat, whole grains, fruit, vegetables, fibre and dairy intake also still resulted in a 49% (1.35, 1.64) and 66% (1.48, 1.86) increased risk of CVD mortality and heart disease mortality, respectively [119].

In the European Prospective Investigation into Cancer and Nutrition (EPIC) cohort, each additional standard deviation (SD) increment in UPF intake per day was associated with a 0.12 kg (0.09, 0.15) greater increase in weight over 5 years of follow-up, which was unaltered after further adjusting for Mediterranean diet score (β: 0.12 kg/5 years (0.09, 0.15)) [111]. In sensitivity analyses of fully adjusted models including Mediterranean diet adherence, UPF intake was associated with a higher risk of overweight/obesity (relative risk (RR): 1.05 (1.04, 1.06)) and obesity (RR: 1.05 (1.03, 1.07)) per 1SD increase in UPF per day. This corresponded to a 15% (1.11, 1.19) higher risk of overweight/obesity in participants with normal weight and a 16% (1.09, 1.23) higher risk of obesity in participants with overweight at baseline, when comparing the highest vs. lowest quintiles of UPF intake [111].

In the China Nutrition and Health Survey (CNHS), consuming $\geq$50 g of UPF per day was associated with an increased risk of overweight/obesity (OR: 1.85 (1.58, 2.17)) and central obesity (OR: 2.04 (1.79, 2.33)), when compared to no UPF intake. Adjustment for traditional and modern dietary patterns did not alter the increased risks (overweight/obesity, OR: 1.45 (1.21, 1.74), central obesity, OR: 1.50 (1.29, 1.74)) [108].

In the Seniors Study on Nutrition and Cardiovascular Risk in Spain (Seniors-ENRICA-1), Sandoval-Insausti et al. found an increased risk of abdominal obesity (OR: 1.62 (1.04, 2.54) in the highest vs. lowest tertile of UPF intake, which was unaltered after adjustment for Mediterranean diet adherence and fibre and omega-3 fatty acid intake (OR: 1.61 (1.01, 2.56)) [110].

Donat-Vargas et al. identified an increased risk of hypertriglyceridaemia (OR: 2.00 (1.04, 3.85)) and low-HDL cholesterol (OR: 2.04 (1.22, 3.41)), as well as a significant increase in blood triglycerides (β: 6.11 mg/dL (1.30, 10.91)) when comparing the highest vs. lowest tertile of UPF intake [136]. Adjustment for unprocessed or minimally processed food intake did not alter the increased risk of hypertriglyceridaemia (OR: 2.66 (1.20, 5.90)), low-HDL cholesterol (OR: 2.23 (1.22, 4.05)) or change in blood triglycerides (β: 6.87 mg/dL (1.48, 12.27)) [136].

In the Pelotas-Brazil 2004 Birth Cohort, Costa et al. found a 0.09 kg/m^2 (0.07, 0.10) greater gain in FMI from ages 6 to 11, per 100 g daily increase in UPF intake [116]. Adjustment for other NOVA food groups (minimally processed and processed food, and processed culinary ingredients intake) significantly increased the associated FMI gain to 0.14 kg/m^2 (0.13, 0.15) from age 6 to 11, per 100 g daily increase in UPF intake [116].

In the Avon Longitudinal Study of Parents and Children (ALSPAC) cohort, the highest vs. lowest quintile of UPF intake was associated with a 0.06 kg/m^2 (0.04, 0.08) and 0.03 kg/m^2 (0.01, 0.05) greater yearly increase in BMI and FMI, respectively, from the age of 7 to 24 [115]. Adjustment for saturated fat, sugar, sodium and fibre intake did not alter the association between UPF intake and increases in BMI (β: 0.07 kg/m^2/year (0.04, 0.08)) or FMI (β: 0.03 kg/m^2/year (0.01, 0.05)) [115].

Koniecnzna et al. conducted a prospective analysis of the PREDIMED-Plus trial over the course of 12 months. Each 10% increment in UPF intake was associated with increases in total (β: 0.09 (0.06, 0.13)) and visceral (β: 0.09 (0.05, 0.13)) fat mass z-scores. Adjusting for overall repeated measures of saturated and trans fat, sodium, glycaemic index, alcohol and fibre intake across the 12 month study did not alter the significant association between UPF intake and increases in total (β: 0.06 (0.03, 0.09) and visceral (β: 0.06 (0.01, 0.10)) fat mass z-scores per 10% increase in daily UPF intake [109]. Adjusting instead for overall repeated measures of Mediterranean diet pattern adherence across the 12 month study also did not alter the association between each 10% increment in UPF intake and increases in total (β: 0.06 (0.02, 0.09)) and visceral (β: 0.06 (0.01, 0.10)) fat mass z-scores [109].

In the Tianjin Chronic Low-grade Systemic Inflammation and Health (TCLSIH) Cohort Study, the highest vs. lowest quartile of UPF intake was associated with a 17% (1.07, 1.29) higher risk of non-alcoholic fatty liver disease (NAFLD) in the age, sex and BMI adjusted model. After adjustment for other confounders including for a healthy diet score based on fruit, vegetable, red meat and fish intake, the increased risk associated with the highest vs. lowest quintile of UPF intake was 19% (1.08, 1.31) [125].

Zhang et al. found a 21% (1.10, 1.33) increased risk of hyperuricaemia in the highest vs. lowest quartile of UPF intake, which was still associated with a 17% (1.06, 1.30) increased risk of hyperuricaemia after adjustment for dietary pattern [134].

In a separate study, Zhang et al. reported that each 10% increment in UPF in the diet was associated with a -0.30 kg (-0.50, -0.09) and -0.0043 kg/kg weight (-0.0073, -0.0014) yearly reduction in absolute and weight-adjusted grip strength, respectively [138].

Adjustment for further covariates including a healthy diet score (based on fruit, vegetable, unprocessed red meat and fish intake), dietary supplement use and protein and milk intake did not alter the association, with each 10% increment in UPF intake still associated with −0.32 kg (−0.53, −0.11) and −0.0046 kg/kg weight (−0.0076, −0.0016) yearly reductions in absolute and weight-adjusted grip strength, respectively [138].

In a combined analysis of the Nurses' Health Study, the Nurses' Health Study II and the Health Professionals Follow-up Study, Lo et al. found a 75% (1.29, 2.35) increased risk of Crohn's disease in the highest vs. lowest quartile of UPF intake after adjusting for age, cohort and calendar year. The increased risk was unchanged after further covariate adjustments, including for diet quality defined by the AHEI-2010 (HR: 1.70 (1.23, 2.35)) [130].

In the Prospective Urban Rural Epidemiology (PURE) cohort, Narula et al. identified an 82% (1.22, 2.72) increased risk of inflammatory bowel disease (IBD) and a 450% (1.67, 12.13) increased risk of Crohn's disease in those consuming five or more UPF servings per day, compared with those consuming less than one serving per day. Adjustment for AHEI-2010 still resulted in a 92% (1.28, 2.90) increased risk of IBD and a 490% (1.78, 13.45) increased risk of Crohn's disease [128].

In the Norwegian Mother, Father and Child Cohort Study, Borge et al. reported that each 1 SD increase in maternal UPF intake was associated with an increase in absolute (0.38 (0.27, 0.49)) and relative (4.5% (3.3, 4.9)) measures of child attention deficit hyperactivity disorder (ADHD) symptoms at age 8, using the Parent Rating Scale for Disruptive Behaviour Disorders [137]. Adjustment for child Diet Quality Index (based on diet diversity, diet quality and diet equilibrium [139]) did not alter the associated increase in absolute (0.25 (0.13, 0.38)) or relative (3.0% (1.5, 4.5)) ADHD symptoms [137].

Three studies have considered the impact of diet quality and dietary pattern using alternative methods. In the ATTICA cohort, each additional weekly serving of UPF was associated with a 10% (1.02, 1.21) increased risk of CVD. Kouvari et al. then performed sub-group analysis based on Mediterranean diet pattern adherence. Participants with moderate to high adherence to the Mediterranean diet had an attenuated (8% (0.98, 1.19)) risk of CVD per weekly serving of UPF, whereas participants with low adherence to the Mediterranean diet had an even greater risk of 19% (1.12, 1.25), per weekly serving of UPF [140].

Bonaccio et al. identified that diet quality (defined by the FSA-NPS-DI) was only significantly associated with all-cause mortality in high UPF consumers (HR per 1 SD increase in FSA-NPS-DI: 1.14 (1.05, 1.25), but not in low UPF consumers (HR: 1.00 (0.93, 1.07) (*p* for interaction = 0.034) in the Moli-sani cohort [141]. The interaction between diet quality and UPF intake was not significant for CVD mortality.

In the ENRICA study, the highest vs. lowest quartile of UPF intake had a 44% (1.01, 2.07) increased risk of all-cause mortality [91]. Instead of dietary adjustment, Blanco-Rojo et al. compared the highest vs. lowest intakes of nutrients from UPF intake, including total, saturated and trans fat, carbohydrates, sugar, sodium and fibre [91]. The nutrient content of UPFs was not associated with an increased mortality risk, except for trans fat (HR highest vs. lowest quartile: 1.39 (1.00, 1.92), *p* = 0.047).

6.2. Adjustment for Fat, Sodium, Carbohydrate and Dietary Pattern

Two studies have simultaneously adjusted for fat, sodium and carbohydrate intake and for dietary pattern, which are reported in Supplementary Table S1. For cancer outcomes, Fiolet et al. adjusted for both intakes of lipids (including fat), sodium, and carbohydrates and Western dietary pattern, resulting in a 13% (1.07, 1.18) and 11% (1.01, 1.21) increased risk of all cancer and breast cancer per 10% increase in UPF in the diet [126].

Adjibade et al. identified a 21% (1.15, 1.27) higher risk of depressive symptoms per 10% increase in UPF in the diet in the NutriNet-Santé cohort [131]. After adjusting for intakes of lipids (including fat), sodium, and carbohydrates and for healthy and Western dietary patterns, the risk of depressive symptoms per 10% increase in UPF in the diet was still 22% (1.16, 1.29) [131].

6.3. Adjustment for Fat And/or Sugar and/or Sodium

Some studies have adjusted for one or two components of fat and/or sugar and/or sodium intake, rather than all three components. One study adjusted for carbohydrate intake, rather than sugar intake [114]. These adjustments are reported in Supplementary Table S2. The significant associations between UPF intake and all-cause mortality, overweight or obesity, central obesity, T2DM, hypertension, gestational weight gain, neonatal anthropometrics and blood lipid profiles were unchanged following these dietary adjustments [103,108,113,114,120,123,135,136].

6.4. Adjustment for Other Dietary Components

Other measures used for dietary adjustment are provided in Supplementary Table S3. Other dietary adjustments include for fried foods, fruit and vegetables, UPF soft drinks, multivitamin use and excluding bacon, sausage and processed meats from ultra-processed food intake. These adjustments had no impact on the association between higher intakes of UPF and risk of all-cause mortality, cancer, overweight/obesity, increased total and visceral fat mass, increased BMI and FMI, NAFLD, weight and waist circumference gain, adverse blood lipid profiles, grip strength decline, incident hypertension and renal function decline [103,104,106,107,109,111,112,115,121,122,125,126,133,136,138].

6.5. Dietary Adjustments That Explain the Association between UPF Intake and Health-Related Outcomes

To date, only two studies have performed dietary adjustments that explain the association between higher UPF intakes and adverse health-related outcomes. In the PREDIMED-Plus study, each 10% increase in UPF in the diet was associated with a 5% (0.00, 0.09, $p = 0.031$) increase in android:gynoid fat ratio z-score during 12 months of follow-up [109]. Adjusting for repeated measures of sodium, saturated and trans fat, alcohol, fibre and glycaemic index, or adjusting for repeated measures of Mediterranean Diet adherence during the 12-month follow-up period resulted in a non-significant association between UPF intake and android:gynoid fat ratio z-score [109].

In the Moli-sani cohort, the highest vs. lowest quartile of UPF intake had a 36% (1.01, 1.83) higher risk of other cause mortality (any mortality, excluding CVD and cancer). However, after adjusting for Mediterranean diet score, this became non-significant (1.26 (0.94, 1.69)) [105]. As noted in Section 6.1, the increased risk of IHD/cerebrovascular mortality also became non-significant after adjusting for saturated fat, sugar, sodium and dietary cholesterol.

6.6. Adjustment for Total Energy Intake

An ultra-processed diet has been shown to increase energy intake in comparison with a minimally processed diet [49]. Energy intake may be a mediator of both nutritional aspects (high energy density and palatability), and of some ultra-processing aspects (a degraded food matrix influencing oro-sensory exposure and satiety) of UPFs. Adjustment for total daily energy intake is not only useful to control for measurement error in epidemiological dietary assessment to improve risk estimation of other dietary measures [142,143], but it can also provide information on the associated risk between UPF intake and adverse health outcomes, independent of energy intake [144].

Adjustment for energy intake can be achieved using several methods [145,146]. However, it has typically been performed by energy-adjusting the UPF independent variable, either via the residual method (regressing UPF intake onto total energy intake to produce residuals) or via the nutrient density method (usually as 'energy intake from UPFs/total energy intake', though 'total weight of UPFs/total food weight' has also been used to capture the non-nutritive aspects of UPFs) [146]. Total energy intake is then included as a covariate in the model [146]. Some studies instead use absolute UPF intake as the independent variable, and then include total energy intake in the model.

Table 5 presents the prospective cohort studies performing adjustments for total energy intake. Forty-seven studies have performed some form of energy adjustment across 131 models. 80 models demonstrate a significant association between energy-adjusted UPF intake and a health-related outcome. 6/6 models were significantly associated with all-cause mortality, 12/15 models were significantly associated with any CVD outcome, 3/3 models were significantly associated with T2DM, 15/17 models were significantly associated with adult weight gain/overweight/obesity, and 15/25 models with gestational or child anthropometrics. Twenty-one non-significant models with energy adjustment were from multiple models for child appetitive traits (eight; Vedovato et al. [99]), childhood anthropometrics and glucose profiles (six; Costa et al. [98]), child asthma and wheezing (four; Machado Azeredo et al. [101]) and childhood lipid profiles (three; Rauber et al. [95]). Four studies provided insufficient detail on energy adjustments [87,96,140,147].

6.7. Prospective Studies Reporting Mediation Analyses

Besides being included as a covariate within models, formal mediation analysis can be used to determine whether dietary components mediate the association between UPF intake and adverse health-related outcomes [149,150]. Few studies to date have performed mediation analyses between UPF intake, dietary components and health-related outcomes.

Bonaccio et al. examined the mediating role of nutrients and energy content on all-cause mortality, CVD mortality and IHD/cerebrovascular mortality [105]. All dietary factors combined (sugar, saturated fat, dietary cholesterol, dietary sodium and energy content) significantly accounted for 41.3% ((11.9%, 78.5%), $p < 0.001$) of IHD/cerebrovascular mortality risk, but did not account for all-cause mortality (12.8% (1.6%, 56.5%), $p = 0.14$) or CVD mortality (11.5% (1.5%, 53.3%), $p = 0.15$) risk. Sugar content alone accounted for 23.2% ((9.7%, 45.9%), $p < 0.001$), 18.0% ((7.2%, 38.4%), $p = 0.003$) and 36.3% ((13.8%, 67.0%), $p < 0.001$)) of the associated risk between UPF intake and all-cause mortality, CVD mortality and IHD/cerebrovascular mortality, respectively. Saturated fat or sodium content did not account for any of the associated risks.

Fiolet et al. performed mediation analyses for sodium, total lipids, saturated fat, monounsaturated fat, polyunsaturated fat, carbohydrate and for Western dietary pattern, with all mediation effects for the association between UPF intake and overall cancer being less than 2% (all $p > 0.05$) [126].

Koniecnzna et al. found that repeated measures of saturated fat, trans fat and fibre explained 11–30% of the associations between UPF intake and increases in measures of central and overall adiposity over 12 months of follow-up [109]. Repeated measures of sodium, total energy intake and glycaemic index did not mediate any of the associations.

Costa et al. identified that 58.2% (0.07 kg/m^2 (0.05, 0.10)) of the association between UPF intake and the increase in FMI from age 6 to 11 in children was mediated by energy content, with the remaining 41.8% being either a direct effect of ultra-processing, or as a result of unmeasured variables [116].

Vedovato et al. showed that energy intake was a mediator between UPF intake at 4 years of age and the appetite traits, 'satiety responsiveness' and 'food fussiness', but not with 'food responsiveness', at age 7 [99].

Gomes et al. showed that the percentage of total energy derived from UPFs in the third trimester was associated with total energy intake in the third trimester, which was also associated with gestational weight gain in the third trimester [93].

Table 5. Prospective cohort studies adjusting for total energy intake.

Author, Year	Outcome	Method of Analysis	Energy Adjustment	Effect
Schnabel 2019 [102]	All-cause mortality	HR per 10% increase in UPF	UPF as % weight + adjusted for TEI	1.15 (1.04, 1.27) [1]
Rico-Campa 2019 [103]	All-cause mortality	HR 1st vs. 4th quartile	Energy-adjusted UPF + adjusted for TEI	1.62 (1.13, 2.33) [2]
	Cardiovascular deaths	HR 1st vs. 4th quartile	Energy-adjusted UPF + adjusted for TEI	2.16 (0.92, 5.06) [2]
	Cancer deaths	HR 1st vs. 4th quartile	Energy-adjusted UPF + adjusted for TEI	1.22 (0.70, 2.12) [2]
Blanco-Rojo 2019 [91]	All-cause mortality	HR 1st vs. 4th quartile	UPF as % TEI	1.44 (1.01, 2.07) [3]
Kim 2019 [104]	All-cause mortality	HR 1st vs. 4th quartile	UPF servings/day + adjusted for TEI	1.31 (1.09, 1.58) [4]
	CVD mortality	HR 1st vs. 4th quartile	UPF servings/day + adjusted for TEI	1.10 (0.74, 1.67) [4]
Romero Ferreiro 2021 [89]	All-cause mortality	HR per 10% increase in UPF	UPF as % TEI + adjusted for TEI	1.16 (1.06, 1.26) [5]
Bonaccio 2021 [105]	All-cause mortality	HR 1st vs. 4th quartile	UPF as % weight + adjusted for TEI and energy content of UPFs	1.35 (1.15, 1.58) [6]
	CVD mortality	HR 1st vs. 4th quartile	UPF as % weight + adjusted for TEI and energy content of UPFs	1.66 (1.28, 2.16) [6]
	IHD/cerebrovascular mortality	HR 1st vs. 4th quartile	UPF as % weight + adjusted for TEI and energy content of UPFs	1.48 (1.05, 2.09) [6]
	Cancer mortality	HR 1st vs. 4th quartile	UPF as % weight + adjusted for TEI	1.00 (0.80, 1.26) [6]
	Other cause mortality	HR 1st vs. 4th quartile	UPF as % weight + adjusted for TEI	1.36 (1.01, 1.83) [6]
Beslay 2020 [106]	BMI change (kg/m^2)	Beta per 10% increase in UPF	UPF as % weight + adjusted for TEI	0.02 (0.01, 0.02) [7]
	Overweight	HR per 10% increase in UPF	UPF as % weight + adjusted for TEI	1.11 (1.08, 1.14) [7]
	Obesity	HR per 10% increase in UPF	UPF as % weight + adjusted for TEI	1.09 (1.05, 1.13) [7]
Mendonça 2016 [107]	Overweight/obesity	HR 1st vs. 4th quartile	UPF servings/day + adjusted for TEI	1.27 (1.09, 1.49) [8]
Li 2021 [108]	Overweight/obesity	OR none vs. $\geq$50 g/day	Absolute UPF g/day + adjusted for TEI	1.85 (1.58, 2.17) [9]
	Central obesity	OR none vs. $\geq$50 g/day	Absolute UPF g/day + adjusted for TEI	2.04 (1.79, 2.33) [9]
Koniecnzna 2021 [109]	Total fat mass (z-score)	Beta per 10% increase in UPF	UPF as % weight + adjusted for TEI	0.09 (0.06, 0.12) [10]
	Visceral fat mass (z-score)	Beta per 10% increase in UPF	UPF as % weight + adjusted for TEI	0.09 (0.04, 0.13) [10]
	Android:Gynoid fat ratio (z-score)	Beta per 10% increase in UPF	UPF as % weight + adjusted for TEI	0.04 (0.00, 0.08) $p = 0.055$ [10]
Sandoval-Insausti 2020 [110]	Abdominal obesity	OR 1st vs. 3rd tertile	UPF as % TEI + adjusted for TEI	2.55 (1.04, 6.27) [11]
Cordova 2021 [111]	Weight gain (kg)	Beta per 1SD increase in UPF/day	Energy-adjusted UPF	0.118 (0.085, 0.151) [12]
	Overweight/obesity	RR per 1SD increase in UPF/day	Energy-adjusted UPF	1·05 (1.04, 1.06) [12]
	Obesity	RR per 1SD increase in UPF/day	Energy-adjusted UPF	1·05 (1.03, 1.07) [12]

Table 5. *Cont.*

Author, Year	Outcome	Method of Analysis	Energy Adjustment	Effect
Canhada 2020 [112]	Large weight gain (≥90th percentile: ≥1.68 kg/year)	RR 1st vs. 4th quartile	UPF as % TEI + adjusted for TEI	1.27 (1.07, 1.51) [13]
	Large WC gain (≥90th percentile: ≥2.42 cm/year)	RR 1st vs. 4th quartile	UPF as % TEI + adjusted for TEI	1.36 (1.14, 1.61) [13]
	Incident overweight/obesity	RR 1st vs. 4th quartile	UPF as % TEI + adjusted for TEI	1.22 (1.04, 1.42) [13]
	Incident obesity	RR 1st vs. 4th quartile	UPF as % TEI + adjusted for TEI	1.02 (0.85, 1.21) [13]
Rohatgi 2017 [113]	Gestational weight gain (kg)	Beta per 1% increase in UPF intake	UPF as % TEI + adjusted for TEI	1.3 (0.3, 2.4) [14]
	Neonate thigh skinfold thickness (mm)	Beta per 1% increase in UPF intake	UPF as % TEI + adjusted for TEI	0.20 (0.005, 0.40) [14]
	Neonate subscapular skinfold thickness (mm)	Beta per 1% increase in UPF intake	UPF as % TEI + adjusted for TEI	0.10 (0.02, 0.30) [14]
	Neonate body fat percentage (%)	Beta per 1% increase in UPF intake	UPF as % TEI + adjusted for TEI	0.60 (0.04, 1.20) [14]
Gomes 2021 [93]	Gestational weight gain 3rd trimester (kg)	Beta per 1% increase in UPF intake during 3rd trimester	UPF as % TEI	4.17 (0.55, 7.79) [15]
	Gestational weight gain 2nd trimester (kg)	Beta per 1% increase in UPF intake in 2nd trimester	UPF as % TEI	−1.50 (−5.08, 2.08) [15]
Leone 2021 [114]	Gestational diabetes pooled	OR 1st vs. 3rd tertile	Energy-adjusted UPF + adjusted for TEI	1.10 (0.74, 1.64) [16]
	Gestational diabetes <30	OR 1st vs. 3rd tertile	Energy-adjusted UPF + adjusted for TEI	0.89 (0.54, 1.46) [16]
	Gestational diabetes ≥30	OR 1st vs. 3rd tertile	Energy-adjusted UPF + adjusted for TEI	2.05 (1.03, 4.07) [16]
Chang 2021 [115]	BMI (kg/m^2)/year	Beta 1st vs. 5th quintile	UPF as % weight + adjusted for child's TEI	0.06 (0.04, 0.08) [17]
	Fat mass index (kg/m^2)/year	Beta 1st vs. 5th quintile	UPF as % weight + adjusted for child's TEI	0.03 (0.01, 0.05) [17]
	Lean mass index (kg/m^2)/year	Beta 1st vs. 5th quintile	UPF as % weight + adjusted for child's TEI	0.004 (−0.007, 0.01) [17]
	Body fat percentage (%)/year	Beta 1st vs. 5th quintile	UPF as % weight + adjusted for child's TEI	0.004 (−0.05, 0.06) [17]
	Weight (kg/year)	Beta 1st vs. 5th quintile	UPF as % weight + adjusted for child's TEI	0.20 (0.11, 0.28) [17]
	Waist circumference (cm/year)	Beta 1st vs. 5th quintile	UPF as % weight + adjusted for child's TEI	0.17 (0.11, 0.22) [17]
	BMI z-score	Beta 1st vs. 5th quintile	UPF as % weight + adjusted for child's TEI	0.01 (0.003, 0.01) [17]
	Fat mass (kg/year)	Beta 1st vs. 5th quintile	UPF as % weight + adjusted for child's TEI	0.15 (0.08, 0.21) [17]
	Lean mass (kg/year)	Beta 1st vs. 5th quintile	UPF as % weight + adjusted for child's TEI	-0.04 (-0.11, 0.02) [17]

Table 5. *Cont.*

Author, Year	Outcome	Method of Analysis	Energy Adjustment	Effect
Costa 2021 [116]	Fat mass index (kg/m^2)	Beta/100 g increase in UPF intake	Absolute UPF g/day + adjusted for energy intake/expenditure ratio + TEI	0.05 (0.04, 0.06) [18]
Vedovato 2021 [99]	BMI z-score age 10	Beta per 1 kcal/100 kcal/d increase in energy from UPF at age 4	UPF as % TEI at age 4	0.028 (0.006, 0.051) [19]
	BMI z-score age 10	Beta per 1 kcal/100 kcal/d increase in energy from UPF at age 7	UPF as % TEI at age 7	0.014 (−0.007, 0.036) [19]
	Enjoyment of food at age 7	Beta per 1 kcal/100 kcal/d increase in energy from UPF at age 4	UPF as % TEI at age 4	−0.002 (−0.021, 0.016) [19]
	Food responsiveness at age 7	Beta per 1 kcal/100 kcal/d increase in energy from UPF at age 4	UPF as % TEI at age 4	0.017 (−0.001, 0.035) [19]
	Emotional overeating at age 7	Beta per 1 kcal/100 kcal/d increase in energy from UPF at age 4	UPF as % TEI at age 4	0.010 (−0.006, 0.026) [19]
	Emotional undereating at age 7	Beta per 1 kcal/100 kcal/d increase in energy from UPF at age 4	UPF as % TEI at age 4	0.007 (−0.012, 0.027) [19]
	Satiety Responsiveness at age 7	Beta per 1 kcal/100 kcal/d increase in energy from UPF at age 4	UPF as % TEI at age 4	0.013 (−0.004, 0.029) [19]
	Slowness in eating at age 7	Beta per 1 kcal/100 kcal/d increase in energy from UPF at age 4	UPF as % TEI at age 4	−0.015 (−0.035, 0.006) [19]
	Food Fussiness at age 7	Beta per 1 kcal/100 kcal/d increase in energy from UPF at age 4	UPF as % TEI at age 4	0.026 (0.007, 0.045) [19]
	Desire to Drink at age 7	Beta per 1 kcal/100 kcal/d increase in energy from UPF at age 4	UPF as % TEI at age 4	0.018 (−0.003, 0.039) [19]
Costa 2019 [98]	△BMI age 4 to 8	Beta per 10% increase in UPF intake	UPF as % TEI at age 4	0.00 (−0.02, 0.01) [20]
	△WC age 4 to 8	Beta per 10% increase in UPF intake	UPF as % TEI at age 4	0.07 (0.01, 0.13) [20]
	△WHR age 4 to 8	Beta per 10% increase in UPF intake	UPF as % TEI at age 4	0.00 (0.00, 0.00) [20]
	△Sum skinfolds age 4 to 8	Beta per 10% increase in UPF intake	UPF as % TEI at age 4	0.05 (−0.04, 0.15) [20]
	Glucose (mmol/L)	Beta per 10% increase in UPF intake	UPF as % TEI at age 4	0.00 (−0.01, 0.00) [20]
	Insulin (uU/mL)	Beta per 10% increase in UPF intake	UPF as % TEI at age 4	0.00 (−0.00, 0.01) [20]
	HOMA-IR	Beta per 10% increase in UPF intake	UPF as % TEI at age 4	0.00 (−0.01, 0.01) [20]
Srour 2019 [117]	All CVD	HR per 10% increase in UPF	UPF as % weight + adjusted for TEI	1.12 (1.05, 1.20) [21]
	Coronary heart disease	HR per 10% increase in UPF	UPF as % weight + adjusted for TEI	1.13 (1.02, 1.24) [21]
	Cerebrovascular disease	HR per 10% increase in UPF	UPF as % weight + adjusted for TEI	1.11 (1.01, 1.21) [21]
Du 2021 [86]	Incident CAD	HR 1st vs. 4th quartile	Energy-adjusted UPF + adjusted for TEI	1.21 (1.06, 1.37) [22]

Table 5. *Cont.*

Author, Year	Outcome	Method of Analysis	Energy Adjustment	Effect
Juul 2021 [118]	Overall CVD	HR per serving UPF/day	Energy-adjusted UPF + adjusted for TEI	1.05 (1.02, 1.08) [23]
	CVD mortality	HR per serving UPF/day	Energy-adjusted UPF + adjusted for TEI	1.09 (1.02, 1.16) [23]
	Incident hard CVD	HR per serving UPF/day	Energy-adjusted UPF + adjusted for TEI	1.07 (1.03, 1.12) [23]
	Hard coronary heart disease	HR per serving UPF/day	Energy-adjusted UPF + adjusted for TEI	1.10 (1.04, 1.15) [23]
Zhong 2021 [119]	CVD mortality	HR 1st vs. 5th quartile	Energy-adjusted UPF + adjusted for TEI	1.50 (1.36, 1.64) [24]
	Heart disease mortality	HR 1st vs. 5th quartile	Energy-adjusted UPF + adjusted for TEI	1.68 (1.50, 1.87) [24]
	Cerebrovascular disease mortality	HR 1st vs. 5th quartile	Energy-adjusted UPF + adjusted for TEI	0.94 (0.76, 1.17) [24]
Scaranni 2021 [120]	Incident hypertension	OR 1st vs. 3rd tertile	UPF as % TEI + adjusted for TEI	1.23 (1.06, 1.44) [25]
	Change in SBP	Beta 1st vs. 3rd tertile	UPF as % TEI + adjusted for TEI	−0.54 (−1.23, 0.15) [25]
	Change in DBP	Beta 1st vs. 3rd tertile	UPF as % TEI + adjusted for TEI	0.08 (−0.39, 0.56) [25]
Mendonça 2017 [122]	Hypertension	HR 1st vs. 3rd tertile	Energy-adjusted UPF + adjusted for TEI	1.21 (1.06, 1.37) [26]
Rezende-Alves 2021 [148]	Hypertension	RR 1st vs. 5th quartile	UPF as % TEI	1.35 (1.01, 1.82) [27]
Monge 2021 [121]	Incident hypertension	≤20% vs. >45% of energy from any UPF	UPF as % TEI + adjusted for TEI	0.98 (0.84, 1.14) [28]
	Incident hypertension	≤20% vs. >45% of energy from liquid UPF	UPF as % TEI + adjusted for TEI	1.34 (1.10, 1.65) [28]
	Incident hypertension	≤20% vs. >45% of energy from solid UPF	UPF as % TEI + adjusted for TEI	0.91 (0.82, 1.01) [28]
Llavero-Valero 2021 [123]	T2DM	HR 1st vs. 3rd tertile	Energy-adjusted UPF + adjusted for TEI	1.52 (1.05, 2.22) [29]
Srour 2020 [124]	T2DM	HR per 10% increase in UPF	UPF as % weight + adjusted for TEI	1.15 (1.06, 1.25) [30]
Levy 2021 [88]	T2DM	HR per 10% increase in UPF	UPF as % weight + adjusted for TEI	1.20 (1.12, 1.29) [31]
Zhang 2021 [125]	NAFLD	HR 1st vs. 4th quartile	UPF g/1000kcal + adjusted for TEI	1.19 (1.08, 1.31) [32]
Fiolet 2018 [126]	All cancers	HR per 10% increase in UPF	UPF as % TEI + adjusted for TEI (exc. Alcohol)	1.12 (1.06, 1.18) [33]
	Breast cancer	HR per 10% increase in UPF	UPF as % TEI + adjusted for TEI (exc. Alcohol)	1.11 (1.02, 1.22) [33]
	Prostate cancer	HR per 10% increase in UPF	UPF as % TEI + adjusted for TEI (exc. Alcohol)	0.98 (0.83, 1.16) [33]
	Colorectal cancer	HR per 10% increase in UPF	UPF as % TEI + adjusted for TEI (exc. Alcohol)	1.13 (0.92, 1.38) [33]
Vasseur 2021 [127]	IBD	RR 1st vs. 3rd tertile	UPF as % weight + adjusted for TEI	1.44 (0.70, 2.94) [34]
Narula 2021 [128]	IBD	HR <1 vs. ≥5 servings UPF/day	UPF servings/day + adjusted for TEI	1.82 (1.22, 2.72) [35]
	Crohn's disease	HR <1 vs. ≥5 servings UPF/day	UPF servings/day + adjusted for TEI	4.50 (1.67, 12.13) [35]
	Ulcerative Colitis	HR <1 vs. ≥5 servings UPF/day	UPF servings/day + adjusted for TEI	1.46 (0.93, 2.28) [35]

Table 5. *Cont.*

Author, Year	Outcome	Method of Analysis	Energy Adjustment	Effect
Schnabel 2018 [129]	Irritable bowel syndrome	OR 1st vs. 4th quartile	UPF as % weight + adjusted for TEI	1.24 (1.12, 1.38) [36]
	Functional Constipation	OR 1st vs. 4th quartile	UPF as % weight + adjusted for TEI	1.00 (0.87, 1.15) [36]
	Functional diarrhoea	OR 1st vs. 4th quartile	UPF as % weight + adjusted for TEI	0.94 (0.71, 1.26) [36]
	Functional dyspepsia	OR 1st vs. 4th quartile	UPF as % weight + adjusted for TEI	1.26 (1.07, 1.48) [36]
Lo 2021 [130]	Crohn's disease	HR 1st vs. 4th quartile	UPF as % TEI + adjusted for TEI	1.70 (1.23, 2.35) [37]
	Ulcerative Colitis	HR 1st vs. 4th quartile	UPF as % TEI + adjusted for TEI	1.20 (0.91, 1.58) [37]
Adjibade 2019 [131]	Depressive symptoms	per 10% increase in UPF	UPF as % weight + adjusted for TEI	1.21 (1.15, 1.27) [38]
Gómez-Donoso 2020 [132]	Incident depression	HR 1st vs. 4th quartile	Energy-adjusted UPF + adjusted for TEI	1.33 (1.07, 1.64) [39]
Rey-Garcia 2021 [133]	Renal function	OR 1st vs. 3rd tertile	UPF as % TEI + adjusted for TEI	1.75 (1.16, 2.64) [40]
Zhang 2021 [134]	Hyperuricemia	HR 1st vs. 4th quartile	UPF servings/day + adjusted for TEI	1.17 (1.06, 1.30) [41]
Leffa 2020 [135]	Total cholesterol at age 6	Beta per 10% increase in UPF intake at age 3	UPF as % TEI + adjusted for TEI	0.07 (0.00, 0.14) $p = 0.046$ [42]
	LDL-cholesterol at age 6	Beta per 10% increase in UPF intake at age 3	UPF as % TEI + adjusted for TEI	0.03 (−0.03, 0.09) [42]
	HDL-cholesterol at age 6	Beta per 10% increase in UPF intake at age 3	UPF as % TEI + adjusted for TEI	0.01 (−0.02, 0.05) [42]
	TAG at age 6	Beta per 10% increase in UPF intake at age 3	UPF as % TEI + adjusted for TEI	0.04 (0.01, 0.07) $p = 0.024$ [42]
Rauber 2015 [95]	ΔTotal cholesterol from 3–4 to 7–8	Beta per 1% increase in energy from UPF	UPF as % TEI + adjusted for TEI at age 7–8	0.430 (0.008, 0.853) [43]
	ΔLDL cholesterol from 3–4 to 7–8	Beta per 1% increase in energy from UPF	UPF as % TEI + adjusted for TEI at age 7–8	0.369 (0.005, 0.733) [43]
	ΔnHDL cholesterol from 3–4 to 7–8	Beta per 1% increase in energy from UPF	UPF as % TEI + adjusted for TEI at age 7–8	0.319 (−0.059, 0.697) [43]
	ΔTriglycerides from 3–4 to 7–8	Beta per 1% increase in energy from UPF	UPF as % TEI + adjusted for TEI at age 7–8	−0.465 (−0.955, 0.025) [43]
	ΔHDL cholesterol from 3–4 to 7–8	Beta per 1% increase in energy from UPF	UPF as % TEI + adjusted for TEI at age 7–8	0.125 (−0.026, 0.277) [43]
Donat-Vargas 2021 [136]	Incident hypertriglyceridemia (≥150 mg/dL)	OR 1st vs. 3rd tertile	UPF as % TEI + adjusted for TEI	2.21 (1.09, 4.49) [44]
	Low HDL-cholesterol (<40 in men or <50 mg/dL in women)	OR 1st vs. 3rd tertile	UPF as % TEI + adjusted for TEI	2.04 (1.18, 3.53) [44]
	High LDL-cholesterol (>129 mg/dL)	OR 1st vs. 3rd tertile	UPF as % TEI + adjusted for TEI	1.13 (0.52, 2.46) [44]
	Δtriglycerides (mg/dL)	Beta 1st vs. 3rd tertile	UPF as % TEI + adjusted for TEI	6.23 (1.26, 11.21) [44]
	ΔHDL cholesterol (mg/dL)	Beta 1st vs. 3rd tertile	UPF as % TEI + adjusted for TEI	0.02 (−1.45, 1.49) [44]
	ΔLDL cholesterol (mg/dL)	Beta 1st vs. 3rd tertile	UPF as % TEI + adjusted for TEI	−3.43 (−8.60, 1.74) [44]

Table 5. *Cont.*

Author, Year	Outcome	Method of Analysis	Energy Adjustment	Effect
Machado Azeredo 2020 [101]	Wheeze at age 11	OR 1st vs. 5th quintile of UPF at age 6	UPF as % TEI + adjusted for TEI and TEI:EEI	0.78 (0.51, 1.19) [45]
	Asthma at age 11	OR 1st vs. 5th quintile of UPF at age 6	UPF as % TEI + adjusted for TEI and TEI:EEI	0.83 (0.59, 1.17) [45]
	Mild/moderate asthma at age 11	OR 1st vs. 5th quintile of UPF at age 6	UPF as % TEI + adjusted for TEI and TEI:EEI	0.63 (0.34, 1.17) [45]
	Severe Asthma at age 11	OR 1st vs. 5th quintile of UPF at age 6	UPF as % TEI + adjusted for TEI and TEI:EEI	0.94 (0.54, 1.65) [45]
Borge 2021 [137]	ADHD diagnosis at 8 years	RR per 1 SD increase in UPF	UPF as % TEI	1.07 (0.99, 1.18) [46]
	ADHD symptoms (absolute) at 8 years	Beta per 1 SD increase in UPF	UPF as % TEI	0.25 (0.13, 0.38) [46,*]
	ADHD symptoms (relative) at 8 years	Beta per 1 SD increase in UPF	UPF as % TEI	3.0 (1.5, 4.5) [46,*]
Zhang 2021 [138]	Change in grip strength (kg/year)	Beta per 10% increase in UPF	UPF as % weight + adjusted for TEI	−0.3207 (−0.5281, −0.1133) [47]
	Change in weight-adjusted grip strength (kg/kg/year)	Beta per 10% increase in UPF	UPF as % weight + adjusted for TEI	−0.0046 (−0.0076, −0.0016) [47]

Energy-adjusted UPF via the residual method. TEI, total energy intake; TEI:EEI, energy intake:expenditure ratio; OR, odds ratio, HR, hazard ratio; RR, relative risk; ADHD, attention deficit hyperactivity disorder; CVD, cardiovascular disease; IHD, ischemic heart disease; HOMA-IR, Homeostatic Model Assessment for Insulin Resistance; BMI, body mass index; T2DM, type 2 diabetes mellitus; IBD, inflammatory bowel disease; NAFLD, non-alcoholic fatty liver disease; UPF, ultra-processed food; SBP, systolic blood pressure; DBP, diastolic blood pressure; LDL, low-density lipoprotein; HDL, high-density lipoprotein; nHDL, non-high-density lipoprotein; TAG, triacylglycerol; WC, waist circumference; WHR, waist-to-hip ratio. [1] Adjusted for sex, age, income level, education level, marital status, residence, BMI, physical activity level, smoking status, energy intake, alcohol intake, season of food records, first-degree family history of cancer or cardiovascular diseases and number of food records. [2] Adjusted for age, sex, marital status, physical activity, smoking status, snacking, special diet at baseline, BMI, total energy intake, alcohol consumption, family history of CVD, diabetes or hypertension at baseline, self-reported hypercholesterolaemia, baseline CVD, cancer or depression, education level and lifelong smoking stratified by recruitment period, deciles of age, sedentary index (sum of hours each day spent watching television, using a computer and driving) and television viewing. [3] Adjusted for age, sex, education level, living alone, smoking status, former drinker, physical activity, time watching television, time devoted to other sedentary activities, number of medications per day and specific chronic conditions diagnosed by a physician (chronic respiratory disease, coronary heart disease, stroke, heart failure, osteoarthritis, cancer and depression). [4] Adjusted for age, sex, race/ethnicity, total energy intake, poverty level, education level, smoking status, physical activity and alcohol intake. [5] Adjusted for age, sex, BMI, physical activity, alcohol intake, smoking status and total energy intake. [6] Adjusted for sex, age, energy intake, educational level, housing tenure, smoking, BMI, leisure-time physical activity, history of cancer, CVD, diabetes, hypertension and hyperlipidaemia and residence. [7] Adjusted for age, sex, marital status, educational level, physical activity, smoking status, alcohol consumption, energy intake and number of dietary records (overweight and obesity outcomes further adjusted for baseline BMI). [8] Adjusted for age, sex, marital status, educational status, physical activity, energy intake, television watching, siesta sleep, smoking status, snacking between meals, following a special diet at baseline, baseline BMI and consumption of fruit and vegetables. [9] Adjusted for age, sex and energy intake. [10] Adjusted for age, sex, study arm, follow-up time, educational level, marital status, smoking habits, type 2 diabetes prevalence, height and repeated measures of total energy intake, physical activity and sedentary behaviour. [11] Adjusted for age, sex, educational level, marital status, smoking, ex-drinker status, physical activity in the household and at leisure time, number of medications consumed per day, number of chronic diseases diagnosed by a doctor (chronic obstructive pulmonary disease/asthma, coronary heart disease, stroke, heart failure, osteoarthritis or depression) and total energy intake, Mediterranean dietary pattern, fibre and very long chain omega-3 fatty acid intake. [12] Adjusted for age, sex, baseline BMI, educational level, physical activity, baseline alcohol intake, baseline smoking status and plausibility of dietary energy reporting and for the modified relative Mediterranean diet score. Overweight and obesity outcomes further adjusted for country/centre, follow-up time in years and smoking status at follow-up (instead of baseline smoking status). [13] Adjusted for age, sex, colour/race, centre, income, school achievement, smoking, physical activity and energy intake. Additionally for incident overweight/obesity and weight gain: baseline BMI. Additionally for waist gain: waist circumference at baseline. [14] Adjusted for maternal age, race, socioeconomic status, weight status, average daily energy intake, time spent in moderate physical activity and fat intake. [15] Adjusted for cohort and potential confounding variables, with $p < 0.25$. [16] Adjusted for age, BMI, education, smoking status, physical activity, family history of diabetes, recruitment year, time between recruitment and the first pregnancy or gestational diabetes, number of pregnancies during follow-up, parity, multiple pregnancies, time spent watching TV, hypertension, following a nutritional therapy, and energy intake. [17] Adjusted for age, baseline UPF, age*baseline UPF interaction term, child sex, race, birth weight, physical activity, quintiles of Index of Multiple Deprivation, the mother's prepregnancy BMI, marital status, highest educational attainment, socioeconomic status and the child's baseline total energy intake. [18] Adjusted for skin colour, maternal age and schooling, birthweight and sex, screen time, total energy intake and energy intake/expenditure ratio. [19] Adjusted for maternal age, maternal education and BMI before pregnancy, exclusive breast-feeding for the first 6 months, child physical exercise and daily screen time. For appetitive behaviours, additionally parental concerns, and for BMI, additionally child BMI z-score at 4 years old. [20] Adjusted for sex, group status in the early phase (intervention and control), pre-pregnancy BMI, birth weight, breastfeeding, familyincome, maternal schooling and total screen duration. [21] Adjusted for age, sex, energy intake, number of 24-h dietary records, smoking status, educational level, physical activity, BMI, alcohol intake, and family history of CVD. [22] Adjusted for age, sex, race, centre and total energy intake. [23] Adjusted for age, sex, education, smoking status, alcohol intake, total energy intake and physical activity. [24] Adjusted for age, sex, race, educational level, marital status, study centre, aspirin use, history of hypertension or diabetes, smoking status, alcohol consumption, BMI, physical activity and energy intake. [25] Adjusted for age, sex, colour or race, education, time since baseline, SBP/DBP/ hypertension, physical activity, smoking, alcohol consumption, sodium intake and total energy intake. [26] Adjusted for age, sex, physical activity, hours of TV watching, baseline BMI, smoking status, use of analgesics, following a special diet at baseline, family history of hypertension, hypercholesterolemia, alcohol consumption, total energy intake, olive oil intake and consumption of fruits and vegetables. [27] Adjusted for age, marital status, skin colour, per capita income, physical activity, smoking, obesity, family history of hypertension and previous diagnosis of T2DM, hypercholesterolaemia and hypertriglyceridaemia. [28] Adjusted for age, indigenous, internet access, insurance, family history of hypertension, menopausal status, smoking status, physical activity, total energy intake and multivitamin intake. [29] Adjusted for age, sex, BMI, educational status, family history of diabetes, smoking status, snacking between meals, active and sedentary lifestyle score, following a special diet at baseline and total energy intake. [30] Adjusted for age, sex, educational level, baseline BMI, physical activity, smoking status, alcohol intake, number of 24-h dietary records, total energy intake, Food Standards Agency nutrient profiling system dietary index score, and family history of T2DM. [31] Adjusted for age, sex, ethnicity, family history of T2DM, Index of Multiple Deprivation, physical activity level, current smoking status and total energy intake. [32] Adjusted for age, sex, BMI, healthy diet score, total energy intake, smoking status, alcohol drinking status, educational level, occupation, monthly household income, physical activity, family history of disease (including cardiovascular disease, hypertension, hyperlipidaemia and diabetes) and depressive symptoms. [33] Adjusted for age, sex, total energy intake without alcohol, number of 24-h dietary records, smoking status, educational level, physical activity, height, BMI, alcohol intake, family history of cancers (and for breast cancer outcome, additionally adjusted for menopausal status, hormonal treatment for menopause, oral contraception and number of children). [34] Adjusted for age, sex, income level, education level, marital status, residence, BMI, physical activity, smoking status, hormonal contraception, number of 24-h dietary records, healthy dietary pattern and total energy intake. [35] Adjusted for age, sex, geographical region, education, alcohol intake, smoking status, BMI, total energy intake, and location. [36] Adjusted for age, sex, income level, education level, marital status, residence, BMI, physical activity, smoking status, total energy intake, season of food records and time between food record and functional gastrointestinal disorders questionnaire. [37] Adjusted for age, cohort, calendar year, AHEI-2010, race, family history of IBD, smoking, BMI, physical activity, total energy intake, regular NSAIDs use, oral contraceptives use and menopausal hormone therapy. [38] Adjusted for age, sex, BMI, marital status, educational level, occupational categories, household income per consumption unit, residential area, number of 24-h dietary records, inclusion month, total energy intake without alcohol, alcohol intake, smoking status and physical activity. [39] Adjusted for age, sex, year of entrance to the cohort, Mediterranean diet, baseline BMI, total energy intake, physical activity, smoking status, marital status, living alone, employment status, working hours per week, health-related career, years of education and baseline self-perception of competitiveness, anxiety and dependence levels. [40] Adjusted for age, sex and total energy intake. [41] Adjusted for age, sex, BMI, smoking status, alcohol consumption status, education levels, employment status, household income, physical activity, depression symptoms, family history of disease (including cardiovascular disease, hypertension, hyperlipidaemia and diabetes), hypertension, hyperlipidaemia, diabetes, metabolic syndrome, total energy intake and dietary patterns (sweet, animal and healthy patterns). [42] Adjusted for sex, group status in the early phase (intervention and control), family income, pre-pregnancy BMI, childbirth weight, BMI z-scores at 3 years, total energy and total fat intake at 3 years. [43] Adjusted for sex, group, birth weight, family income, maternal schooling, BMI-for-age z-scores and total energy intake at 7–8 years. [44] Adjusted for age, sex, fibre intake, total energy intake, educational level, marital status, smoking status, BMI, physical activity, alcohol consumption, number of medications and number of chronic conditions. [45] Adjusted for TEI and TEI:EEI. [46] Adjusted for child diet quality score using Diet Quality Index at 3 years, maternal pre-pregnancy BMI, maternal education, smoking and alcohol intake during pregnancy, maternal symptoms of depression and ADHD, maternal age, parity, child sex and child birth quarter. [47] Adjusted for baseline age, sex, BMI, smoking status, alcohol drinking status, education level, employment, monthly household income, physical activity, family history of disease (including CVD, hypertension, hyperlipidaemia, and diabetes), depressive symptoms, hypertension, hyperlipidaemia, diabetes, total energy intake, healthy diet score, dietary supplement use, total protein intake and milk intake. * 95% credible intervals.

7. Discussion

This review provides novel insights into the relative impact of nutrient content and dietary patterns vs. ultra-processing on obesity and adverse health-related outcomes. The analyses reported here from prospective cohort studies have been largely unexplored to date. Consistent across many studies, adjustment for fat, sugar and sodium intake, or adjustment for adherence to a range of healthy or unhealthy dietary patterns has a minimal impact on the adverse associations between UPF intake and a diverse range of health-related outcomes. These findings strongly point towards aspects of ultra-processing as being important factors that impact health, and question the ability to conclude that the adverse outcomes from UPFs can be solely attributed to their nutritional quality.

A meta-analysis of nationally representative samples demonstrates that diets high in UPF tend to contain greater intakes of energy, free sugars, total and saturated fat, and lower intakes of fibre, protein and some micronutrients [30]. The NOVA classification therefore captures important aspects of nutrient quality, despite this not being a core aspect of the UPF definition [28]. It is unsurprising therefore, that the detrimental associations between UPF intake and obesity, CVD and all-cause mortality have been largely attributed to the poor nutritional quality of high UPF diets [71]. If this were the case, then adjustment for aspects of dietary quality should explain the associations between UPFs and poor health outcomes, or at least, explain a significant proportion of the association. However, the majority of the models from prospective studies retain a significantly increased risk of poor health from UPF intake, and are also largely unaltered in magnitude, following dietary adjustment. The findings from this review are in alignment with the results from a metabolic ward cross-over study, the only randomised controlled trial comparing diets of differing levels processing [49]. Participants consumed ad libitum, minimally processed or ultra-processed diets, matched for energy and nutrient content, for two weeks each. The ultra-processed diet resulted in greater energy intake (+508 ± 106 kcal/day), leading to weight gain (+0.9 ± 0.3 kg). In contrast, the minimally processed diet resulted in weight loss (-0.9 ± 0.3 kg), despite diets being matched for energy and nutrient content [49].

The Mediterranean diet, considered to be one of the healthiest dietary patterns for reducing CVD risk [151], consists predominantly of whole grains, fruits, vegetables, beans, pulses and legumes, of which, their consumption is inversely associated with UPF intake [30]. Therefore, the impact of UPFs on health could just be that they displace more healthful foods, or that they overlap with pre-established unhealthy dietary patterns. However, adjustment for Mediterranean diet adherence, for the Western dietary pattern or for other dietary pattern indices, did not alter the majority of the significant associations between UPF intake and health-related outcomes, including the increased risk of weight gain or obesity.

High UPF diets are also characterised by the displacement of minimally processed foods, as defined by NOVA [30]. Few studies have performed dietary adjustment for other NOVA food groups. However, in those that have, adjustment for other NOVA food groups not only did not explain, but in fact, increased the risk associated with UPF intake and FMI gain from age 6 to 11 [116], and adjustment for unprocessed or minimally processed food intake did not alter the increased risk of T2DM [124].

Although limited at this stage, these adjustments would suggest that UPF intake has a direct effect on health-related outcomes, rather than simply displacing healthy foods intake. This may indicate the importance of considering the nature and extent of food processing as an important dimension of dietary pattern analysis.

Discussions over the relative importance of nutrient content vs. ultra-processing continue [24,81]. However, recent reports have not taken into account the dietary adjustments from prospective studies reported in this review [24,81,152]. The aspects of ultra-processing that lead to adverse health outcomes are poorly understood, and the findings from this review highlight the need for research into mechanisms of ultra-processing as being a priority, in order to determine the long-term potential for UPF reformulation, or need for elimination to address the growing obesity pandemic. On a case-by-case basis, choosing

UPF reformulations over high fat, salt or sugar alternatives can be beneficial to reduce intakes of nutrients to limit, which are known to be associated with poor health [153]. However, given the high prevalence of UPFs within diets [30], if ultra-processing itself directly results in poor health, then scaling up the case-by-case reformulation approach to the whole diet still leaves an ultra-processed dietary pattern that displaces minimally processed foods, and thus will not sufficiently address current health risks. The nutritional quality of food is important, but is not the sole determinant of the healthiness of a diet [10]. The importance of dietary patterns, food groups and foods as a whole, rather than specific individual nutrients, has previously been highlighted [8,154]. Indeed, the diversity of chemicals and nutrients consumed in human diets is vast, yet current nutrient profiling methods only consider a fraction of the 26,000 or so biochemicals in food [155].

Current dietary policies vary across nations and health organisations. The public health implications regarding whether UPFs should be reformulated based on nutrient content or removed from the diet are important. The UK government and Cancer Research UK currently adopt a reformulation approach to high fat, salt or sugar foods, and do not consider the nature and extent of processing in their dietary recommendations [156–158]. However, advice to avoid UPFs is becoming increasingly more prevalent. The American Heart Association now recommends limiting UPF intake [12], and the World Health Organisation and UNICEF recognise the importance of UPF consumption for ending childhood obesity [159,160]. UPFs are also recognised by the Pan American Health Organisation as important for reducing health risk, as part of their nutrient profiling model [161]. Some national dietary guidelines now encourage limiting UPF intake including Brazil [162], Uruguay [163] and Israel [164]. France is also planning to reduce UPF consumption by 20% from 2018 to 2021 [165].

This review discusses the results from over 1,000,000 participants across more than 20 different prospective cohorts, covering many countries, demographic profiles and age groups. The studies in this review utilise dietary assessment methods including 24-hour dietary records and food frequency questionnaires, that are not designed specifically for the application of NOVA classification. Similar foods can be classed as ultra-processed (e.g., pre-packaged bread) or processed (e.g., artisanal bread), which may result in the misclassification of foods with the dietary assessment methods used. Furthermore, other important dietary aspects may have not been captured or suitably adjusted for. Most studies have adjusted for fat, sugar and sodium intake or for overall dietary patterns, which are important dietary factors for health, and proposed to be explanatory factors for the associations between UPF intake and health outcomes.

8. Conclusions

Experts for and against the NOVA classification have often focussed on nutrient quality as an important explanatory link between UPFs, obesity and adverse health-related outcomes. However, many of the prospective studies published to date have performed analyses adjusting for nutrient content and overall dietary patterns. These adjustments do not explain the association between UPFs, obesity and adverse health-related outcomes, with estimates remaining significant. These findings raise important questions regarding current policy and future research needs, suggesting that the nature and extent of processing is an important dietary dimension, and whether UPF reformulations can sufficiently address the growing transition towards high UPF diets and the associated risk of obesity and poor health.

Supplementary Materials: The following are available online at https://www.mdpi.com/article/10.3390/nu14010023/s1, Table S1: Prospective studies adjusting for fat, sodium and carbohydrate intake and dietary pattern, Table S2: Prospective studies adjusting for components of fat, sodium and carbohydrates, Table S3: Prospective studies adjusting for other dietary components.

Author Contributions: S.J.D., writing—first draft; R.L.B., writing—editing and reviewing, supervision. All authors have read and agreed to the published version of the manuscript.

Funding: S.J.D. is funded by a Medical Research Council grant (MR/N013867/1). R.L.B. is funded by the National Institute for Health Research, Sir Jules Thorn Charitable Trust and Rosetrees Trust.

Institutional Review Board Statement: Not applicable.

Informed Consent Statement: Not applicable.

Data Availability Statement: Not applicable.

Acknowledgments: The authors would like to thank Chloe Firman for her review of the final draft.

Conflicts of Interest: The authors declare no conflict of interest.

References

1. WHO. Obesity. Available online: https://www.who.int/westernpacific/health-topics/obesity (accessed on 7 July 2021).
2. Steel, N.; Ford, J.A.; Newton, J.N.; Davis, A.C.J.; Vos, T.; Naghavi, M.; Glenn, S.; Hughes, A.; Dalton, A.M.; Stockton, D.; et al. Changes in Health in the Countries of the UK and 150 English Local Authority Areas 1990-2016: A Systematic Analysis for the Global Burden of Disease Study 2016. *Lancet* **2018**, *392*, 1647–1661. [CrossRef]
3. Kolotkin, R.L.; Andersen, J.R. A Systematic Review of Reviews: Exploring the Relationship between Obesity, Weight Loss and Health-Related Quality of Life. *Clin. Obes.* **2017**, *7*, 273–289. [CrossRef] [PubMed]
4. Di Angelantonio, E.; Bhupathiraju, S.N.; Wormser, D.; Gao, P.; Kaptoge, S.; De Gonzalez, A.B.; Cairns, B.J.; Huxley, R.; Jackson, C.L.; Joshy, G.; et al. The Global BMI Mortality Collaboration Body-Mass Index and All-Cause Mortality: Individual-Participant-Data Meta-Analysis of 239 Prospective Studies in Four Continents. *Lancet* **2016**, *388*, 776–786. [CrossRef]
5. Afshin, A.; Sur, P.J.; Fay, K.A.; Cornaby, L.; Ferrara, G.; Salama, J.S.; Mullany, E.C.; Abate, K.H.; Abbafati, C.; Abebe, Z.; et al. Health Effects of Dietary Risks in 195 Countries, 1990–2017: A Systematic Analysis for the Global Burden of Disease Study 2017. *Lancet* **2019**, *393*, 1958–1972. [CrossRef]
6. Stanaway, J.D.; Afshin, A.; Gakidou, E.; Lim, S.S.; Abate, D.; Abate, K.H.; Abbafati, C.; Abbasi, N.; Abbastabar, H.; Abd-Allah, F.; et al. Global, Regional, and National Comparative Risk Assessment of 84 Behavioural, Environmental and Occupational, and Metabolic Risks or Clusters of Risks for 195 Countries and Territories, 1990–2017: A Systematic Analysis for the Global Burden of Disease Study 2017. *Lancet* **2018**, *392*, 1923–1994. [CrossRef]
7. Mozaffarian, D.; Rosenberg, I.; Uauy, R. History of Modern Nutrition Science—Implications for Current Research, Dietary Guidelines, and Food Policy. *BMJ* **2018**, *361*, k2392. [CrossRef] [PubMed]
8. Mozaffarian, D. Dietary and Policy Priorities for Cardiovascular Disease, Diabetes, and Obesity: A Comprehensive Review. *Circulation* **2016**, *133*, 187–225. [CrossRef]
9. Micha, R.; Shulkin, M.L.; Peñalvo, J.L.; Khatibzadeh, S.; Singh, G.M.; Rao, M.; Fahimi, S.; Powles, J.; Mozaffarian, D. Etiologic Effects and Optimal Intakes of Foods and Nutrients for Risk of Cardiovascular Diseases and Diabetes: Systematic Reviews and Meta-Analyses from the Nutrition and Chronic Diseases Expert Group (NutriCoDE). *PLoS ONE* **2017**, *12*, e0175149. [CrossRef]
10. Katz, D.L.; Meller, S. Can We Say What Diet Is Best for Health? *Annu. Rev. Public Health* **2014**, *35*, 83–103. [CrossRef] [PubMed]
11. English, L.K.; Ard, J.D.; Bailey, R.L.; Bates, M.; Bazzano, L.A.; Boushey, C.J.; Brown, C.; Butera, G.; Callahan, E.H.; de Jesus, J.; et al. Evaluation of Dietary Patterns and All-Cause Mortality: A Systematic Review. *JAMA Netw. Open* **2021**, *4*, e2122277. [CrossRef]
12. Lichtenstein, A.H.; Appel, L.J.; Vadiveloo, M.; Hu, F.B.; Kris-Etherton, P.M.; Rebholz, C.M.; Sacks, F.M.; Thorndike, A.N.; Van Horn, L.; Wylie-Rosett, J.; et al. 2021 Dietary Guidance to Improve Cardiovascular Health: A Scientific Statement from the American Heart Association. *Circulation* **2021**, *144*, e472–e487. [CrossRef]
13. Schlesinger, S.; Neuenschwander, M.; Schwedhelm, C.; Hoffmann, G.; Bechthold, A.; Boeing, H.; Schwingshackl, L. Food Groups and Risk of Overweight, Obesity, and Weight Gain: A Systematic Review and Dose-Response Meta-Analysis of Prospective Studies. *Adv. Nutr.* **2019**, *10*, 205–218. [CrossRef] [PubMed]
14. Schwingshackl, L.; Schwedhelm, C.; Hoffmann, G.; Lampousi, A.-M.; Knüppel, S.; Iqbal, K.; Bechthold, A.; Schlesinger, S.; Boeing, H. Food Groups and Risk of All-Cause Mortality: A Systematic Review and Meta-Analysis of Prospective Studies. *Am. J. Clin. Nutr.* **2017**, *105*, 1462–1473. [CrossRef]
15. Tapsell, L.C.; Neale, E.P.; Satija, A.; Hu, F.B. Foods, Nutrients, and Dietary Patterns: Interconnections and Implications for Dietary Guidelines. *Adv. Nutr.* **2016**, *7*, 445–454. [CrossRef] [PubMed]
16. Wirfält, E.; Drake, I.; Wallström, P. What Do Review Papers Conclude about Food and Dietary Patterns? *Food Nutr. Res.* **2013**, *57*, 20523. [CrossRef]
17. Fabiani, R.; Naldini, G.; Chiavarini, M. Dietary Patterns and Metabolic Syndrome in Adult Subjects: A Systematic Review and Meta-Analysis. *Nutrients* **2019**, *11*, 2056. [CrossRef]
18. Srour, B.; Touvier, M. Ultra-Processed Foods and Human Health: What Do We Already Know and What Will Further Research Tell Us? *EClinicalMedicine* **2021**, *32*, 100747. [CrossRef]
19. Popkin, B. Ultra-Processed Foods' Impacts on Health. 2030—Food, Agriculture and Rural Development in Latin America and the Caribbean, No. 34. Santiago de Chile. FAO. 2019. Available online: https://www.fao.org/documents/card/en/c/ca7349en/ (accessed on 14 October 2021).

20. Zobel, E.H.; Hansen, T.W.; Rossing, P.; von Scholten, B.J. Global Changes in Food Supply and the Obesity Epidemic. *Curr. Obes. Rep.* **2016**, *5*, 449–455. [CrossRef] [PubMed]
21. Popkin, B.M.; Adair, L.S.; Ng, S.W. The Global Nutrition Transition: The Pandemic of Obesity in Developing Countries. *Nutr. Rev.* **2012**, *70*, 3–21. [CrossRef] [PubMed]
22. Ng, M.; Fleming, T.; Robinson, M.; Thomson, B.; Graetz, N.; Margono, C.; Mullany, E.C.; Biryukov, S.; Abbafati, C.; Abera, S.F.; et al. Global, Regional, and National Prevalence of Overweight and Obesity in Children and Adults during 1980–2013: A Systematic Analysis for the Global Burden of Disease Study 2013. *Lancet* **2014**, *384*, 766–781. [CrossRef]
23. Mozaffarian, D.; Ludwig, D.S. Dietary Guidelines in the 21st Century—A Time for Food. *JAMA* **2010**, *304*, 681–682. [CrossRef] [PubMed]
24. Tobias, D.K.; Hall, K.D. Eliminate or Reformulate Ultra-Processed Foods? Biological Mechanisms Matter. *Cell Metab.* **2021**, *33*, 2314–2315. [CrossRef] [PubMed]
25. Crino, M.; Barakat, T.; Trevena, H.; Neal, B. Systematic Review and Comparison of Classification Frameworks Describing the Degree of Food Processing. *Nutr. Food Technol. Open Access* **2017**, *3*. [CrossRef]
26. Monteiro, C.A.; Cannon, G.; Moubarac, J.-C.; Levy, R.B.; Louzada, M.L.C.; Jaime, P.C. The UN Decade of Nutrition, the NOVA Food Classification and the Trouble with Ultra-Processing. *Public Health Nutr.* **2018**, *21*, 5–17. [CrossRef]
27. Monteiro, C.A.; Cannon, G.; Levy, R.; Moubarac, J.-C.; Jaime, P.; Martins, A.P.; Canella, D.; Louzada, M.; Parra, D. NOVA. The Star Shines Bright. *World Nutr. J.* **2016**, *7*, 28–38.
28. Monteiro, C.A.; Cannon, G.; Levy, R.B.; Moubarac, J.-C.; Louzada, M.L.; Rauber, F.; Khandpur, N.; Cediel, G.; Neri, D.; Martinez-Steele, E.; et al. Ultra-Processed Foods: What They Are and How to Identify Them. *Public Health Nutr.* **2019**, *22*, 936–941. [CrossRef] [PubMed]
29. Baker, P.; Machado, P.; Santos, T.; Sievert, K.; Backholer, K.; Hadjikakou, M.; Russell, C.; Huse, O.; Bell, C.; Scrinis, G.; et al. Ultra-Processed Foods and the Nutrition Transition: Global, Regional and National Trends, Food Systems Transformations and Political Economy Drivers. *Obes. Rev.* **2020**, *21*, e13126. [CrossRef]
30. Martini, D.; Godos, J.; Bonaccio, M.; Vitaglione, P.; Grosso, G. Ultra-Processed Foods and Nutritional Dietary Profile: A Meta-Analysis of Nationally Representative Samples. *Nutrients* **2021**, *13*, 3390. [CrossRef]
31. Wang, L.; Martínez Steele, E.; Du, M.; Pomeranz, J.L.; O'Connor, L.E.; Herrick, K.A.; Luo, H.; Zhang, X.; Mozaffarian, D.; Zhang, F.F. Trends in Consumption of Ultraprocessed Foods Among US Youths Aged 2-19 Years, 1999-2018. *JAMA* **2021**, *326*, 519–530. [CrossRef]
32. de Araújo, C.R.B.; Ribeiro, K.D.d.S.; de Oliveira, A.F.; de Morais, I.L.; Breda, J.; Padrão, P.; Moreira, P. Degree of Processing and Nutritional Value of Children's Food Products. *Public Health Nutr.* **2021**, *24*, 5977–5984. [CrossRef]
33. Chen, X.; Zhang, Z.; Yang, H.; Qiu, P.; Wang, H.; Wang, F.; Zhao, Q.; Fang, J.; Nie, J. Consumption of Ultra-Processed Foods and Health Outcomes: A Systematic Review of Epidemiological Studies. *Nutr. J.* **2020**, *19*, 86. [CrossRef] [PubMed]
34. Elizabeth, L.; Machado, P.; Zinöcker, M.; Baker, P.; Lawrence, M. Ultra-Processed Foods and Health Outcomes: A Narrative Review. *Nutrients* **2020**, *12*, 1955. [CrossRef]
35. Lane, M.M.; Davis, J.A.; Beattie, S.; Gómez-Donoso, C.; Loughman, A.; O'Neil, A.; Jacka, F.; Berk, M.; Page, R.; Marx, W.; et al. Ultraprocessed Food and Chronic Noncommunicable Diseases: A Systematic Review and Meta-Analysis of 43 Observational Studies. *Obes. Rev.* **2021**, *22*, e13146. [CrossRef]
36. Askari, M.; Heshmati, J.; Shahinfar, H.; Tripathi, N.; Daneshzad, E. Ultra-Processed Food and the Risk of Overweight and Obesity: A Systematic Review and Meta-Analysis of Observational Studies. *Int. J. Obes.* **2020**, *44*, 2080–2091. [CrossRef] [PubMed]
37. Moradi, S.; Entezari, M.H.; Mohammadi, H.; Jayedi, A.; Lazaridi, A.-V.; Kermani, M.A.H.; Miraghajani, M. Ultra-Processed Food Consumption and Adult Obesity Risk: A Systematic Review and Dose-Response Meta-Analysis. *Crit. Rev. Food Sci. Nutr.* **2021**. [CrossRef]
38. Jardim, M.Z.; Costa, B.V.d.L.; Pessoa, M.C.; Duarte, C.K. Ultra-Processed Foods Increase Noncommunicable Chronic Disease Risk. *Nutr. Res.* **2021**, *95*, 19–34. [CrossRef]
39. Silva Meneguelli, T.; Viana Hinkelmann, J.; Hermsdorff, H.H.M.; Zulet, M.Á.; Martínez, J.A.; Bressan, J. Food Consumption by Degree of Processing and Cardiometabolic Risk: A Systematic Review. *Int. J. Food Sci. Nutr.* **2020**, *71*, 678–692. [CrossRef]
40. Pagliai, G.; Dinu, M.; Madarena, M.P.; Bonaccio, M.; Iacoviello, L.; Sofi, F. Consumption of Ultra-Processed Foods and Health Status: A Systematic Review and Meta-Analysis. *Br. J. Nutr.* **2021**, *125*, 308–318. [CrossRef] [PubMed]
41. Monteiro, C.A.; Moubarac, J.-C.; Levy, R.B.; Canella, D.S.; Louzada, M.L.d.C.; Cannon, G. Household Availability of Ultra-Processed Foods and Obesity in Nineteen European Countries. *Public Health Nutr.* **2018**, *21*, 18–26. [CrossRef]
42. Vandevijvere, S.; Jaacks, L.M.; Monteiro, C.A.; Moubarac, J.-C.; Girling-Butcher, M.; Lee, A.C.; Pan, A.; Bentham, J.; Swinburn, B. Global Trends in Ultraprocessed Food and Drink Product Sales and Their Association with Adult Body Mass Index Trajectories. *Obes. Rev.* **2019**, *20* (Suppl. 2), 10–19. [CrossRef]
43. Poti, J.M.; Braga, B.; Qin, B. Ultra-Processed Food Intake and Obesity: What Really Matters for Health—Processing or Nutrient Content? *Curr. Obes. Rep.* **2017**, *6*, 420–431. [CrossRef]
44. Rolls, B.; Cunningham, P.; Diktas, H. Properties of Ultraprocessed Foods That Can Drive Excess Intake. *Nutr. Today* **2020**, *55*, 109–115. [CrossRef]
45. Juul, F.; Vaidean, G.; Parekh, N. Ultra-Processed Foods and Cardiovascular Diseases: Potential Mechanisms of Action. *Adv. Nutr.* **2021**, *12*, 1673–1680. [CrossRef] [PubMed]

46. Gupta, S.; Hawk, T.; Aggarwal, A.; Drewnowski, A. Characterizing Ultra-Processed Foods by Energy Density, Nutrient Density, and Cost. *Front. Nutr.* **2019**, *6*, 70. [CrossRef] [PubMed]
47. Rauber, F.; da Costa Louzada, M.L.; Steele, E.M.; Millett, C.; Monteiro, C.A.; Levy, R.B. Ultra-Processed Food Consumption and Chronic Non-Communicable Diseases-Related Dietary Nutrient Profile in the UK (2008–2014). *Nutrients* **2018**, *10*, 587. [CrossRef] [PubMed]
48. Micek, A.; Godos, J.; Grosso, G. Nutrient and Energy Contribution of Ultra-Processed Foods in the Diet of Nations: A Meta-Analysis. *Eur. J. Public Health* **2021**, *31*, 159–160. [CrossRef]
49. Hall, K.D.; Ayuketah, A.; Brychta, R.; Cai, H.; Cassimatis, T.; Chen, K.Y.; Chung, S.T.; Costa, E.; Courville, A.; Darcey, V.; et al. Ultra-Processed Diets Cause Excess Calorie Intake and Weight Gain: An Inpatient Randomized Controlled Trial of Ad Libitum Food Intake. *Cell Metab.* **2019**, *30*, 67–77.e3. [CrossRef]
50. Fardet, A.; Rock, E. Ultra-Processed Foods: A New Holistic Paradigm? *Trends Food Sci. Technol.* **2019**, *93*, 174–184. [CrossRef]
51. Forde, C.G.; Mars, M.; de Graaf, K. Ultra-Processing or Oral Processing? A Role for Energy Density and Eating Rate in Moderating Energy Intake from Processed Foods. *Curr. Dev. Nutr.* **2020**, *4*, 1–6. [CrossRef]
52. Bolhuis, D.P.; Forde, C.G. Application of Food Texture to Moderate Oral Processing Behaviors and Energy Intake. *Trends Food Sci. Technol.* **2020**, *106*, 445–456. [CrossRef]
53. Fardet, A.; Méjean, C.; Labouré, H.; Andreeva, V.A.; Feron, G. The Degree of Processing of Foods Which Are Most Widely Consumed by the French Elderly Population Is Associated with Satiety and Glycemic Potentials and Nutrient Profiles. *Food Funct.* **2017**, *8*, 651–658. [CrossRef] [PubMed]
54. Fardet, A. Minimally Processed Foods Are More Satiating and Less Hyperglycemic than Ultra-Processed Foods: A Preliminary Study with 98 Ready-to-Eat Foods. *Food Funct.* **2016**, *7*, 2338–2346. [CrossRef]
55. Slyper, A. Oral Processing, Satiation and Obesity: Overview and Hypotheses. *Diabetes Metab. Syndr. Obes.* **2021**, *14*, 3399–3415. [CrossRef] [PubMed]
56. Mariath, A.B.; Machado, A.D.; Ferreira, L.d.N.M.; Ribeiro, S.M.L. The Possible Role of Increased Consumption of Ultra-Processed Food Products in the Development of Frailty: A Threat for Healthy Ageing? *Br. J. Nutr.* **2021**, 1–6. [CrossRef]
57. Trakman, G.L.; Fehily, S.; Basnayake, C.; Hamilton, A.L.; Russell, E.; Wilson-O'Brien, A.; Kamm, M.A. Diet and Gut Microbiome in Gastrointestinal Disease. *J. Gastroenterol. Hepatol.* **2021**, 1. [CrossRef]
58. Swinburn, B.A.; Sacks, G.; Hall, K.D.; McPherson, K.; Finegood, D.T.; Moodie, M.L.; Gortmaker, S.L. The Global Obesity Pandemic: Shaped by Global Drivers and Local Environments. *Lancet* **2011**, *378*, 804–814. [CrossRef]
59. Mendes, C.; Miranda, L.; Claro, R.; Horta, P. Food Marketing in Supermarket Circulars in Brazil: An Obstacle to Healthy Eating. *Prev. Med. Rep.* **2021**, *21*, 101304. [CrossRef]
60. Vandevijvere, S.; Van Dam, I. The Nature of Food Promotions over One Year in Circulars from Leading Belgian Supermarket Chains. *Arch. Public Health* **2021**, *79*, 84. [CrossRef] [PubMed]
61. Luiten, C.M.; Steenhuis, I.H.; Eyles, H.; Mhurchu, C.N.; Waterlander, W.E. Ultra-Processed Foods Have the Worst Nutrient Profile, yet They Are the Most Available Packaged Products in a Sample of New Zealand Supermarkets. *Public Health Nutr.* **2016**, *19*, 530–538. [CrossRef]
62. Hollands, G.J.; Shemilt, I.; Marteau, T.M.; Jebb, S.A.; Lewis, H.B.; Wei, Y.; Higgins, J.P.T.; Ogilvie, D. Portion, Package or Tableware Size for Changing Selection and Consumption of Food, Alcohol and Tobacco. *Cochrane Database Syst. Rev.* **2015**, *9*, CD011045. [CrossRef]
63. Juul, F.; Lin, Y.; Deierlein, A.L.; Vaidean, G.; Parekh, N. Trends in Food Consumption by Degree of Processing and Diet Quality over 17 Years: Results from the Framingham Offspring Study. *Br. J. Nutr.* **2021**, *126*, 1861–1871. [CrossRef]
64. Gibney, M.J. Ultraprocessed Foods and Their Application to Nutrition Policy. *Nutr. Today* **2020**, *55*, 16–21. [CrossRef]
65. Derbyshire, E. Are All 'Ultra-Processed' Foods Nutritional Demons? A Commentary and Nutritional Profiling Analysis. *Trends Food Sci. Technol.* **2019**, *94*, 98–104. [CrossRef]
66. Marino, M.; Puppo, F.; Del Bo', C.; Vinelli, V.; Riso, P.; Porrini, M.; Martini, D. A Systematic Review of Worldwide Consumption of Ultra-Processed Foods: Findings and Criticisms. *Nutrients* **2021**, *13*, 2778. [CrossRef]
67. Steele, E.M.; Batis, C.; Cediel, G.; Louzada, M.L.d.C.; Khandpur, N.; Machado, P.; Moubarac, J.-C.; Rauber, F.; Jedlicki, M.R.; Levy, R.B.; et al. The Burden of Excessive Saturated Fatty Acid Intake Attributed to Ultra-Processed Food Consumption: A Study Conducted with Nationally Representative Cross-Sectional Studies from Eight Countries. *J. Nutr. Sci.* **2021**, *10*, e43. [CrossRef] [PubMed]
68. Moreira, P.V.L.; Baraldi, L.G.; Moubarac, J.-C.; Monteiro, C.A.; Newton, A.; Capewell, S.; O'Flaherty, M. Comparing Different Policy Scenarios to Reduce the Consumption of Ultra-Processed Foods in UK: Impact on Cardiovascular Disease Mortality Using a Modelling Approach. *PLoS ONE* **2015**, *10*, e0118353. [CrossRef]
69. Starck, C.S.; Blumfield, M.; Keighley, T.; Marshall, S.; Petocz, P.; Inan-Eroglu, E.; Abbott, K.; Cassettari, T.; Ali, A.; Wham, C.; et al. Nutrient Dense, Low-Cost Foods Can Improve the Affordability and Quality of the New Zealand Diet-A Substitution Modeling Study. *Int. J. Environ. Res. Public Health* **2021**, *18*, 7950. [CrossRef]
70. Verly-Jr, E.; Pereira, A.d.S.; Marques, E.S.; Horta, P.M.; Canella, D.S.; Cunha, D.B. Reducing Ultra-Processed Foods and Increasing Diet Quality in Affordable and Culturally Acceptable Diets: A Study Case from Brazil Using Linear Programming. *Br. J. Nutr.* **2021**, *126*, 572–581. [CrossRef] [PubMed]
71. Gibney, M.J. Ultra-Processed Foods: Definitions and Policy Issues. *Curr. Dev. Nutr.* **2019**, *3*, nzy077. [CrossRef]

72. Gibney, M.J.; Forde, C.G.; Mullally, D.; Gibney, E.R. Ultra-Processed Foods in Human Health: A Critical Appraisal. *Am. J. Clin. Nutr.* **2017**, *106*, 717–724. [CrossRef]
73. Drewnowski, A.; Gupta, S.; Darmon, N. An Overlap Between "Ultraprocessed" Foods and the Preexisting Nutrient Rich Foods Index? *Nutr. Today* **2020**, *55*, 75–81. [CrossRef]
74. Romero Ferreiro, C.; Lora Pablos, D.; Gómez de la Cámara, A. Two Dimensions of Nutritional Value: Nutri-Score and NOVA. *Nutrients* **2021**, *13*, 2783. [CrossRef] [PubMed]
75. European Commission. Nutrition Claims. Available online: https://ec.europa.eu/food/safety/labelling-and-nutrition/nutrition-and-health-claims/nutrition-claims_en (accessed on 7 November 2021).
76. Jones, J.M. Food Processing: Criteria for Dietary Guidance and Public Health? *Proc. Nutr. Soc.* **2019**, *78*, 4–18. [CrossRef] [PubMed]
77. Estell, M.L.; Barrett, E.M.; Kissock, K.R.; Grafenauer, S.J.; Jones, J.M.; Beck, E.J. Fortification of Grain Foods and NOVA: The Potential for Altered Nutrient Intakes While Avoiding Ultra-Processed Foods. *Eur. J. Nutr.* **2021**. [CrossRef] [PubMed]
78. Vergeer, L.; Veira, P.; Bernstein, J.T.; Weippert, M.; L'Abbé, M.R. The Calorie and Nutrient Density of More- Versus Less-Processed Packaged Food and Beverage Products in the Canadian Food Supply. *Nutrients* **2019**, *11*, 2782. [CrossRef]
79. Monteiro, C.; Cannon, G. The Food System. Ultra-Processing. Product Reformulation Will Not Improve Public Health. *World Nutr.* **2012**, *3*, 406–434.
80. Fardet, A. Complex Foods versus Functional Foods, Nutraceuticals and Dietary Supplements: Differential Health Impact (Part 1). *Agro Food Ind. Hi Tech* **2015**, *26*, 6.
81. Adams, J.; Hofman, K.; Moubarac, J.-C.; Thow, A.M. Public Health Response to Ultra-Processed Food and Drinks. *BMJ* **2020**, *369*, m2391. [CrossRef]
82. Scrinis, G.; Monteiro, C.A. Ultra-Processed Foods and the Limits of Product Reformulation. *Public Health Nutr.* **2018**, *21*, 247–252. [CrossRef]
83. Fardet, A.; Rock, E.; Bassama, J.; Bohuon, P.; Prabhasankar, P.; Monteiro, C.; Moubarac, J.-C.; Achir, N. Current Food Classifications in Epidemiological Studies Do Not Enable Solid Nutritional Recommendations for Preventing Diet-Related Chronic Diseases: The Impact of Food Processing. *Adv. Nutr.* **2015**, *6*, 629–638. [CrossRef]
84. Zinöcker, M.K.; Lindseth, I.A. The Western Diet–Microbiome-Host Interaction and Its Role in Metabolic Disease. *Nutrients* **2018**, *10*, 365. [CrossRef] [PubMed]
85. Kwon, T.W.; Hong, J.H.; Moon, G.S.; Song, Y.S.; Kim, J.I.; Kim, J.C.; Kim, M.J. Food Technology: Challenge for Health Promotion. *BioFactors* **2004**, *22*, 279–287. [CrossRef]
86. Du, S.; Kim, H.; Rebholz, C.M. Higher Ultra-Processed Food Consumption Is Associated with Increased Risk of Incident Coronary Artery Disease in the Atherosclerosis Risk in Communities Study. *J. Nutr.* **2021**, *151*, 3746–3754. [CrossRef] [PubMed]
87. Sandoval-Insausti, H.; Blanco-Rojo, R.; Graciani, A.; López-García, E.; Moreno-Franco, B.; Laclaustra, M.; Donat-Vargas, C.; Ordovás, J.M.; Rodríguez-Artalejo, F.; Guallar-Castillón, P. Ultra-Processed Food Consumption and Incident Frailty: A Prospective Cohort Study of Older Adults. *J. Gerontol. A Biol. Sci. Med. Sci.* **2020**, *75*, 1126–1133. [CrossRef]
88. Levy, R.B.; Rauber, F.; Chang, K.; Louzada, M.L.d.C.; Monteiro, C.A.; Millett, C.; Vamos, E.P. Ultra-Processed Food Consumption and Type 2 Diabetes Incidence: A Prospective Cohort Study. *Clin. Nutr.* **2021**, *40*, 3608–3614. [CrossRef]
89. Romero Ferreiro, C.; Martín-Arriscado Arroba, C.; Cancelas Navia, P.; Lora Pablos, D.; Gómez de la Cámara, A. Ultra-Processed Food Intake and All-Cause Mortality: DRECE Cohort Study. *Public Health Nutr.* **2021**, 1–28. [CrossRef]
90. Rauber, F.; Chang, K.; Vamos, E.P.; da Costa Louzada, M.L.; Monteiro, C.A.; Millett, C.; Levy, R.B. Ultra-Processed Food Consumption and Risk of Obesity: A Prospective Cohort Study of UK Biobank. *Eur. J. Nutr.* **2021**, *60*, 2169–2180. [CrossRef] [PubMed]
91. Blanco-Rojo, R.; Sandoval-Insausti, H.; López-Garcia, E.; Graciani, A.; Ordovás, J.M.; Banegas, J.R.; Rodríguez-Artalejo, F.; Guallar-Castillón, P. Consumption of Ultra-Processed Foods and Mortality: A National Prospective Cohort in Spain. *Mayo Clin. Proc.* **2019**, *94*, 2178–2188. [CrossRef]
92. Silva, C.F.M.; Saunders, C.; Peres, W.; Folino, B.; Kamel, T.; Dos Santos, M.S.; Padilha, P. Effect of Ultra-Processed Foods Consumption on Glycemic Control and Gestational Weight Gain in Pregnant with Pregestational Diabetes Mellitus Using Carbohydrate Counting. *PeerJ* **2021**, *9*, e10514. [CrossRef]
93. Gomes, C.d.B.; Malta, M.B.; Benício, M.H.D.; Carvalhaes, M.A.d.B.L. Consumption of Ultra-Processed Foods in the Third Gestational Trimester and Increased Weight Gain: A Brazilian Cohort Study. *Public Health Nutr.* **2021**, *24*, 3304–3312. [CrossRef]
94. Gadelha, P.C.F.P.; de Arruda, I.K.G.; Coelho, P.B.P.; Queiroz, P.M.A.; Maio, R.; da Silva Diniz, A. Consumption of Ultraprocessed Foods, Nutritional Status, and Dyslipidemia in Schoolchildren: A Cohort Study. *Eur. J. Clin. Nutr.* **2019**, *73*, 1194–1199. [CrossRef]
95. Rauber, F.; Campagnolo, P.D.B.; Hoffman, D.J.; Vitolo, M.R. Consumption of Ultra-Processed Food Products and Its Effects on Children's Lipid Profiles: A Longitudinal Study. *Nutr. Metab. Cardiovasc. Dis.* **2015**, *25*, 116–122. [CrossRef]
96. Cunha, D.B.; da Costa, T.H.M.; da Veiga, G.V.; Pereira, R.A.; Sichieri, R. Ultra-Processed Food Consumption and Adiposity Trajectories in a Brazilian Cohort of Adolescents: ELANA Study. *Nutr. Diabetes* **2018**, *8*, 28. [CrossRef]
97. Bawaked, R.A.; Fernández-Barrés, S.; Navarrete-Muñoz, E.M.; González-Palacios, S.; Guxens, M.; Irizar, A.; Lertxundi, A.; Sunyer, J.; Vioque, J.; Schröder, H.; et al. Impact of Lifestyle Behaviors in Early Childhood on Obesity and Cardiometabolic Risk in Children: Results from the Spanish INMA Birth Cohort Study. *Pediatr. Obes.* **2020**, *15*, e12590. [CrossRef]

98. Costa, C.S.; Rauber, F.; Leffa, P.S.; Sangalli, C.N.; Campagnolo, P.D.B.; Vitolo, M.R. Ultra-Processed Food Consumption and Its Effects on Anthropometric and Glucose Profile: A Longitudinal Study during Childhood. *Nutr. Metab. Cardiovasc. Dis.* **2019**, *29*, 177–184. [CrossRef] [PubMed]

99. Vedovato, G.M.; Vilela, S.; Severo, M.; Rodrigues, S.; Lopes, C.; Oliveira, A. Ultra-Processed Food Consumption, Appetitive Traits and BMI in Children: A Prospective Study. *Br. J. Nutr.* **2021**, *125*, 1427–1436. [CrossRef] [PubMed]

100. de Melo, J.M.M.; Dourado, B.L.L.F.S.; de Menezes, R.C.E.; Longo-Silva, G.; da Silveira, J.A.C. Early Onset of Overweight among Children from Low-Income Families: The Role of Exclusive Breastfeeding and Maternal Intake of Ultra-Processed Food. *Pediatr. Obes.* **2021**, *16*, e12825. [CrossRef]

101. Machado Azeredo, C.; Cortese, M.; Costa, C.d.S.; Bjornevik, K.; Barros, A.J.D.; Barros, F.C.; Santos, I.S.; Matijasevich, A. Ultra-Processed Food Consumption during Childhood and Asthma in Adolescence: Data from the 2004 Pelotas Birth Cohort Study. *Pediatr. Allergy Immunol.* **2020**, *31*, 27–37. [CrossRef] [PubMed]

102. Schnabel, L.; Kesse-Guyot, E.; Allès, B.; Touvier, M.; Srour, B.; Hercberg, S.; Buscail, C.; Julia, C. Association Between Ultraprocessed Food Consumption and Risk of Mortality Among Middle-Aged Adults in France. *JAMA Intern. Med.* **2019**, *179*, 490–498. [CrossRef]

103. Rico-Campà, A.; Martínez-González, M.A.; Alvarez-Alvarez, I.; Mendonça, R.d.D.; de la Fuente-Arrillaga, C.; Gómez-Donoso, C.; Bes-Rastrollo, M. Association between Consumption of Ultra-Processed Foods and All Cause Mortality: SUN Prospective Cohort Study. *BMJ* **2019**, *365*, l1949. [CrossRef]

104. Kim, H.; Hu, E.A.; Rebholz, C.M. Ultra-Processed Food Intake and Mortality in the United States: Results from the Third National Health and Nutrition Examination Survey (NHANES III 1988-1994). *Public Health Nutr.* **2019**, *22*, 1777–1785. [CrossRef]

105. Bonaccio, M.; Di Castelnuovo, A.; Costanzo, S.; De Curtis, A.; Persichillo, M.; Sofi, F.; Cerletti, C.; Donati, M.B.; de Gaetano, G.; Iacoviello, L.; et al. Ultra-Processed Food Consumption Is Associated with Increased Risk of All-Cause and Cardiovascular Mortality in the Moli-Sani Study. *Am. J. Clin. Nutr.* **2021**, *113*, 446–455. [CrossRef]

106. Beslay, M.; Srour, B.; Méjean, C.; Allès, B.; Fiolet, T.; Debras, C.; Chazelas, E.; Deschasaux, M.; Wendeu-Foyet, M.G.; Hercberg, S.; et al. Ultra-Processed Food Intake in Association with BMI Change and Risk of Overweight and Obesity: A Prospective Analysis of the French NutriNet-Santé Cohort. *PLoS Med.* **2020**, *17*, e1003256. [CrossRef] [PubMed]

107. Mendonça, R.d.D.; Pimenta, A.M.; Gea, A.; de la Fuente-Arrillaga, C.; Martinez-Gonzalez, M.A.; Lopes, A.C.S.; Bes-Rastrollo, M. Ultraprocessed Food Consumption and Risk of Overweight and Obesity: The University of Navarra Follow-Up (SUN) Cohort Study. *Am. J. Clin. Nutr.* **2016**, *104*, 1433–1440. [CrossRef] [PubMed]

108. Li, M.; Shi, Z. Ultra-Processed Food Consumption Associated with Overweight/Obesity among Chinese Adults—Results from China Health and Nutrition Survey 1997–2011. *Nutrients* **2021**, *13*, 2796. [CrossRef]

109. Konieczna, J.; Morey, M.; Abete, I.; Bes-Rastrollo, M.; Ruiz-Canela, M.; Vioque, J.; Gonzalez-Palacios, S.; Daimiel, L.; Salas-Salvadó, J.; Fiol, M.; et al. Contribution of Ultra-Processed Foods in Visceral Fat Deposition and Other Adiposity Indicators: Prospective Analysis Nested in the PREDIMED-Plus Trial. *Clin. Nutr.* **2021**, *40*, 4290–4300. [CrossRef]

110. Sandoval-Insausti, H.; Jiménez-Onsurbe, M.; Donat-Vargas, C.; Rey-García, J.; Banegas, J.R.; Rodríguez-Artalejo, F.; Guallar-Castillón, P. Ultra-Processed Food Consumption Is Associated with Abdominal Obesity: A Prospective Cohort Study in Older Adults. *Nutrients* **2020**, *12*, 2368. [CrossRef]

111. Cordova, R.; Kliemann, N.; Huybrechts, I.; Rauber, F.; Vamos, E.P.; Levy, R.B.; Wagner, K.-H.; Viallon, V.; Casagrande, C.; Nicolas, G.; et al. Consumption of Ultra-Processed Foods Associated with Weight Gain and Obesity in Adults: A Multi-National Cohort Study. *Clin. Nutr.* **2021**, *40*, 5079–5088. [CrossRef] [PubMed]

112. Canhada, S.L.; Luft, V.C.; Giatti, L.; Duncan, B.B.; Chor, D.; da Fonseca, M.d.J.M.; Matos, S.M.A.; Molina, M.d.C.B.; Barreto, S.M.; Levy, R.B.; et al. Ultra-Processed Foods, Incident Overweight and Obesity, and Longitudinal Changes in Weight and Waist Circumference: The Brazilian Longitudinal Study of Adult Health (ELSA-Brasil). *Public Health Nutr.* **2020**, *23*, 1076–1086. [CrossRef]

113. Rohatgi, K.W.; Tinius, R.A.; Cade, W.T.; Steele, E.M.; Cahill, A.G.; Parra, D.C. Relationships between Consumption of Ultra-Processed Foods, Gestational Weight Gain and Neonatal Outcomes in a Sample of US Pregnant Women. *PeerJ* **2017**, *5*, e4091. [CrossRef]

114. Leone, A.; Martínez-González, M.Á.; Craig, W.; Fresán, U.; Gómez-Donoso, C.; Bes-Rastrollo, M. Pre-Gestational Consumption of Ultra-Processed Foods and Risk of Gestational Diabetes in a Mediterranean Cohort. The SUN Project. *Nutrients* **2021**, *13*, 2202. [CrossRef] [PubMed]

115. Chang, K.; Khandpur, N.; Neri, D.; Touvier, M.; Huybrechts, I.; Millett, C.; Vamos, E.P. Association Between Childhood Consumption of Ultraprocessed Food and Adiposity Trajectories in the Avon Longitudinal Study of Parents and Children Birth Cohort. *JAMA Pediatr.* **2021**, *175*, e211573. [CrossRef]

116. Costa, C.d.S.; Assunção, M.C.F.; Loret de Mola, C.; Cardoso, J.d.S.; Matijasevich, A.; Barros, A.J.D.; Santos, I.S. Role of Ultra-Processed Food in Fat Mass Index between 6 and 11 Years of Age: A Cohort Study. *Int. J. Epidemiol.* **2021**, *50*, 256–265. [CrossRef] [PubMed]

117. Srour, B.; Fezeu, L.K.; Kesse-Guyot, E.; Allès, B.; Méjean, C.; Andrianasolo, R.M.; Chazelas, E.; Deschasaux, M.; Hercberg, S.; Galan, P.; et al. Ultra-Processed Food Intake and Risk of Cardiovascular Disease: Prospective Cohort Study (NutriNet-Santé). *BMJ* **2019**, *365*, l1451. [CrossRef]

118. Juul, F.; Vaidean, G.; Lin, Y.; Deierlein, A.L.; Parekh, N. Ultra-Processed Foods and Incident Cardiovascular Disease in the Framingham Offspring Study. *J. Am. Coll. Cardiol.* **2021**, *77*, 1520–1531. [CrossRef] [PubMed]
119. Zhong, G.-C.; Gu, H.-T.; Peng, Y.; Wang, K.; Wu, Y.-Q.-L.; Hu, T.-Y.; Jing, F.-C.; Hao, F.-B. Association of Ultra-Processed Food Consumption with Cardiovascular Mortality in the US Population: Long-Term Results from a Large Prospective Multicenter Study. *Int. J. Behav. Nutr. Phys. Act.* **2021**, *18*, 21. [CrossRef]
120. da Silva Scaranni, P.D.O.; de Oliveira Cardoso, L.; Chor, D.; Melo, E.C.P.; Matos, S.M.A.; Giatti, L.; Barreto, S.M.; da Fonseca, M.D.J.M. Ultra-Processed Foods, Changes in Blood Pressure and Incidence of Hypertension: The Brazilian Longitudinal Study of Adult Health (ELSA-Brasil). *Public Health Nutr.* **2021**, *24*, 3352–3360. [CrossRef]
121. Monge, A.; Silva Canella, D.; López-Olmedo, N.; Lajous, M.; Cortés-Valencia, A.; Stern, D. Ultraprocessed Beverages and Processed Meats Increase the Incidence of Hypertension in Mexican Women. *Br. J. Nutr.* **2021**, *126*, 600–611. [CrossRef]
122. de Mendonça, R.D.; Lopes, A.C.S.; Pimenta, A.M.; Gea, A.; Martinez-Gonzalez, M.A.; Bes-Rastrollo, M. Ultra-Processed Food Consumption and the Incidence of Hypertension in a Mediterranean Cohort: The Seguimiento Universidad de Navarra Project. *Am. J. Hypertens.* **2017**, *30*, 358–366. [CrossRef]
123. Llavero-Valero, M.; Martín, J.E.-S.; Martínez-González, M.A.; Basterra-Gortari, F.J.; de la Fuente-Arrillaga, C.; Bes-Rastrollo, M. Ultra-Processed Foods and Type-2 Diabetes Risk in the SUN Project: A Prospective Cohort Study. *Clin. Nutr.* **2021**, *40*, 2817–2824. [CrossRef]
124. Srour, B.; Fezeu, L.K.; Kesse-Guyot, E.; Allès, B.; Debras, C.; Druesne-Pecollo, N.; Chazelas, E.; Deschasaux, M.; Hercberg, S.; Galan, P.; et al. Ultraprocessed Food Consumption and Risk of Type 2 Diabetes Among Participants of the NutriNet-Santé Prospective Cohort. *JAMA Intern. Med.* **2020**, *180*, 283–291. [CrossRef] [PubMed]
125. Zhang, S.; Gan, S.; Zhang, Q.; Liu, L.; Meng, G.; Yao, Z.; Wu, H.; Gu, Y.; Wang, Y.; Zhang, T.; et al. Ultra-Processed Food Consumption and the Risk of Non-Alcoholic Fatty Liver Disease in the Tianjin Chronic Low-Grade Systemic Inflammation and Health Cohort Study. *Int. J. Epidemiol.* **2021**, dyab174. [CrossRef] [PubMed]
126. Fiolet, T.; Srour, B.; Sellem, L.; Kesse-Guyot, E.; Allès, B.; Méjean, C.; Deschasaux, M.; Fassier, P.; Latino-Martel, P.; Beslay, M.; et al. Consumption of Ultra-Processed Foods and Cancer Risk: Results from NutriNet-Santé Prospective Cohort. *BMJ* **2018**, *360*, k322. [CrossRef]
127. Vasseur, P.; Dugelay, E.; Benamouzig, R.; Savoye, G.; Lan, A.; Srour, B.; Hercberg, S.; Touvier, M.; Hugot, J.-P.; Julia, C.; et al. Dietary Patterns, Ultra-Processed Food, and the Risk of Inflammatory Bowel Diseases in the NutriNet-Santé Cohort. *Inflamm. Bowel Dis.* **2021**, *27*, 65–73. [CrossRef]
128. Narula, N.; Wong, E.C.L.; Dehghan, M.; Mente, A.; Rangarajan, S.; Lanas, F.; Lopez-Jaramillo, P.; Rohatgi, P.; Lakshmi, P.V.M.; Varma, R.P.; et al. Association of Ultra-Processed Food Intake with Risk of Inflammatory Bowel Disease: Prospective Cohort Study. *BMJ* **2021**, *374*, n1554. [CrossRef]
129. Schnabel, L.; Buscail, C.; Sabate, J.-M.; Bouchoucha, M.; Kesse-Guyot, E.; Allès, B.; Touvier, M.; Monteiro, C.A.; Hercberg, S.; Benamouzig, R.; et al. Association Between Ultra-Processed Food Consumption and Functional Gastrointestinal Disorders: Results from the French NutriNet-Santé Cohort. *Am. J. Gastroenterol.* **2018**, *113*, 1217–1228. [CrossRef] [PubMed]
130. Lo, C.-H.; Khandpur, N.; Rossato, S.; Lochhead, P.; Lopes, E.; Burke, K.; Richter, J.; Song, M.; Korat, A.; Sun, Q.; et al. Ultra-Processed Foods and Risk of Crohn's Disease and Ulcerative Colitis: A Prospective Cohort Study. *Clin. Gastroenterol. Hepatol.* **2021**. [CrossRef]
131. Adjibade, M.; Julia, C.; Allès, B.; Touvier, M.; Lemogne, C.; Srour, B.; Hercberg, S.; Galan, P.; Assmann, K.E.; Kesse-Guyot, E. Prospective Association between Ultra-Processed Food Consumption and Incident Depressive Symptoms in the French NutriNet-Santé Cohort. *BMC Med.* **2019**, *17*, 78. [CrossRef]
132. Gómez-Donoso, C.; Sánchez-Villegas, A.; Martínez-González, M.A.; Gea, A.; de Mendonça, R.D.; Lahortiga-Ramos, F.; Bes-Rastrollo, M. Ultra-Processed Food Consumption and the Incidence of Depression in a Mediterranean Cohort: The SUN Project. *Eur. J. Nutr.* **2020**, *59*, 1093–1103. [CrossRef]
133. Rey-García, J.; Donat-Vargas, C.; Sandoval-Insausti, H.; Bayan-Bravo, A.; Moreno-Franco, B.; Banegas, J.R.; Rodríguez-Artalejo, F.; Guallar-Castillón, P. Ultra-Processed Food Consumption Is Associated with Renal Function Decline in Older Adults: A Prospective Cohort Study. *Nutrients* **2021**, *13*, 428. [CrossRef]
134. Zhang, T.; Gan, S.; Ye, M.; Meng, G.; Zhang, Q.; Liu, L.; Wu, H.; Gu, Y.; Zhang, S.; Wang, Y.; et al. Association between Consumption of Ultra-Processed Foods and Hyperuricemia: TCLSIH Prospective Cohort Study. *Nutr. Metab. Cardiovasc. Dis.* **2021**, *31*, 1993–2003. [CrossRef] [PubMed]
135. Leffa, P.S.; Hoffman, D.J.; Rauber, F.; Sangalli, C.N.; Valmórbida, J.L.; Vitolo, M.R. Longitudinal Associations between Ultra-Processed Foods and Blood Lipids in Childhood. *Br. J. Nutr.* **2020**, *124*, 341–348. [CrossRef]
136. Donat-Vargas, C.; Sandoval-Insausti, H.; Rey-García, J.; Moreno-Franco, B.; Åkesson, A.; Banegas, J.R.; Rodríguez-Artalejo, F.; Guallar-Castillón, P. High Consumption of Ultra-Processed Food Is Associated with Incident Dyslipidemia: A Prospective Study of Older Adults. *J. Nutr.* **2021**, *151*, 2390–2398. [CrossRef]
137. Borge, T.C.; Biele, G.; Papadopoulou, E.; Andersen, L.F.; Jacka, F.; Eggesbø, M.; Caspersen, I.H.; Aase, H.; Meltzer, H.M.; Brantsæter, A.L. The Associations between Maternal and Child Diet Quality and Child ADHD—Findings from a Large Norwegian Pregnancy Cohort Study. *BMC Psychiatry* **2021**, *21*, 139. [CrossRef]

138. Zhang, S.; Gu, Y.; Rayamajhi, S.; Thapa, A.; Meng, G.; Zhang, Q.; Liu, L.; Wu, H.; Zhang, T.; Wang, X.; et al. Ultra-Processed Food Intake Is Associated with Grip Strength Decline in Middle-Aged and Older Adults: A Prospective Analysis of the TCLSIH Study. *Eur. J. Nutr.* **2021**. [CrossRef] [PubMed]

139. Huybrechts, I.; Vereecken, C.; De Bacquer, D.; Vandevijvere, S.; Van Oyen, H.; Maes, L.; Vanhauwaert, E.; Temme, L.; De Backer, G.; De Henauw, S. Reproducibility and Validity of a Diet Quality Index for Children Assessed Using a FFQ. *Br. J. Nutr.* **2010**, *104*, 135–144. [CrossRef]

140. Kouvari, M.; Chrysohoou, C.; Georgousopoulou, E.; Skoumas, J.; Pitsavos, C.; Panagiotakos, D.B. Ultra-Processed Foods and Ten-Year Cardiovascular Disease Incidence in a Mediterranean Population: Results from a Population-Based Cohort Study. *Eur. Heart J.* **2021**, *42*, 2433. [CrossRef]

141. Bonaccio, M.; Di Castelnuovo, A.; Ruggiero, E.; Costanzo, S.; Esposito, S.; Persichillo, M.; Cerletti, C.; Donati, M.; de Gaetano, G.; Iacoviello, L. Ultra-Processed Food Consumption Modifies the Association of Nutri-Score with All-Cause Mortality. *Eur. J. Public Health* **2021**, *31*, 169–170. [CrossRef]

142. Kipnis, V.; Subar, A.F.; Midthune, D.; Freedman, L.S.; Ballard-Barbash, R.; Troiano, R.P.; Bingham, S.; Schoeller, D.A.; Schatzkin, A.; Carroll, R.J. Structure of Dietary Measurement Error: Results of the OPEN Biomarker Study. *Am. J. Epidemiol.* **2003**, *158*, 14–21. [CrossRef] [PubMed]

143. Subar, A.F.; Freedman, L.S.; Tooze, J.A.; Kirkpatrick, S.I.; Boushey, C.; Neuhouser, M.L.; Thompson, F.E.; Potischman, N.; Guenther, P.M.; Tarasuk, V.; et al. Addressing Current Criticism Regarding the Value of Self-Report Dietary Data. *J. Nutr.* **2015**, *145*, 2639–2645. [CrossRef] [PubMed]

144. Willett, W. *Monographs in Epidemiology and Biostatistics. Nutritional Epidemiology*, 3rd ed.; Oxford University Press: New York, NY, USA, 2012; ISBN 978-0-19-975403-8.

145. Hu, F.B.; Stampfer, M.J.; Rimm, E.; Ascherio, A.; Rosner, B.A.; Spiegelman, D.; Willett, W.C. Dietary Fat and Coronary Heart Disease: A Comparison of Approaches for Adjusting for Total Energy Intake and Modeling Repeated Dietary Measurements. *Am. J. Epidemiol.* **1999**, *149*, 531–540. [CrossRef] [PubMed]

146. Willett, W.C.; Howe, G.R.; Kushi, L.H. Adjustment for Total Energy Intake in Epidemiologic Studies. *Am. J. Clin. Nutr.* **1997**, *65*, 1220S–1228S. [CrossRef]

147. Padilha, P.; Felizardo, C.; Saunders, C.; Cunha, L.; Pinheiro, A.; Belfort, G.; Santos, K.; Ferreira, N. Consumption of Ultraprocessed Foods by Pregnant Women with Diabetes Mellitus (P11-021-19). *Curr. Dev. Nutr.* **2019**, *3*, nzz048-P11. [CrossRef]

148. Rezende-Alves, K.; Hermsdorff, H.; Miranda, A.; Lopes, A.; Bressan, J.; Pimenta, A. Food processing and risk of hypertension: Cohort of Universities of Minas Gerais, Brazil (CUME Project). *Public Health Nutr.* **2021**, *24*, 4071–4079. [CrossRef]

149. Lange, T.; Vansteelandt, S.; Bekaert, M. A Simple Unified Approach for Estimating Natural Direct and Indirect Effects. *Am. J. Epidemiol.* **2012**, *176*, 190–195. [CrossRef]

150. Lockwood, C.M.; DeFrancesco, C.A.; Elliot, D.L.; Beresford, S.A.A.; Toobert, D.J. Mediation Analyses: Applications in Nutrition Research and Reading the Literature. *J. Am. Diet. Assoc.* **2010**, *110*, 753–762. [CrossRef]

151. Widmer, R.J.; Flammer, A.J.; Lerman, L.O.; Lerman, A. The Mediterranean Diet, Its Components, and Cardiovascular Disease. *Am. J. Med.* **2015**, *128*, 229–238. [CrossRef]

152. Crimarco, A.; Landry, M.J.; Gardner, C.D. Ultra-Processed Foods, Weight Gain, and Co-Morbidity Risk. *Curr. Obes. Rep.* **2021**, 1–13. [CrossRef] [PubMed]

153. Bandy, L.K.; Scarborough, P.; Harrington, R.A.; Rayner, M.; Jebb, S.A. Reductions in Sugar Sales from Soft Drinks in the UK from 2015 to 2018. *BMC Med.* **2020**, *18*, 20. [CrossRef] [PubMed]

154. Jacobs, D.R.; Tapsell, L.C. Food, Not Nutrients, Is the Fundamental Unit in Nutrition. *Nutr. Rev.* **2007**, *65*, 439–450. [CrossRef]

155. Barabási, A.-L.; Menichetti, G.; Loscalzo, J. The Unmapped Chemical Complexity of Our Diet. *Nat. Food* **2020**, *1*, 33–37. [CrossRef]

156. Cancer Research UK. What Is a Healthy Diet? Available online: https://www.cancerresearchuk.org/about-cancer/causes-of-cancer/diet-and-cancer/what-is-a-healthy-diet (accessed on 11 October 2021).

157. NHS. The Eatwell Guide. Available online: https://www.nhs.uk/live-well/eat-well/the-eatwell-guide/ (accessed on 11 October 2021).

158. GOV.UK. Government Recommendations for Energy and Nutrients for Males and Females Aged 1–18 Years and 19+ Years. Available online: https://assets.publishing.service.gov.uk/government/uploads/system/uploads/attachment_data/file/618167/government_dietary_recommendations.pdf (accessed on 14 October 2021).

159. UNICEF. Prevention of Overweight and Obesity in Children and Adolescents. Available online: https://www.unicef.org/media/92336/file/Programming-Guidance-Overweight-Prevention.pdf (accessed on 3 November 2021).

160. World Health Organisation. Report of the Commission on Ending Childhood Obesity. Available online: http://apps.who.int/iris/bitstream/handle/10665/204176/9789241510066_eng.pdf (accessed on 3 November 2021).

161. Pan American Health Organisation. Nutrient Profile Model. Available online: https://iris.paho.org/bitstream/handle/10665.2/18621/9789275118733_eng.pdf (accessed on 19 October 2021).

162. Ministry of Health of Brazil. Dietary Guidelines for the Brazilian Population. Available online: https://bvsms.saude.gov.br/bvs/publicacoes/dietary_guidelines_brazilian_population.pdf (accessed on 14 October 2021).
163. Food and Agriculture Organization of the United Nations. Dietary Guidelines for the Uruguayan Population: For a Healthy, Shared and Enjoyable Diet. Available online: http://www.fao.org/nutrition/education/food-dietary-guidelines/regions/uruguay/en/ (accessed on 14 October 2021).
164. The Israeli Ministry of Health. Nutritional Recommendations. Available online: https://www.health.gov.il/PublicationsFiles/dietary%20guidelines%20EN.pdf (accessed on 18 October 2021).
165. High Council of Public Health. Relatif Aux Objectifs de Santé Publique Quantifiés Pour La Politique Nutritionnelle de Santé Publique (PNNS) 2018–2022 [Quantified Public Health Objectives for Public Health Nutrition Policy (PNNS) 2018–2022]. Available online: https://www.hcsp.fr/explore.cgi/avisrapportsdomaine?clefr=648 (accessed on 13 November 2021).

nutrients

MDPI

Article

Effects of Nutrition Intervention on Blood Glucose, Body Composition, and Phase Angle in Obese and Overweight Patients with Diabetic Foot Ulcers

Raedeh Basiri [1,2,3,4,*], Maria T. Spicer [3], Thomas Ledermann [5] and Bahram H. Arjmandi [3,4,*]

[1] Department of Nutrition and Food Studies, George Mason University, Fairfax, VA 22030, USA
[2] Institute for Biohealth Innovation, George Mason University, Fairfax, VA 22030, USA
[3] Department of Nutrition and Integrative Physiology, Florida State University, Tallahassee, FL 32306, USA
[4] Center for Advancing Exercise and Nutrition Research on Aging, Florida State University, Tallahassee, FL 32306, USA
[5] Department of Family and Child Sciences, Florida State University, Tallahassee, FL 32306, USA
* Correspondence: rbasiri@gmu.edu (R.B.); barjmandi@fsu.edu (B.H.A.)

Abstract: Nutrition can play an important role in the treatment of chronic wounds such as diabetic foot ulcers (DFUs); however, diet therapy is not currently part of the standard care for DFUs. There are numerous controversies about dietary recommendations, especially regarding calories and macronutrients, for overweight and obese patients with DFUs. This study examined the effects of nutrition education and supplementation on body composition in overweight and obese patients with DFUs. Twenty-nine patients with DFUs between the ages of 30 and 70 years were randomly assigned to either the treatment group (nutritional supplements, diet education, and standard care) or the control group (standard care). At baseline, the mean body mass index (BMI) was 33.5 kg/m^2 for the treatment group and 34.1 kg/m^2 for the control group. HbA1c decreased in both groups, with no significant difference between the groups. On average, patients in the treatment group lost less lean body mass and gained less fat than the control group ((3.8 kg vs. 4.9 kg) and (0.9 kg vs. 3.6 kg), respectively). While the interaction between group and time did not reach statistical significance for any of the study variables after adjustments for confounding variables, the observed changes are clinically relevant.

Keywords: nutrition education; DFU; nutrition supplementation; body composition; phase angle; protein; macronutrients; micronutrients; chronic wounds; wound healing; diabetic foot ulcer

Citation: Basiri, R.; Spicer, M.T.; Ledermann, T.; Arjmandi, B.H. Effects of Nutrition Intervention on Blood Glucose, Body Composition, and Phase Angle in Obese and Overweight Patients with Diabetic Foot Ulcers. *Nutrients* **2022**, *14*, 3564. https://doi.org/10.3390/nu14173564

Academic Editor: Javier Gómez-Ambrosi

Received: 11 August 2022
Accepted: 26 August 2022
Published: 30 August 2022

Publisher's Note: MDPI stays neutral with regard to jurisdictional claims in published maps and institutional affiliations.

1. Introduction

Diabetic foot ulcers (DFUs) are among the most common complications of uncontrolled diabetes [1]. It has been reported that 25% of patients with diabetes develop DFUs during their lifetime [2]. DFUs significantly affect the patient's health and socioeconomic well-being and negatively affect the quality of life of the patients and their family members [3]. Nutrition can play a key role in the prevention and improvement of the clinical outcomes of DFUs [4]. In patients with chronic wounds, cellular activity and inflammation in the healing wound increase metabolic needs; therefore, they require more energy and a higher nutrient intake. In DFUs, the hypermetabolism nature of the wound, as well as a decreased sensitivity to insulin, increased counter-regulatory hormones such as cortisol, catecholamines, and glucagon due to a high level of stress, and not enough energy intake results in the body utilizing muscle proteins as a source of energy [5,6]. Apart from the wound itself, the kinetics of whole-body protein metabolism are elevated in diabetes and the net balance is diminished [7]. Increased protein catabolism and negative nitrogen balance have been reported in patients with uncontrolled diabetes [8]. Additionally, it has been reported that energy expenditure is significantly higher in patients with type 2 diabetes, in comparison with non-diabetic individuals [9,10]. Despite the higher need for energy sources and essential nutrients during wound healing, diabetic patients are usually recommended to follow low-calorie/low-carbohydrate diets to better manage their glycemic indices and related complications, particularly if they are overweight or obese [11]. However, restrictive diets could result in an inadequate intake of essential micronutrients, as well as the energy sources and protein that are vital for wound healing. It has been reported that patients with DFUs have a significantly lower intake of energy, protein, and micronutrients compared to dietary reference intakes (DRI), which are designed for a healthy population [12–14]. A significantly low dietary intake of energy and nutrients has also been shown in overweight and obese patients with DFUs [15]. Insufficient energy intake can result in muscle wasting, the loss of subcutaneous tissue, and consequently, poor wound healing [16]. Protein is responsible for cell proliferation, collagen, and connective tissue synthesis, as well as the antibody synthesis needed for immune system function [17]; therefore, adequate protein intake supplemented by non-protein energy sources promotes a positive nitrogen balance, which is crucial for improving wound healing in DFUs. Although an increased need for energy and nutrient intake has been established in patients with chronic wounds [18] and an inadequate intake of energy sources and nutrients has repeatedly been reported in patients with DFUs, there are still controversies about the nutritional recommendations for DFU patients, particularly for those who are overweight or obese. This study examined the effects of nutrition education and supplementation on long-term blood glucose control, body composition, and phase angle as an indicator of cellular health and cell membrane integrity in patients with DFUs.

2. Materials and Methods

2.1. Screening and Recruitment

This study was approved by the Institutional Review Board (IRB) of Tallahassee Memorial HealthCare (TMH, Tallahassee, FL, USA) and Florida State University, and is registered at clinicaltrials.gov NCT04055064. The study was advertised at the Tallahassee Memorial Wound Healing Center; interested participants were prescreened by one of the medical staff or a nurse at the clinic, based on the inclusion/exclusion criteria. The potential participants were then scheduled for a screening visit with the researcher. Patients were included in the study if they were between the ages of 30 and 70 years, had a body mass index (BMI) $\geq$ 25 kg/m^2, had at least one diabetic foot ulcer of grade 1A [19], and were receiving medications for glycemic control.

Patients were excluded from the study if they were pregnant or lactating, had used bioengineered tissue within the four weeks before baseline, had high concentrations of hemoglobin A1C (HbA1c) of > 12%, known immunosuppression, liver failure/cirrhosis, active malignancy, myocardial infarction or heart failure in the past three months, chronic kidney disease, underwent radiation therapy for the treatment of their wounds, excessive use of alcohol according to the standards of the World Health Organization, or were subject to any physiological or mental condition that might affect the study regimen.

After screening, eligible patients were provided with details of the study and were asked to sign the consent form if they were interested in participating in the study. Participants were then randomly assigned to either the control or the treatment group.

2.2. Study Intervention

Standard wound care was provided to all the participants at the TMH Wound Healing Center. Additionally, patients in the treatment group were asked to consume more low-fat protein sources with high bioavailability, vegetables, high-fiber carbohydrates, and a lower amount of simple carbohydrates. Participants were also educated about the different food groups and were given examples of healthier food items in each group. The nutrition education was conducted by the researcher (nutritionist) for at least 10 min at baseline and was then repeated every four weeks for each patient in the treatment group. The treatment group was also provided with two servings of Boost Glucose Control nutritional formula (Nestlé Health Science, NJ, USA) and instructed to consume one supplement in the morning and one in the afternoon, during the study. Consuming two servings of supplements provided patients with extra energy, protein, and essential vitamins and minerals. Table 1 shows the energy and macronutrient contents of the supplement. The complete nutrient content of the supplement has been published elsewhere [15].

Table 1. Energy and macronutrient contents of one serving (237 mL) of the nutritional supplement.

Nutrient	Amount
Calories (kcal)	250
Calories from Fat (kcal)	110
Total Fat (g)	12
Saturated Fat (g)	1.5
Total Carbohydrate (g)	23
Dietary Fiber (g)	3
Sugars (g)	6
Protein [1] (g)	14

[1] Includes protein from caseinate and L-arginine.

The supplement used in this study was designed for diabetic patients and contained a slow-release carbohydrate source called tapioca dextrin, which is digested slowly. Tapioca is resistant to amylase and prevents a sudden increase in blood glucose [20]. This study aimed to provide patients with adequate supplements that could help the patients to meet their extra need for energy and protein and to support them with at least 50% of the Recommended Dietary Allowance (RDA) recommendations for the essential vitamins and minerals for wound healing. We anticipated that the combination of nutrition education and supplements could significantly improve the dietary intake of participants and support them in meeting their nutrient recommendations.

2.3. Anthropometric and Body Composition Measurement

Height was self-reported, and weight was measured at baseline and every four weeks during the study, using a stand-on scale (Seca Mechanical Column Scale; Hamburg, Germany). BMI was calculated from weight and height using the $BMI = weight (kg)/Height^2 (m^2)$ formula. Body composition was evaluated using a bioelectrical impedance analyzer (BIA) 310 e (Biodynamics Corporation, Seattle, WA, USA) which has been shown to accurately estimate body cell mass and lean body mass [21]. The within-day and between-day coefficients of variation (CVs)% for hand-to-foot (whole-body) model impedance have been reported as 0.2% to 0.7% and 0.9% to 1.8%, respectively [22]. Patients were instructed to remove their right shoe and sock and lay down on their backs. Their feet were 12 to 18 inches apart, while their hands were placed palm-down at 6 to 12 inches from the torso. Two adhesive sensor pads were placed on the right wrist/hand and two on the right ankle/foot. A small current (800 µA at 50 kHz) was passed through the electrodes to measure fat body weight (kg), lean body mass weight (kg), and reactance and resistance; the measurements were used for calculating the phase angle (PA). Fat body weight is the total amount of stored lipids in the body and consists of subcutaneous and visceral fat. Lean body mass weight was calculated by subtracting body fat (kg) from total body weight (kg). PA indicates a relationship between electric resistance (R) and reactance (Rc) and is an indicator of cellular health and function [15,16]. Lower phase angles are correlated with the duration of disease, inflammation, malnutrition, and mortality in diabetic patients [23–26]. PA was calculated directly from reactance and resistance, using the $PA = arctangent\ reactance\ (ohm)/resistance\ (ohm) \times 180°/\pi$ formula [27].

2.4. HbA1c Measurement

To assess the effects of the intervention on long-term glucose homeostasis, HbA1c was evaluated at baseline and the end of the study, using the HbA1c Now+ test (Polymer Technology Systems, Indianapolis, IN, USA).

2.5. Dietary Assessment

The dietary intake of participants was estimated using 24-hour recall forms. Participants were asked about all the foods and beverages consumed during the last 24 h as well as any prescribed or voluntary use of nutritional supplements. Nutrient intake was estimated using the Food Processor SQL, version 11.1.480 (ESHA's Food Processor®, Salem, OR, USA).

2.6. Statistical Analysis

The Statistical Package for Social Science (SPSS), version 25.0 (SPSS, Inc., Chicago, IL, USA), was utilized to analyze our data. For all tests, $p < 0.05$ was set as the statistically significant level. Population characteristics were evaluated at baseline using descriptive statistics. An independent sample t-test was used to compare the means of potential confounding variables between groups at baseline; if the effect was significant, they were set as covariates in the model. The independent sample t-test was also used for the comparison of HbA1c concentrations between groups at baseline and the end of the study. All the other variables were analyzed using multilevel modeling (mixed model), while Bonferroni's post hoc test was utilized for pairwise comparisons if the F-statistic was significant.

3. Results

Out of 95 patients who were screened for the study, 42 met the inclusion criteria and were interested in participating in this study. Thirteen patients were then excluded from the study, due to a change in their clinic or because of missing their next two appointments. Laboratory, clinical, and statistical analyses were performed for a total of 29 patients.

3.1. General Characteristics

Descriptive data of the relevant characteristics of the participants at baseline are outlined in Table 2. There were no statistically significant differences in ethnicity, age, BMI, HbA1C, duration of diabetes, wound area, or wound age estimation among the participants of the two groups. Gender distribution differed in the groups; however, the effects of gender on each variable were evaluated and, if the effect was significant, it was added to the model as a covariate. Participants in the treatment group had a longer duration of diabetes, in comparison with the ones in the control group (14.4 ± 8 years vs. 11.7 ± 6 years, respectively), but this difference did not reach statistical significance. No significant differences were observed between the two groups in terms of the indicators of socioeconomic status (SES) and other factors that could affect the nutritional status of patients, such as appetite problems, previous unintentional weight loss, and cultural and religious dietary restrictions. Living alone, having financial concerns, being employed, and having food needs were considered indicators of SES. Referrals to registered dietitians (RDs) were not part of the standard care for patients with DFUs and only 38% of the patients had visited RDs at least once in the past.

3.2. Dietary Intake of Participants at Baseline and during the Follow-Up

The mean dietary intake of participants in terms of energy and protein was 50% and 48.7% of the recommendations when compared to the minimum recommendations for energy (30 kcal/kg) and protein (1.2 g/kg) made by the National Pressure Ulcer Advisory Panel (NPUAP). The mean dietary intake of essential micronutrients for wound healing was also alarmingly lower than DRI in this population. Details about the change in energy, protein, and micronutrient intake of the treatment and control group during the study have been published elsewhere [15]. In summary, the dietary intake of energy did not change significantly during the study for either the treatment or control group, and the interaction between time and group was not statistically significant. Compared to NPUAP recommendations, energy intake increased from 52.0% to 68.0% in the treatment group and from 43.7% to 57.8% in the control group. The increase in the protein intake of the treatment group was higher (from 54.5% to 84.9%) than the control group (from 43.1% to 54.7%) when it was compared with NPUAP recommendations. The interaction between group and time was not statistically significant; however, the change in dietary intake of protein in the treatment group is clinically relevant. Although the treatment group was provided with an extra 500 kcal of energy and an additional 28 g of protein, they still could not meet the NPUAP recommendations for energy and protein intake. The dietary intake of copper, zinc, vitamin A, vitamin C, and vitamin E significantly increased in the treatment group; however, no significant changes in the dietary intake of the participants in the control group were observed during the follow-up.

Table 2. Baseline characteristics of participants, according to group.

Variable	Treatment (*n* = 15)	Control (*n* = 14)	*p*-Value
Women/men	7/8	3/11	0.08
Age (year) Means ± SD	52.9 ± 9.74	53.8 ± 12.8	0.84
Ethnicity African American/white	4/11	3/11	0.75
BMI [1] (kg/m^2) Means ± SD	33.5 ± 7.98	34.1 ± 6.04	0.84
HbA1C [2] Means ± SD	7.95 ± 2.06	8.40 ± 2.16	0.57
Duration of diabetes Means ± SD	14.40 ± 8.03	11.7 ± 6.17	0.32
Wound age (months) Means ± SD	10.97 ± 15.09	10.58 ± 18.27	0.95
Smoking (yes/no)	3/12	3/11	1.00

[1] BMI: body mass index. [2] HbA1C: Hemoglobin A1C.

3.3. Hemoglobin A1c

The concentration of HbA1c decreased at a similar rate in both the treatment and control groups (0.31% and 0.39%, respectively). At the end of the study, there were no significant differences between the concentration of HbA1c among the groups; therefore, supplementation with extra energy sources and nutrients did not have any negative effect on long-term blood glucose control in patients with DFUs.

3.4. Anthropometrics and Body Composition

3.4.1. Changes in Body Mass Index

The mean BMI at baseline for patients in the treatment and control groups were 33.5 kg/m^2 and 34.1 kg/m^2, respectively. We examined the potential effects of gender and age on BMI; however, since their effects were not significant, we did not include them in the statistical model. The interaction between time and group did not reach statistical significance for BMI. Therefore, the intervention did not have any negative effects on the BMI of the patients in the treatment group.

3.4.2. Changes in Lean Body Mass

The mean LBM of the treatment group was slightly lower than that of the control group (68.2 kg vs. 69.1 kg, respectively) at baseline; however, the difference was not statistically significant. The effects of gender (*p* < 0.001), age (*p* = 0.001), wound age estimation (*p* < 0.001), and duration of diabetes (*p* < 0.001) on the LBM change were significant; therefore, we added these factors as covariates to the statistical model. Although the mean LBM of both groups was lowered during the study, patients in the treatment group lost less LBM than the patients in the control group (3.8 kg vs. 4.9 kg, respectively). The interaction between time and group did not reach statistical significance for LBM after adjustments for the confounding variables; however, the difference is clinically relevant.

3.4.3. Changes in Body Fat Weight

At baseline, there was no significant difference in body fat weight between the two groups. We examined the potential effects of confounding variables, such as gender (p = 0.04), duration of diabetes ($p < 0.001$), and wound age estimation ($p < 0.001$). All the significant confounding variables were included as covariates in the model. The body fat weight changed in the treatment group from 31.7 kg to 32.1 kg and in the control group from 34.3 kg to 35.9 kg. While the interaction between time and group did not reach statistical significance for body fat weight after adjustments for confounding variables, these changes are clinically relevant.

3.5. Changes in Phase Angle

The PA for the participants in the treatment group was slightly higher than that in the control group at baseline; however, the difference was not statistically significant (7.0° vs. 6.8°, respectively). We examined the potential effects of the confounding variables age, gender, BMI, duration of diabetes, and wound age estimation. Age was the only significant variable (p = 0.001); therefore, it was kept in the model as a covariate. After adjustments for the effects of age, PA was decreased by 0.3° in the treatment group and 0.6° in the control group during the study period. Even though the difference between the two groups did not reach statical significance (p = 0.09) for PA, these changes are clinically relevant.

4. Discussion

The findings of this study showed that supplementing the diet with extra energy sources and nutrients did not have any negative effects on long-term blood glucose control or the body composition of overweight and obese patients with DFUs when combined with nutrition education. Additionally, our results showed that our intervention had some positive effects on the body composition and PA of DFU patients in the treatment group. As is similar to the findings in other research [16,28–30], LBM decreased in our participants during the study; however, our treatment group could maintain LBM better than the control group. Although the effect was not statistically significant, preventing the loss of LBM leads to a better nitrogen balance, which is important for the faster healing of chronic wounds [4,28,31,32]. This could also be one of the reasons that DFUs in our treatment group healed 12.85-fold faster than in the control group [15].

While we had no intentions to decrease body fat during the healing time, this might have happened as a result of educating our treatment group about consuming nutrient-dense foods. The observed increase in protein intake along with receiving an adequate quantity of micronutrients [12] could have positive effects on regulating patients' appetite and decreasing cravings for sugar and sweets in the treatment group. This could be confirmed by our data, which showed lower sugar intake and no significant increase in energy intake in the treatment group during the study period, despite their receiving extra energy (500 kcal/d) through supplements. Our results showed that the mean sugar intake decreased by 20 g/d in the treatment group and 8 g/d in the control group. This highlights the importance of nutrition education in changing eating behaviors in this population.

We previously showed that our intervention also significantly decreased inflammation in the treatment group [33]. Rieu et al. [34] reported that the reduction of low-grade inflammation (lower IL6 and IL1β) decreased muscle mass loss and increased muscle protein synthesis by 24.8% ($p < 0.05$). Therefore, the better preservation of lean body mass that occurred in the treatment group might have happened partly as a result of decreased inflammation in the treatment group, due to an increased intake of micronutrients and antioxidants. Nonetheless, further studies are needed to replicate our findings.

Although the change in PA did not reach statistical significance for the duration of the study, scientific evidence has shown that one grade increase in PA is associated with a 33% reduction in the risk of mortality or renal death in diabetic patients [35]; therefore, the observed lower decrease in PA in the treatment group relative to the control group is clinically relevant. Our results showed that HbA1c decreased in both groups during the study and there were no significant differences between the two groups during the study. Therefore, supplementing DFU patients with additional energy sources and nutrients did not have negative effects on long-term glucose control when it was combined with nutrition education. This confirms that overweight and obese patients with DFUs could benefit from a generalized diet when it is combined with nutrition education. Nutrition education for DFU patients should prioritize dietary needs for wound healing. Consuming nutrient-dense foods should be emphasized for meeting energy and nutrient needs during wound healing. When the wound is fully healed, calorie restrictions with better nutritional support may be applied if necessary. Other studies should be conducted using a larger population, applying a similar approach, to confirm our results. If our observations are confirmed in similar studies, our method would add new approaches to the treatment of patients with DFUs.

To our knowledge, this is the first study that evaluates the effects of nutrition education and supplementation with extra calories, protein, and micronutrients on long-term blood glucose control and body composition in overweight and obese patients with DFUs. The strength of our study was that the patients were educated to choose nutrient-dense foods in addition to receiving supplements, which made it easier for them to meet their dietary requirements. It is also important to note that the supplement was tolerated well, and participants did not report any adverse effects related to the use of the supplement. One of the limitations of this study was that due to the small population in the area, the effects of nutrition education or supplementation on the outcome variables could not be evaluated independently. Additionally, the participants of this study were not provided with individualized dietary recommendations. Different medications might have different effects on blood glucose concentrations and wound outcomes; however, we did not collect data about the medications used by our participants. Currently, there are no recommendations regarding the dietary intake of energy sources and protein for patients with DFUs; therefore, the assessment of the dietary intake of our participants was conducted based on the recommendations from NPUAP, which are for patients with pressure ulcers. The supplement used in this study was not designed for patients with DFUs; however, it was the most appropriate manufactured formula that could help our participants to meet most of their dietary needs. Future research should identify the optimum amounts of energy sources and nutrients for faster wound healing in diabetic patients with foot ulcers. Routine visits with a dietitian are essential for assessing the dietary needs of patients and designing individualized nutrition therapy, which can result in effective clinical outcomes. Identifying the adequate dietary intake of macro- and micronutrients in diabetic patients with foot ulcers, especially for those who are overweight or obese, is critical for expediting the wound-healing process and can make a substantial difference to medical expenses and quality of life in this population.

5. Conclusions

Dietary recommendations for overweight and obese individuals with DFUs should prioritize proper wound healing by recommending that patients consume adequate energy sources and essential nutrients.

Author Contributions: Conceptualization, R.B. and M.T.S.; methodology, R.B., M.T.S. and B.H.A.; software, R.B.; formal analysis, R.B. and T.L.; investigation, R.B.; resources, R.B. and M.T.S.; data curation, R.B.; writing—original draft preparation, R.B.; writing—R.B., B.H.A. and M.T.S.; visualization, R.B.; supervision, B.H.A. and M.T.S.; project administration, R.B.; funding acquisition, R.B. and M.T.S. All authors have read and agreed to the published version of the manuscript.

Funding: This research received no external funding.

Institutional Review Board Statement: The study was conducted according to the guidelines of the Declaration of Helsinki and was approved by the Institutional Review Board of Florida State University (2016.18608) and Tallahassee Memorial HealthCare (2016-10).

Informed Consent Statement: Informed consent was obtained from all subjects involved in the study.

Data Availability Statement: The datasets generated from this study are available from the corresponding author upon reasonable request.

Acknowledgments: We sincerely thank the nurses, physicians, and staff at the Tallahassee Memorial Wound Healing Center and the participants.

Conflicts of Interest: The authors declare no conflict of interest.

References

1. Deshpande, A.D.; Harris-Hayes, M.; Schootman, M. Epidemiology of Diabetes and Diabetes-Related Complications. *Phys Ther* **2008**, *88*, 1254–1264. [CrossRef] [PubMed]
2. Rice, J.B.; Desai, U.; Cummings, A.K.G.; Birnbaum, H.G.; Skornicki, M.; Parsons, N.B. Burden of Diabetic Foot Ulcers for Medicare and Private Insurers. *Diabetes Care* **2014**, *37*, 651–658. [CrossRef]
3. Polikandrioti, M.; Vasilopoulos, G.; Koutelekos, I.; Panoutsopoulos, G.; Gerogianni, G.; Babatsikou, F.; Zartaloudi, A.; Toulia, G. Quality of Life in Diabetic Foot Ulcer: Associated Factors and the Impact of Anxiety/Depression and Adherence to Self-Care. *Int. J. Low. Extrem. Wounds* **2020**, *19*, 165–179. [CrossRef]
4. Molnar, J.A.; Vlad, L.G.; Gumus, T. Nutrition and Chronic Wounds: Improving Clinical Outcomes. *Plast. Reconstr. Surg.* **2016**, *138*, 71S–81S. [CrossRef]
5. Bassil, M.S.; Gougeon, R. Muscle Protein Anabolism in Type 2 Diabetes. *Curr. Opin. Clin. Nutr. Metab. Care* **2013**, *16*, 83–88. [CrossRef]
6. Joseph, J.J.; Golden, S.H. Cortisol Dysregulation: The Bidirectional Link between Stress, Depression, and Type 2 Diabetes Mellitus. *Ann. N. Y. Acad. Sci. USA* **2017**, *1391*, 20–34. [CrossRef]
7. Gougeon, R.; Morais, J.A.; Chevalier, S.; Pereira, S. Determinants of Whole-Body Protein Metabolism in Subjects With and Without Type 2 Diabetes. *Diabetes Care* **2008**, *31*, 128–133. [CrossRef]
8. Marchesini, G.; Forlani, G.; Zoli, M.; Vannini, P.; Pisi, E. Muscle Protein Breakdown in Uncontrolled Diabetes as Assessed by Urinary 3-Methylhistidine Excretion. *Diabetologia* **1982**, *23*, 456–458. [CrossRef]
9. Bitz, C.; Toubro, S.; Larsen, T.M.; Harder, H.; Rennie, K.L.; Jebb, S.A.; Astrup, A. Increased 24-h Energy Expenditure in Type 2 Diabetes. *Diabetes Care* **2004**, *27*, 2416–2421. [CrossRef]
10. Alawad, A.O.; Merghani, T.H.; Ballal, M.A. Resting Metabolic Rate in Obese Diabetic and Obese Non-Diabetic Subjects and Its Relation to Glycaemic Control. *BMC Res. Notes* **2013**, *6*, 382. [CrossRef]
11. Diabetes Diet, Eating, & Physical Activity I NIDDK. Available online: https://www.niddk.nih.gov/health-information/diabetes/overview/diet-eating-physical-activity (accessed on 20 July 2022).
12. Basiri, R.; Spicer, M.; Munoz, J.; Arjmandi, B. Nutritional Intervention Improves the Dietary Intake of Essential Micronutrients in Patients with Diabetic Foot Ulcers. *Curr. Dev. Nutr.* **2020**, *4*, 8. [CrossRef]
13. Maier, H.M.; Ilich-Ernst, J.; Arjmandi, B.; Kim, J.-S.; Spicer, M. Deficiencies in Nutritional Intake in Patients with Diabetic Foot Ulcers. *J. Nutr. Ther.* **2016**, *5*, 85–92. [CrossRef]
14. Collins, R.; Burrows, T.; Donnelly, H.; Tehan, P.E. Macronutrient and Micronutrient Intake of Individuals with Diabetic Foot Ulceration: A Short Report. *J. Hum. Nutr. Diet* **2021**, 1–5. [CrossRef]
15. Basiri, R.; Spicer, M.; Levenson, C.; Ormsbee, M.; Ledermann, T.; Arjmandi, B. Nutritional Supplementation Concurrent with Nutrition Education Accelerates the Wound Healing Process in Patients with Diabetic Foot Ulcers. Available online: https://pubmed.ncbi.nlm.nih.gov/32756299/ (accessed on 1 November 2020).
16. Demling, R.H. Nutrition, Anabolism, and the Wound Healing Process: An Overview. *Eplasty* **2009**, *9*, e9.
17. Posthauer, M.E.; Dorner, B.; Collins, N. Nutrition: A Critical Component of Wound Healing. *Adv. Skin Wound Care* **2010**, *23*, 560–572; quiz 573–574. [CrossRef]
18. National Pressure Ulcer Advisory Panel, European Pressure Ulcer Advisory Panel and Pan Pacific Pressure Injury Alliance. In *Prevention and Treatment of Pressure Ulcers: Quick Reference Guide*; Haesler, E. (Ed.) Cambridge Media: Perth, Australia, 2014.
19. Oyibo, S.O.; Jude, E.B.; Tarawneh, I.; Nguyen, H.C.; Harkless, L.B.; Boulton, A.J.M. A Comparison of Two Diabetic Foot Ulcer Classification Systems: The Wagner and the University of Texas Wound Classification Systems. *Diabetes Care* **2001**, *24*, 84–88. [CrossRef]
20. Knapp, B.K.; Parsons, C.M.; Bauer, L.L.; Swanson, K.S.; Fahey, G.C. Soluble Fiber Dextrins and Pullulans Vary in Extent of Hydrolytic Digestion in Vitro and in Energy Value and Attenuate Glycemic and Insulinemic Responses in Dogs. *J. Agric. Food Chem.* **2010**, *58*, 11355–11363. [CrossRef]

21. Aandstad, A.; Holtberget, K.; Hageberg, R.; Holme, I.; Anderssen, S.A. Validity and Reliability of Bioelectrical Impedance Analysis and Skinfold Thickness in Predicting Body Fat in Military Personnel. *Mil. Med.* **2014**, *179*, 208–217. [CrossRef]
22. Lu, H.-K.; Chiang, L.-M.; Chen, Y.-Y.; Chuang, C.-L.; Chen, K.-T.; Dwyer, G.B.; Hsu, Y.-L.; Chen, C.-H.; Hsieh, K.-C. Hand-to-Hand Model for Bioelectrical Impedance Analysis to Estimate Fat Free Mass in a Healthy Population. *Nutrients* **2016**, *8*, 654. [CrossRef]
23. Stobäus, N.; Pirlich, M.; Valentini, L.; Schulzke, J.D.; Norman, K. Determinants of Bioelectrical Phase Angle in Disease. *Br. J. Nutr.* **2012**, *107*, 1217–1220. [CrossRef]
24. Kumar, S.; Dutt, A.; Hemraj, S.; Bhat, S.; Manipadybhima, B. Phase Angle Measurement in Healthy Human Subjects through Bio-Impedance Analysis. *Iran J. Basic Med. Sci.* **2012**, *15*, 1180–1184. [PubMed]
25. Buscemi, S.; Batsis, J.A.; Parrinello, G.; Massenti, F.M.; Rosafio, G.; Sciascia, V.; Costa, F.; Pollina Addario, S.; Mendola, S.; Barile, A.M.; et al. Nutritional Predictors of Mortality after Discharge in Elderly Patients on a Medical Ward. *Eur. J. Clin. Investig.* **2016**, *46*, 609–618. [CrossRef] [PubMed]
26. Dittmar, M.; Reber, H.; Kahaly, G.J. Bioimpedance Phase Angle Indicates Catabolism in Type 2 Diabetes. *Diabetes Med.* **2015**, *32*, 1177–1185. [CrossRef]
27. Barbosa-Silva, M.C.G.; Barros, A.J.D.; Wang, J.; Heymsfield, S.B.; Pierson, R.N. Bioelectrical Impedance Analysis: Population Reference Values for Phase Angle by Age and Sex. *Am. J. Clin. Nutr.* **2005**, *82*, 49–52. [CrossRef]
28. Wolfe, R.R.; Goodenough, R.D.; Burke, J.F.; Wolfe, M.H. Response of Protein and Urea Kinetics in Burn Patients to Different Levels of Protein Intake. *Ann. Surg.* **1983**, *197*, 163–171. [CrossRef]
29. Zhang, X.-J.; Chinkes, D.L.; Doyle, D.; Wolfe, R.R. Metabolism of Skin and Muscle Protein Is Regulated Differently in Response to Nutrition. *Am. J. Physiol. -Endocrinol. Metab.* **1998**, *274*, E484–E492. [CrossRef]
30. Iizaka, S.; Kaitani, T.; Nakagami, G.; Sugama, J.; Sanada, H. Clinical Validity of the Estimated Energy Requirement and the Average Protein Requirement for Nutritional Status Change and Wound Healing in Older Patients with Pressure Ulcers: A Multicenter Prospective Cohort Study. *Geriatr. Gerontol. Int.* **2015**, *15*, 1201–1209. [CrossRef]
31. Langemo, D.; Anderson, J.; Hanson, D.; Hunter, S.; Thompson, P.; Posthauer, M.E. Nutritional Considerations in Wound Care. *Adv. Skin Wound Care* **2006**, *19*, 297–298, 300, 303. [CrossRef]
32. Cheng, Q.; Hu, J.; Yang, P.; Cao, X.; Deng, X.; Yang, Q.; Liu, Z.; Yang, S.; Goswami, R.; Wang, Y.; et al. Sarcopenia Is Independently Associated with Diabetic Foot Disease. *Sci. Rep.* **2017**, *7*, 8372. [CrossRef]
33. Basiri, R.; Spicer, M.; Levenson, C.; Ledermann, T.; Akhavan, N.; Arjmandi, B. Improving Dietary Intake of Essential Nutrients Can Ameliorate Inflammation in Patients with Diabetic Foot Ulcers. *Nutrients* **2022**, *14*, 2393. [CrossRef]
34. Rieu, I.; Magne, H.; Savary-Auzeloux, I.; Averous, J.; Bos, C.; Peyron, M.A.; Combaret, L.; Dardevet, D. Reduction of Low Grade Inflammation Restores Blunting of Postprandial Muscle Anabolism and Limits Sarcopenia in Old Rats. *J. Physiol.* **2009**, *587*, 5483–5492. [CrossRef] [PubMed]
35. D'Alessandro, C.; Barsotti, M.; Cianchi, C.; Mannucci, C.; Morganti, R.; Tassi, S.; Cupisti, A. Nutritional Aspects in Diabetic CKD Patients on Tertiary Care. *Medicina* **2019**, *55*, 427. [CrossRef] [PubMed]

Article

Ketogenic Diet Applied in Weight Reduction of Overweight and Obese Individuals with Progress Prediction by Use of the Modified Wishnofsky Equation

Gordana Markovikj [1], Vesna Knights [1] and Jasenka Gajdoš Kljusurić [2,*]

1   Faculty of Technology and Technical Sciences-Veles, University St. Kliment Ohridski-Bitola, Dimitar Vlahov bb, Veles 1400, North Macedonia
2   Faculty of Food Technology and Biotechnology, University of Zagreb, Pierottijeva 6, Zagreb 10000, Croatia
*   Correspondence: jasenka.gajdos@pbf.unizg.hr; Tel.: +385-1-4605-025

Abstract: Ketogenic diet is often used as diet therapy for certain diseases, among other things, its positive effect related to weight loss is highlighted. Precisely because of the suggestion that KD can help with weight loss, visceral obesity, and appetite control, 100 respondents joined the weight loss program (of which 31% were men and 69% were women). The aforementioned respondents were interviewed in order to determine their eating habits, the amount of food consumed, and the time when they consume meals. Basic anthropometric data (body height, body mass, chest, waist, hips, biceps, and thigh circumferences) were also collected, in order to be able to monitor their progress during the different phases of the ketogenic diet. Important information is the expected body mass during the time frame of a certain keto diet phase. This information is important for the nutritionist, medical doctor, as well as for the participant in the reduced diet program; therefore, the model was developed that modified the original equation according to Wishnofsky. The results show that women lost an average of 22.7 kg (average number of days in the program 79.5), and for men the average weight loss was slightly higher, 29.7 kg (with an average of 76.8 days in the program). The prediction of expected body mass by the modified Wishnofsky's equation was extremely well aligned with the experimental values, as shown by the Bland-Altman graph (bias for women 0.021 kg and −0.697 kg for men) and the coefficient of determination of 0.9903. The modification of the Wishnofsky equation further shed light on the importance of controlled energy reduction during the dietetic options of the ketogenic diet.

Keywords: ketogenic diet; overweight; obesity; modified Wishnofsky equation; modelling

Citation: Markovikj, G.; Knights, V.; Kljusurić, J.G. Ketogenic Diet Applied in Weight Reduction of Overweight and Obese Individuals with Progress Prediction by Use of the Modified Wishnofsky Equation. *Nutrients* **2023**, *15*, 927. https://doi.org/10.3390/nu15040927

Academic Editor: Javier Gómez-Ambrosi

Received: 30 December 2022
Revised: 6 February 2023
Accepted: 8 February 2023
Published: 12 February 2023

1. Introduction

In recent years, obesity became a serious global health crisis with prevalence increasing nearly threefold from 1975 to 2016 [1]. Research indicates the connection between obesity and numerous diseases and health complications, such as cardiovascular diseases, various types of cancer, type 2 diabetes, hypertension, polycystic ovary syndrome (PCOS), and many others [2,3]. It is important to emphasize that obesity can be prevented by establishing a balanced diet, adequate physical activity, and changes in behaviour and lifestyle [4]. Understanding the principles of energy balance is crucial [5] in approaching the global problem of the western countries: obesity. The concept of energy balance is based on the law of conservation of energy (energy conservation law: energy state of the organism = entered energy–expended energy), which states that energy cannot disappear or be created from nothing, but can only change its forms [6]. The source of energy in human diet are foods and drinks, with the main energy donors: carbohydrates, proteins, fats, and alcohol and the energy consumption varies throughout the day, but also throughout the lifespan [7].

Our organism strives for a state of energy balance and possesses regulatory mechanisms for this purpose. Regulation implies a complex physiological control system that includes neuronal and hormonal signals from the gastrointestinal tract, pancreas, and adipose tissue that reach the hypothalamus and the autonomic nervous system that innervates muscles, organs, and adipose tissue [5]. It was proven that this integrated regulatory system has stronger protection mechanisms for the loss of body mass than for the prevention of excess energy accumulation, and therefore there is a greater chance for the success of increasing body mass than reducing it [8]. The reduction in body mass is the result of a negative energy balance, i.e., increased energy consumption compared to intake [6]; however, sole reduction in energy intake does not result in continuous (infinite) and proportional loss of body mass. Reduction requires temporary changes in diet and physical activity, while long-term maintenance requires permanent changes, which seems to be more difficult [9] because studies show that 35 to 80% of individuals, who reduced at least 10% of their initial body mass, fail to maintain the reduced body mass for more than a year [10].

As successful reduction in body mass is classified, intentional loss of at least 10% of the original body mass is maintained at that level for at least one year. The criterion of 10% was set because already then the risk of diabetes and cardiovascular diseases was significantly reduced [11]. So, with the aim of a better understanding of an observed problem, models are developed, among which mathematical models were developed to try to understand the non-linearity of body mass loss during energy reduction as one of approaches in dealing with obesity. Numerous mathematical models were designed for the purpose of predicting body mass loss, which differ from each other according to the concept of how energy is stored and consumed [12]. The first such model, which combined all the knowledge about calories and energy metabolism developed for predicting the expected body mass based on the timeframe of energy intake reduction is the Wishnofski model from 1958 [13]. Doctor Max Wishnofsky researched energy from food, how it is stored in the body, and by what amount it is necessary to reduce energy intake in order to lose 1 kg of body mass [14]. He designed a regression model that was supposed to serve as a universal measure for assessing body mass change based on an energy intake reduction in a known time frame and with a caloric equivalent of one pound of lost or gained body mass of 3500 kcal (for 1 kg–approximately 7700 kcal) [12]:

$$Weight\ loss\ [lb] = E_s[kcal/day] \cdot \frac{t\ [days]}{3500\ \left[\frac{kcal}{lb}\right]} \tag{1}$$

where:

E_s—imposed daily deficit in energy stores (reduced energy intake or increased exercise generated energy output), [kcal/day];

t—duration of the diet [days].

Studies show that different diet patterns influence diet changes and maintain reduced body weight [2,15], and one of them is the ketogenic diet, which is characterized by a significant reduced intake of carbohydrates (<30 g/day) and standard protein intake (1.2–1.5 g/kg of ideal body weight or 1.0–1.2 g/kg of fat free mass) [16]. This diet is also often used in diet therapy of obesity, type 2 diabetes mellitus, migraines, polycystic ovary syndrome, and even epilepsy [17–22]. There are several types of eating patterns within the keto diet. A standard ketogenic diet implies that fats make up 70% of the daily energy intake (DEI), proteins 20%, and carbohydrates only 10%. In addition to the standard one, the cyclic ketogenic diet includes periods of carbohydrate compensation (after every 5 days the diet is followed by 2 days with increased carbohydrate intake), a targeted ketogenic diet that allows the addition of carbohydrates during periods of intense physical activity (25 to 50 g half an hour before training), and a high-protein ketogenic diet that is similar to the standard diet, but the macronutrient intake ratio is changed (fats: proteins: carbohydrates = 60:35:5) [23].

According to all of the above, the aim of this paper is to demonstrate the usefulness of the Wishnofsky equation based on collected data of people on a ketogenic diet. Several requirements were studied, the most important of which is the accuracy of predicting the course of body mass loss over a certain period of time, as well as different phases during the energy restriction and macronutrient intake based on the ketogenic diet guidelines.

2. Materials and Methods

In the study were included 100 healthy adults (31% of them are males) from Skopje, North Macedonia, enrolled in the program of weight reduction by following keto diet principles. Their anthropometric data (weight, height, circumferences of: chest, waist (two places: (i) narrowest part and (ii) at the navel region), hips, biceps, and thighs), diet habits before the diet, and some basic information related to their food intake were collected in an individual interview with a nutritionist. During the interview were collected such data as frequency of consumption of some nutrition dense food (fruit and vegetables) as well as caloric food (sweets, salty snacks, seeds, and nuts) and beverages (carbonated drinks vs water). The time of meal consumption was also recorded. The measurements were collected since April 2022. All respondents signed the agreement that their data can be used exclusively for scientific purposes, and the principles of the GDPR were respected.

Observed anthropometric parameters of the participants were collected following the recommendation of Casadei and Kiel [24] and they are given in Table 1.

Table 1. Basic anthropometric parameters of the subjects on the first medical examination.

Observed Parameters	Female (N = 69)		Male (N = 31)	
	Mean $\pm$ SD	**[Min–Max]**	**Mean $\pm$ SD**	**[Min–Max]**
Age (years)	37.5 ± 11.1	[18–68]	35 ± 11.8	[18–56]
Body height (cm)	163.8 ± 7.2	[147–184]	175.8 ± 7.4	[162–192]
Body mass (kg), W_0	78.6 ± 17.2	[12.4–152.5]	103.3 ± 25.3	[60.3–237]
BMI (kg/m^2)	$32.9 \pm 8.5*$	[25.3–63.5]	$38.3 \pm 8*$	[26.3–64.3]
	Circumference (cm)			
Chest	88.4 ± 11.7	[64–136]	99.8 ± 14	[74–150]
Waist (narrowest part)	90.5 ± 12.9	[67–138]	101.8 ± 14.7	[74–162]
Waist (at the navel)	100.9 ± 13.5	[75–162]	110.7 ± 15.8	[80–164]
Hips	106.7 ± 13.2	[79–172]	114.9 ± 16.2	[80–179]
Biceps	33.8 ± 5.8	[9–62]	37.1 ± 4.1	[29–50]
Thighs	64.9 ± 9.1	[28–97]	68.5 ± 7.6	[56–97]

*: Statistically significant differences ($p < 0.05$); SD: standard deviation; BMI: body mass index; and W_0: body mass on the beginning of diet program.

In addition to the anthropometric parameters, for each subject, anamnesis was taken about the basic state of health, as well as the number and type of meals and the time of consumption. At the first medical examination, the subjects' body mass, body height, and circumferences of arm, leg, waist, and hips were measured. The initial body mass index (BMI) was expressed as the ratio of the body mass to the square of the body height, and the target body mass for each subject was obtained in the range for the targeted normal BMI (20–25 kg/m^2).

At each follow-up examination, subjects' body mass and circumference were measured to monitor progress. In the case of adequate progress, the allowed energy intake is increased, i.e., the person moves to the next phase of the ketogenic diet. However, if at some point there is a stagnation of progress or an increase in body mass, the subjects are returned to the previous phase and their energy intake is reduced.

The Wishnofsky equation was used because it depends on the phase of the body mass reduction process, and was modified because during the ketogenic diet were included seven different phases (Table 2) and the average energy nutrient composition for the last phase is given in the supplementary Table S1 for menus created by a dietitian and medical doctor.

Table 2. Permissible ranges of energy intake with regard to the phase of the ketogenic diet.

Ketogenic Diet Phase ($i = 1, \ldots, 7$)	Energy Range (EB), kcal (kJ)
I	750–850 (3140–3560)
IIa	850–950 (3560–3975)
IIb	950–1050 (3975–4395)
III	1100–1200 (4605–5025)
IVa	1300–1400 (5440–5860)
IVb	1350–1450 (5650–6070)
V	1500 (6280)

a, b-different energy levels of the same phase of the ketogenic diet.

According to the energy intake of different phases, the Wishnofsky equation (Equation (1)) was modified as follows:

$$W(t) \ = \ W_0 - 0.454 * \Delta EB * \frac{t}{3500} \tag{2}$$

$$Wt_j \ = \ Wt_{j-1} - 0.454 * EB_i * \frac{t_j}{3500} \tag{3}$$

where

W_0—initial body mass [kg]; $W(t)$—expected body mass [kg] after t days where the energy intake was reduced for ΔEB (reduced daily energy intake [kcal/day] compared to the required one);

Wt_{j-1}—initial body mass for the new ketogenic diet phase ($i = 1, \ldots, 7$), the ketogenic diet phase (i) can be repeated several times ($j = 1, \ldots, n$) and the last one ends when $Wt_j = Wd$ (desired body mass). When calculating EB_i, the mean values of the energy range of the different phases of the ketogenic diet were used (Table 2).

The flow chart (Figure 1) presents the ketogenic diet implementation process from the initial body mass (W_0) to the desired one (Wd). Patients start with 800 kcal in phase I until they reach 48% of the difference between initial body mass (W_0) and the desired (Wd). After this phase, the energy intake is increased to 900 kcal (Phase IIa). By reaching 64% of the difference between initial body mass (W_0) and the desired one (Wd), the daily energy intake is upgraded to 1000 kcal (IIb phase). In phase III, the patient reached less than 80% of the difference between W_0 and Wd, and the daily calorie intake is than 1150 kcal. In phase IVa, the patient reached less than 85% of the difference (W_0–Wd) with 1300 kcal per day. Phase IVb, starts when the patient reaches 90% of the difference between initial body mass and desired (W_0–Wd) with 1400 kcal per day. The last phase (phase V) increases the energy intake to 1500 kcal when the achieved body mass is less the 5% from the desired one.

As variables are indicated values for body masses that were recorded for the patient after each examination at a certain diet phase, Wt_j, j is the number of check controls, while other primary parameters are: W_0 as the initial body weight, Wd as the desired body weight, and previously mentioned EB as intake of energy during the day.

Patients who are over-weighted, but still not obese, and where the difference between the initial body mass and the desired one (W_0–Wd) is less than 48%, the diet is directed immediately into the second phase of the diet ($i = 2$: Phase IIa). If this is not the case, the diet plan will start with the first phase. The third phase is approached when the patient reaches 80% of the desired weight loss. In the remaining stages, the patient loses the remaining 20% of body mass.

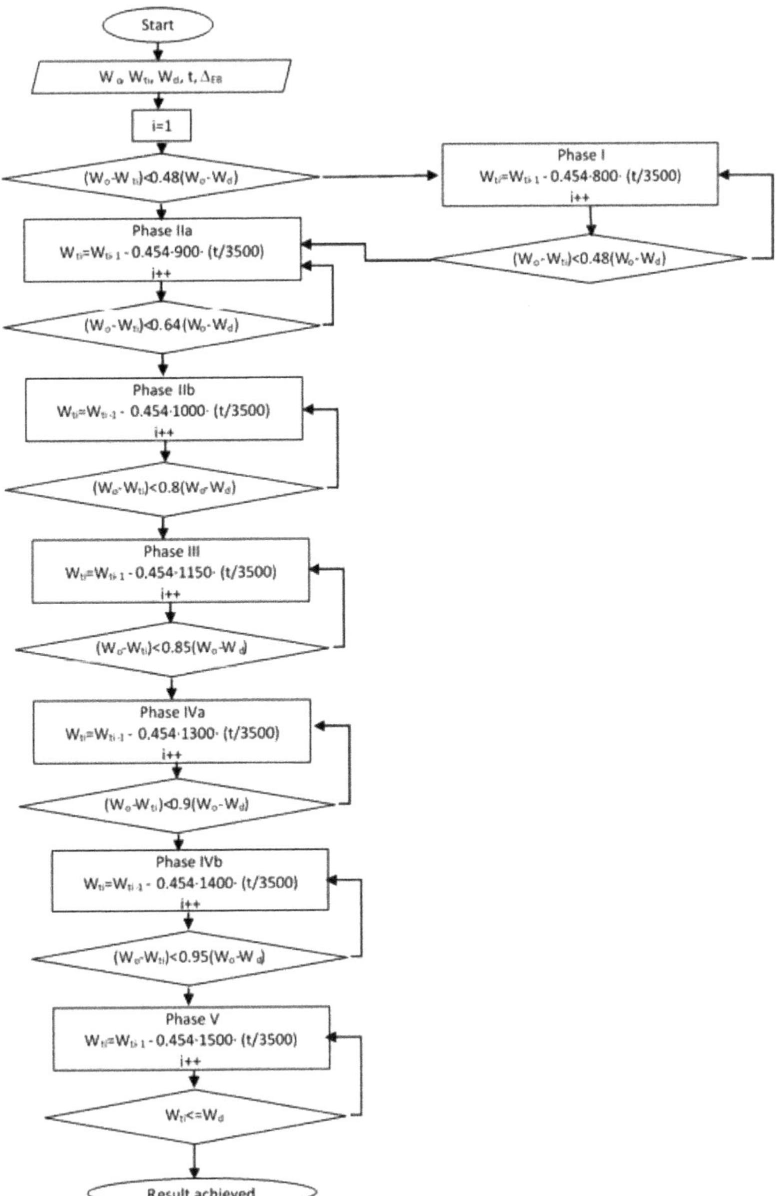

Figure 1. Flow chart presenting ketogenic diet phases.

In addition to all measurements, the expected body mass during each examination was also calculated using the Wishnofsky equation (based on the body mass recorded at the previous examination, the number of days since the previous examination, and the energy intake in that phase). None of three variables used in the calculation are a constant; body mass differs each time, and energy intake also changes analogously, i.e., the phase of the ketogenic diet. The number of days since the previous examination is different for each person.

At the end of the research, the data were statistically processed and the actual situation and progress were compared with the prediction based on the Wishnofsky equation.

All calculations were conducted by use of MS Excel. Calculated were the (i) minimal and maximal values in the observed data set, (ii) standard measure of central tendency (mean, mode), and (iii) standard deviation (SD) as a measure of dispersion. Relative frequencies (as percentage) were used in the display of results related to the eating habits of people involved in the weight loss program. Box-whiskers plot was used to show the progress of body mass loss and the reduced body mass index. The Bland–Altman chart is used to show the effectiveness of predicting body mass using the modified Wishnofsky equation. A simple linear regression was used to show the agreement of body mass data in a certain phase of the ketogenic diet with the exact body mass measured during the regular examination.

3. Results

During the first examination, an interview was conducted (with each subject) in which data were recorded on the frequency of overweight and/or obesity in the family (Figure 2), their eating habits, i.e., the frequency and time of eating (Table 3), and certain types of food (Table 4).

Table 3. Frequency of consumption of certain foods and drinks.

	Consumption Per Week [1] (%)				
Sweets	**non**	**<50 g**	**50–100 g**	**101–200 g**	**>200 g**
Female	5.79 [a]	26.09 [a]	23.19 [a]	37.68 [a]	7.25 [a]
Male	9.68 [a]	38.71 [b]	16.13 [b]	16.13 [b]	19.36 [b]
Chips	**non**	**<50 g**	**50–100 g**	**101–200 g**	**>200 g**
Female	18.84 [a]	33.33 [a]	30.44 [a]	13.04 [a]	4.35 [a]
Male	35.48 [b]	25.81 [b]	19.36 [b]	9.68 [a]	9.68 [a]
Vegetables	**non**	**<500 g**	**500–1000 g**	**1001–1500 g**	**>1500 g**
Female	10.15 [a]	88.41 [a]	1.45 [a]	0 [a]	0 [a]
Male	12.90 [a]	87.10 [a]	0 [a]	0 [a]	0 [a]
Fruits	**non**	**<500 g**	**500–1000 g**	**1001–1500 g**	**>1500 g**
Female	17.39 [a]	49.28 [a]	18.84 [a]	8.70 [a]	5.80 [a]
Male	19.35 [a]	54.84 [a]	16.13 [a]	6.45 [a]	3.23 [a]
Nuts	**non**	**<50 g**	**50–100 g**	**101–200 g**	**200–500 g**
Female	26.09 [a]	31.88 [a]	13.04 a	26.09 [a]	2.90 [a]
Male	22.58 [a]	32.26 [a]	9.68 a	22.6 [a]	12.90 [b]
Seeds	**non**	**<500 g**	**500–1000 g**	**1001–1500 g**	**>1500 g**
Female	56.52 [a]	21.74 [a]	4.35 [a]	17.39 [a]	0 [a]
Male	54.84 [a]	22.58 [a]	6.45 [a]	16.13 [a]	0 [a]
Carb. drinks	**non**	**<0.5 L**	**0.5–1 L**	**1–2 L**	**>2 L**
Female	39.13 [a]	28.99 [a]	20.29 [a]	7.25 [a]	4.35 [a]
Male	25.81 [b]	12.91 [b]	16,13 [a]	38.71 [b]	6.45 [a]
Water [1]	**non**	**<0.5 L**	**0.5–1 L**	**1–2 L**	**>2 L**
Female	56.52 [a]	33.33 [a]	10.14 [a]	0 [a]	0 [a]
Male	74.19 [b]	19.35 [b]	6.45 [a]	0 [a]	0 [a]

[1] Per day; carb. drinks: carbonated drinks; different letters for investigated food group indicate significant differences for the observed consumed amount.

Table 4. Number of meals and their distribution during the day.

	Female (N = 69)		Male (N = 31)	
	Mean ± SD	**Mode [Min–Max]**	**Mean ± SD**	**Mode [Min–Max]**
No. of meals	2.8 ± 0.9	3 [1–6]	2.5 ± 1	2 [1–5]
		Time of meal consumption (h) *		
1st meal	10.6 ± 2.5	10 [4.5–18.5]	11.3 ± 3.9	9 [3–21]
2nd meal	16.8 ± 2.4	17 [11–22]	15.9 ± 2.9	17 [9–21]
3rd meal	19.8 ± 2.6	21 [16–23]	19.5 ± 3.2	21 [13–24]
4th meal	21.3 ± 1.1	20 [20–23]	19.8 ± 0.5	20 [19,20]

* Time is expressed in the form 0–24 h; SD: standard deviation.

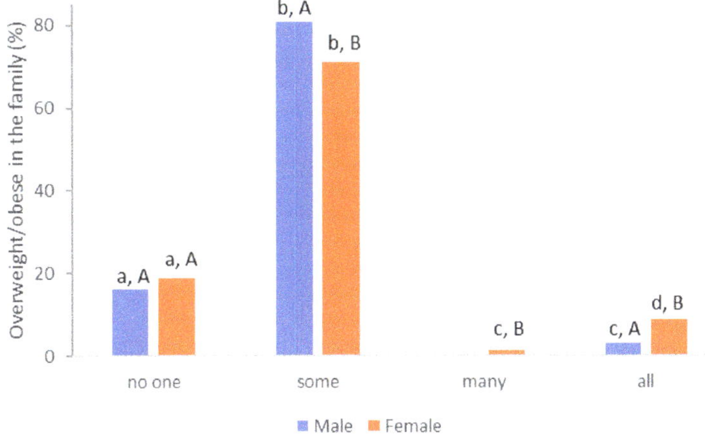

Figure 2. Prevalence of being overweight and/or obesity in the family (different lowercase letters: significant differences in the frequency of overweight or obese family members, within the same gender; different capital letters: significant difference in the frequency of overweight or obese family members within different gender).

From the prevalence of obesity in the family, differences in the answers of the male and female population are visible; however, research by [25] Sattler and associates (2018) shows that it is weight-based stigmatization with motivation to exercise and physical activity in overweight individuals in connection with different genders.

Information on the frequency of consumption of certain foods (sweet, salty, and seeds) and drinks was a source of information on the quality of eating habits (Table 3). Only one third of female and 48.39% female subjects consumed non or less than 50 g of sweets per week, while chips (including other salty snacks) were consumed by over 50% of subjects, regardless the gender is consuming in amounts less than 50 g/week. Unfortunately, it is a devastating fact that the amount of fruit and vegetables consumed during the week is limited to small amounts, indicating that energy-rich food, with low nutritional density, dominates their diet. Higher intake of fruits and vegetables increased weight loss [26]. In the investigated group, the frequency of consuming vegetables was significantly lower, although it can be consumed as a side dish, salad, etc. The following finding is related with the regional consumer habits, including high consumption of nuts and seeds. In the conversation during the interview, it was clear that seeds and nuts are consumed between meals in uncontrolled amounts, although the average caloric contribution in 100 g is in the range of 400–600 kcal [27].

Consumption of beverages shows an exceptional representation of carbonated beverages compared to water, which is consumed most often in the amount of 1–2 L in the male population (38.71%). Carbonated mineral water is also included in CO_2 beverages. Over-weighted and obese individuals have higher demand on fluid intake, and improved hydration is a commonly used strategy by nutritionists to prevent overeating with the goal of promoting a healthy weight among patients [28]. It is a worrying fact that almost 56.52% of women and even 74.19% of men in the group of respondents do not consume water on a daily basis.

However, it is not only the combination of poor nutrition that is related to the problem of excessive body weight or obesity of the respondents, there is also the number of meals and their distribution during the day (Table 4).

In order to additionally determine the frequency of the most common number of meals in the respondents' answers, the mode value was also calculated. Female subjects have more meals (mode value is 3 vs. 2 of the male population, respectively). Late meals dominate (second meal at around 4 pm) while the first meal is extremely late (regardless of gender, around 10 or 11 am) and a lot of them have late night meals (around midnight).

The participants reduced their daily energy intake, guided by the ketogenic diet principles. Successful progress of the subjects can be seen in Table 5.

Table 5. Observed parameters on the end of the diet program.

Observed Parameters	Female (N = 69)		Male (N = 31)	
	Mean $\pm$ SD	[Min–Max]	Mean $\pm$ SD	[Min–Max]
Number of examinations	6.8 $\pm$ 4.6	[1–27]	8 $\pm$ 6.9	[1–38]
No. of days in the program	79.5 $\pm$ 171.6	[41–240]	76.8 $\pm$ 155.5	[10–247]
Planned body mass (kg)	64.3 $\pm$ 8.2	[52–97]	81.6 $\pm$ 12.6	[44–99]
Achieved body mass (kg)	76.7 $\pm$ 16.1	[50.7–146]	100.4 $\pm$ 24	[60.3–221]
Weight loss (kg)	22.7 $\pm$ 13.0	[5.6–55]	29.3 $\pm$ 12.6	[8.1–52.5]
BMI (kg/m^2)	25.2 $\pm$ 3.8 *	[20.1–40.9]	27.8 $\pm$ 3.4 *	[21.6–38.2]
	Circumference reduction (cm)			
Chest	19.9 $\pm$ 8.5	[2–40]	22.4 $\pm$ 10	[5–38]
Waist (narrowest part)	20.7 $\pm$ 9	[4–42]	24 $\pm$ 10	[8–50]
Waist (at the navel)	23 $\pm$ 9.9	[5–51]	26 $\pm$ 9.5	[8–41]
Hips	22.7 $\pm$ 10.9	[3–50]	26.3 $\pm$ 10.2	[8–50]
Biceps	9.3 $\pm$ 3.6	[2–17]	10.1 $\pm$ 2.8	[5–17]
Thighs	13.9 $\pm$ 5.2	[3–24]	11.6 $\pm$ 4.8	[4–24]

*: Statistically significant differences ($p < 0.05$); SD: standard deviation; and BMI: body mass index.

Such an approach in body weight reductions requires numerous examinations (6.8 for females and 8 for male subjects) and a long period of time in the program (79.5 and 76.8 for female and male subjects, respectively).

The reduction in all measured circumferences is dominantly in the waist and hip region for both genders (more than 20 cm reduction).

Although the average body mass that was planned and achieved differs for both genders (Table 5), the success can be seen in Figure 3, indicating the achievement in body mass loss, as well as for the decrease in the body mass index. The first impression is that the male subjects failed to achieve the expected body mass index of normal nutrition. However, the male population more actively accepted physical activity, especially exercise, and therefore their body mass index is slightly higher due to an increase in muscle mass.

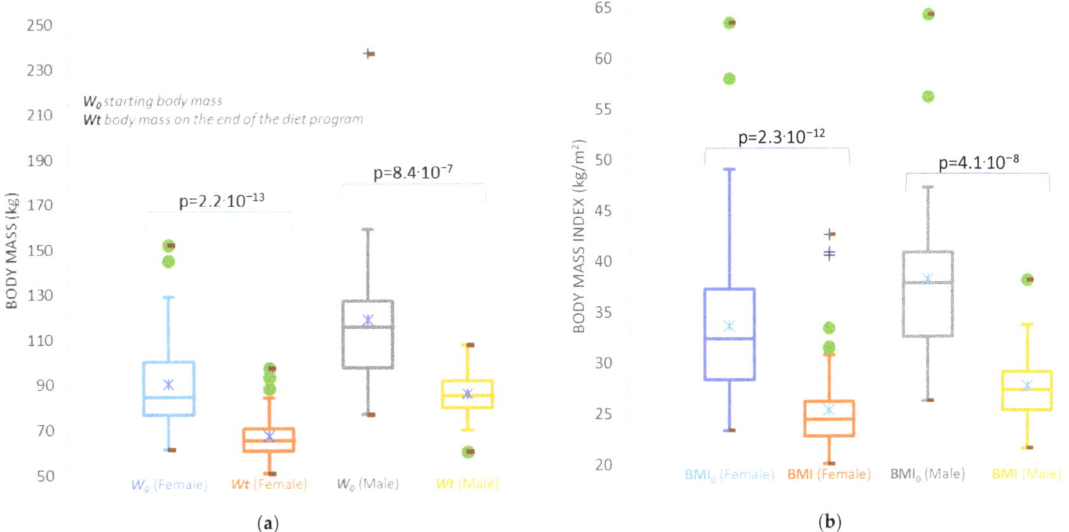

Figure 3. Box-whisker plots of: (**a**) reduced body masses of female and male subjects; (**b**) reduced BMI for female and male subjects enrolled in the ketogenic reduction diet program. Description of what is contained in the second panel.

An accurate perception of the expected body weight after a certain time of reduction in energy intake is necessary for people who are on a weight loss program, but also for nutritionists who lead the program in order to design the appropriate next step of the weight management program [29]. Therefore, last results are devoted to the efficiency of the modified Wishnofsky equation in predicting expected body mass after a certain phase of their diet. It is suggested to use correlations and regressions to assess the agreement between the two quantitative measurement methods, as in our case with the experimental values of body mass, and the predicted one by use of modified equation by Wishnofsky. Correlation will give an insight into the relationship between one variable and another, but will not indicate differences, and thus is not an ideal method for assessing comparability between methods. An alternative is the Bland–Altman graph, which as a basis for quantifying the agreement between two quantitative measurements offers the study of the relationship of the mean difference in the limits of agreement. The Bland–Altman graph defines intervals of agreement, and acceptable limits must be defined in advance, based on the set goals [30]. Our agreement is presented in Figure 4. For both genders, the bias values are very close to zero (−0.697 kg and 0.021 kg) for male and female measures, respectively. The error is 0.0614 in prediction of male body mass and 0.058 in predicting female body mass of a certain phase of the ketogenic diet. A certain proportion of outliers (Figure 4., dots outside the area of limits of agreement ($\pm 1.96 \times$ SD)) is visible, which is dominantly the result of non-adherence to the principles of the keto diet, and precisely the difference between the expected body mass (>5%) vs. the measured body mass during the control examination, which is the indication of relapse. In the supplement material, Figure S1 shows the repetition of ketogenic diet phases for one relapsed participant who started the program from the beginning for three times. The disproportion between the expected body mass (calculated by the modified Wishnofsky's equation) and the measured mass is evident, and greater than $\pm 5\%$. Here, it must be emphasized that none of the input data of the respondents were taken as an outlier (precisely the extreme values, such as the body weight of 237 kg of a male person) that influence the increase in the error. For this reason, the regression line of the experimental values of body masses and those obtained by the modified model is shown (Figure 5).

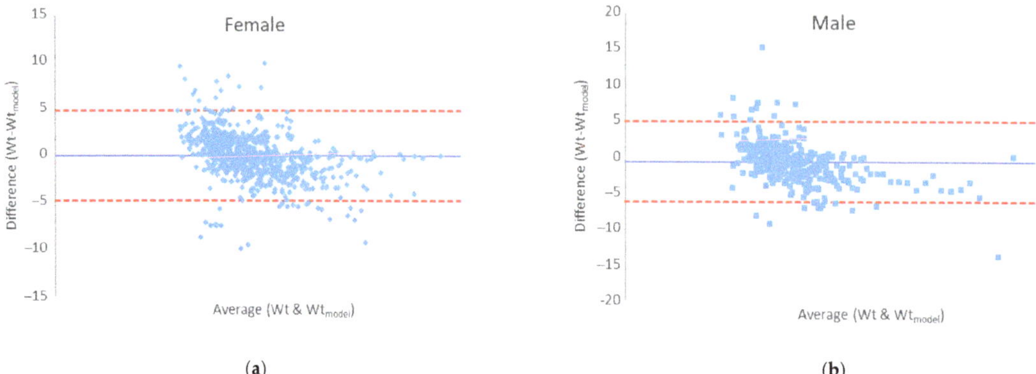

(**a**)

(**b**)

Figure 4. Bland–Altman plot for body mass of different phases of ketogenic diet for (**a**) female and (**b**) male subjects.

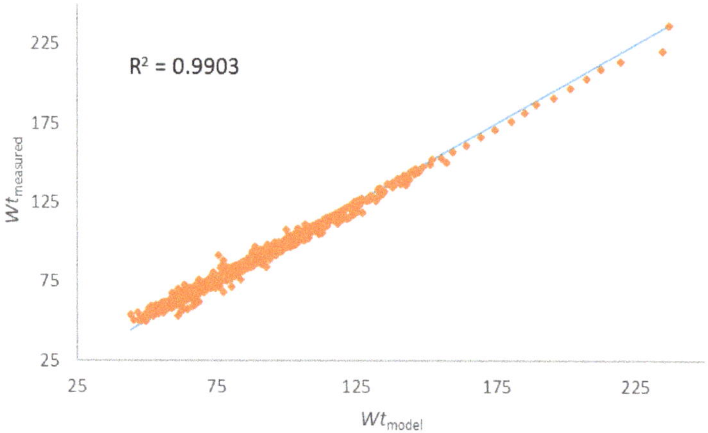

Figure 5. Compatibility of estimated body masses with the modified Wishnofsky equation (*Wt*) during the reduction ketogenic diet and established body masses at the end of the program.

The last efficient test is presented with the regression line of the body masses measured during the examinations and those predicted by the use of the modified equation of Wishnofsky (Figure 5), and it is clear that, even with outliers in the data set, there is still an extremely strong connection between the observed data ($R^2 = 0.9903$).

4. Discussion

In order to avoid the yo-yo effect and preserve weight loss progress, Wing and Phelan [31] defined six key strategies that should be followed: (i) increased level of physical activity (1 h/day), (ii) change in eating habits in the context of avoiding energy-rich foods and foods rich in fats, (iii) regular breakfast (latest in 2 h after waking up), (iv) regular monitoring of body mass, (v) constant eating pattern, and (vi) reacting to minor mistakes by correcting them in a timely manner so that they do not cause a greater return of lost body mass and causing a negative impact on the weight loss progress. Theoretically, thebasic principle of losing weight is quite simple: spend more energy than you take in. However, while the fact is that we have to reduce our calorie intake, it is important to know the exact source and amount of calories eaten, and whether the body can be influenced in the tendency to lose and later to restore the balance. The primary "fuel" of the human body is glucose, i.e., carbohydrates. Therefore, when glucose stores are low, as is the case during a ketogenic diet, the central nervous system must find an alternative source of energy [4]. Then, the energy source becomes ketone bodies–acetoacetate, beta-hydroxybutyrate, and acetone. These molecules are the product of ketogenesis that takes place in the mitochondrial matrix in the liver. Under normal conditions, they are found in the body in very low concentrations (<0.3 mmol/L). Given that they are similar in structure to glucose, they have the ability to use a glucose transporter to cross the blood–brain barrier to be used as an energy source when they reach a concentration of 4 mmol/L in the body. The described state of elevated levels of ketone bodies in the body is called "ketosis" [32].

It is believed that this mechanism forces the body, due to the lack of glucose, i.e., carbohydrates in the diet, to consume fat reserves and thereby reduce the amount of fat tissue and total body mass. In addition, Ketone bodies serve as an alternative energy source for brain metabolism [33]. Bypassing the traditional ways of releasing energy through glycolysis in favour of using ketone bodies has a significant effect on the body, and although the entire mechanism is not fully understood, it is clear that bypassing the metabolism of carbohydrates in the brain can also lead to positive health effects, such as a reduced frequency of epileptic seizures [5,34].

The ketogenic diet guidelines show that the basis of the diet should be fats. Unsaturated fatty acids are allowed, such as nuts, seeds, avocado, tofu, and olive oil, but a higher intake of saturated fatty acids is emphasized, such as butter, animal fat, coconut oil, butter, etc. Proteins are the next macronutrient when considering the share of daily energy intake. There are no big differences in the recommendations of protein sources, but poultry meat, fish, and red meat are recommended in larger quantities than eggs, cheese and milk, and dairy products. In the end, carbohydrates remain [16,23]. As can be seen in our investigated group, the (i) time of consumption and (ii) the number of meals are also important issues related to being overweight. A study conducted among Japanese women showed that those who consumed late dinners or bedtime snacks were more likely to skip breakfast, which explains the late first meal of the investigated subjects. The same study concluded that having a late dinner or bedtime snack is associated with a higher probability of being overweight/obesity [35]. Low-carb vegetables are allowed, i.e., green leafy vegetables, broccoli, cauliflower, Brussels sprouts, asparagus, peppers, onions, garlic, cucumber, mushrooms, etc. In addition to vegetables, fruits contain a high proportion of carbohydrates, and therefore only berries are recommended [36]. Extensive literature overview in the meta-analysis conducted by Arnotti and Bamber [26] investigated the fruit and vegetable consumption in overweight or obese individuals (3719 participants), and it was shown that the effect was large (−2.81 kg; $p < 0.001$). Lipid metabolism, which is a key factor in planning body mass reduction, is an extremely complex process, and models are available that simulate its development with the aim of understanding its biological processes [37]. The models can also be used to optimize and define sustainable diet indicators where the ketogenic diet shows success [38,39]. In order to help both nutritionists and people who are on a weight loss program, a modified model of weight gain from the Wishnofsky equation was proposed. Having a perception of what to expect after a certain period of reduced energy intake is more than encouraging for participants in the weight loss program [25]. Decades after Wishnofsky's equations, different mathematical models were created for predicting expected body mass in a certain time frame based on the law of conservation of energy, and those models differ according to differences in the understanding of what energy consumption entails and what the energy state of the organism entails [40]. The main requirement of a model is its simplicity and acceptable error [37–39]. Our results show that the prediction of expected body mass during the reduction keto diet using the modified Wishnofsky equation is extremely well aligned with the actual progress of people in the weight loss program, regardless of gender. The modification of the equation that includes changes in the phases of energy intake during the ketogenic diet is important because it is not a linear relation of body mass loss, but a non-linear process that is taken into account in this way. An extremely important factor is the time (t) of a certain phase of the diet in which a person establishes control over eating habits and continues with a further constant decrease in body weight. Deviations greater than 5.8% in women and 6.1% in men are an indicator of non-adherence to the basic principles of the diet, and are a corrective factor for the person on a diet and their nutritionist, because one of the goals is certainly the prevention of the "yo-yo" effect in the respondents. Thus, the so-called confidence interval values in the Bland–Altman graph will indicate the above and that the modified Wishnofsky equation did not successfully predict the expected body mass. This effect was confirmed in his research by Thomas and colleagues [13], who state that the use of the Wishnofsky's equation is a rule that is easy to apply, but can lead to an error in predicting body mass loss; however, in the absence of simple and most understandable solutions, it is also an acceptable smaller error [39] in the expected value of body mass during the weight loss program.

As with each method, this model has also a disadvantage: it does not explain the metabolic adaptations that occur in the body, and it also does not take physical activity as an input in the calculations. The average BMI for the male population was slightly higher than 25 kg/m^2, but according to the findings of Weber and co-worker [21], ketogenic diet helped in preserving muscle mass in patients with cancer, and the study of Pasiakos and coworkers [41] conducted on adults varying levels of dietary protein on body composition during energy deficit concluded that consuming dietary protein at levels exceeding recommendations may protect fat-free mass during short-term weight loss. The physical activity in obese people [25] will also affect the increase in muscle mass and consequently affect the BMI, although BMI does not distinguish muscle and fat mass. The focus is exclusively on the energy intake and the time frame of its reduction. Given the limited time period of this research, future research should include physical activity as well as the energy intake during the stabilization of energy intake after the restriction, because the potential of the model was confirmed also for those who had phases in which they returned to increased and inappropriate energy intake.

5. Conclusions

The implementation of supervised body weight loss, with the guidelines of the ketogenic diet, is primarily focused on the reduction in carbohydrates and energy. The proportion of body mass loss is dictated by the sequence of phases of the diet, and with medical supervision, the third phase (of 1300 kcal) of the diet occurs after 60–80% of the target body mass loss (Wt-W_0). For the patient, cognition of the flow and duration of the diet itself is extremely important, and it is necessary to use tools such as prediction equations for body mass loss over a certain period of time. Body mass loss in different phases of the ketogenic diet can be effectively predicted by applying the equation by Wishnofsky, which represents a simple mathematical model that relatively accurately predicts the course of body mass change. It does not require a large number of input variables, which makes it useful for clinical practice, as well to monitor the progress and helping in creating an effective program for body weight loss, especially due to the large problem of obesity in the world.

Given that the modified model of Wishnofksy's equation is proposed for predicting body mass reduction, taking into account that the person adheres to the prescribed guidelines and certain energy intake, the errors that occur are the result of the selection and procedures of the subjects, not the model itself. In this paper, an algorithm for the flow of the phases of the keto diet is generated. Following this algorithm leads to a reliable result of the desired weight. This is a good start for further research, as the next step would be to create a program that generates a variety of food and satisfies this algorithm model.

Supplementary Materials: The following supporting information can be downloaded at: https://www.mdpi.com/article/10.3390/nu15040927/s1, Figure S1: Trend of the expected body mass (BM) and measured body mass for a relapsed individual.; Table S1: Average energy and nutrient content of the 5th phase of the ketogenic diet.

Author Contributions: Conceptualization, V.K. and J.G.K.; methodology, J.G.K..; validation, G.M., V.K. and J.G.K.; formal analysis, J.G.K.; investigation, G.M.; resources, G.M.; data curation, V.K. and J.G.K.; writing—original draft preparation, G.M., V.K. and J.G.K.; writing—review and editing, V.K. and J.G.K.; visualization, V.K. and J.G.K.; supervision, V.K. and J.G.K.; project administration, G.M.; funding acquisition, G.M. All authors have read and agreed to the published version of the manuscript.

Funding: This research received no external funding.

Institutional Review Board Statement: The study was conducted in accordance with the Declaration of Helsinki, and approved by the Institutional Review Board of Универзитет „Св. Климент Охридски"—Битола, Технолошко—технички факултет Велес (protocol code 10-168/1 approved on 28 April 2022).

Informed Consent Statement: Written informed consent was obtained from the patient(s) to publish this paper.

Data Availability Statement: The data that support the findings of this study are available from the corresponding author upon reasonable request.

Acknowledgments: We would like to thank our participants for giving their written consent to the use of their data for processing and publication in this paper.

Conflicts of Interest: The authors declare no conflict of interest.

References

1. WHO. Obesity and Overweight. WHO-World Health Organisation. Available online: https://www.who.int/news-room/fact-sheets/detail/obesity-and-overweight (accessed on 4 July 2022).
2. Ahluwalia, M.K. Chrononutrition—When We Eat Is of the Essence in Tackling Obesity. *Nutrients* **2022**, *14*, 5080. [CrossRef]
3. Vasileva, L.V.; Marchev, A.S.; Georgiev, M.I. Causes and solutions to "globesity": The new FA(S)T alarming global epidemic. *Food Chem. Toxicol.* **2018**, *121*, 173–193. [CrossRef]
4. Di Rosa, C.; Lattanzi, G.; Taylor, S.F.; Manfrini, S.; Khazrai, Y.M. Very low calorie ketogenic diets in overweight and obesity treatment: Effects on anthropometric parameters, body composition, satiety, lipid profile and microbiota. *Obes. Res. Clin. Pract.* **2020**, *14*, 491–503. [CrossRef]
5. Williams, M.S.; Turos, E. The Chemistry of the Ketogenic Diet: Updates and Opportunities in Organic Synthesis. *Int. J. Mol. Sci.* **2021**, *22*, 5230. [CrossRef]
6. Hill, J.O.; Wyatt, R.; Peters, J.C. The Importance of Energy Balance. *Eur. Endocrinol.* **2013**, *9*, 111–115. [CrossRef] [PubMed]
7. Hill, J.O.; Wyatt, H.R.; Peters, J.C. Energy Balance and Obesity. *Circulation* **2012**, *126*, 126–132. [CrossRef] [PubMed]
8. Hafekost, K.; Lawrence, D.; Mitrou, F.; O'Sullivan, T.A.; Zubrick, S.R. Tackling overweight and obesity: Does the public health message match the science? *BMC Med.* **2013**, *11*, 41. [CrossRef]
9. Mann, T.; Tomiyama, A.J.; Westling, E.; Lew, A.M.; Samuels, B.; Chatman, J. Medicare's search for effective obesity treatments: Diets are not the answer. *Am. Psychol.* **2007**, *62*, 220–233. [CrossRef] [PubMed]
10. Hall, K.D.; Kahan, S. Maintenance of Lost Weight and Long-Term Management of Obesity. *Med. Clin. N. Am.* **2018**, *102*, 183–197. [CrossRef]
11. Carey, K.J.; Vitek, W. Weight Cycling in Women: Adaptation or Risk? *Semin. Reprod. Med.* **2020**, *40*, 277–282. [CrossRef] [PubMed]
12. Thomas, D.M.; Martin, C.K.; Heymsfield, S.; Redman, L.M.; Schoeller, D.A.; Levine, J.A. A simple model predicting individual weight change in humans. *J. Biol. Dyn.* **2011**, *6*, 579–599. [CrossRef]
13. Thomas, D.M.; Gonzalez, M.C.; Pereira, A.Z.; Redman, L.M.; Heymsfield, S.B. Time to Correctly Predict the Amount of Weight Loss with Dieting. *J. Acad. Nutr. Diet* **2014**, *114*, 857–861. [CrossRef]
14. Goodman, B. Mysteries of Weight Loss—WebMD. Available online: https://www.webmd.com/special-reports/weight-loss-mysteries/20160104/weight-loss-mysteries (accessed on 4 July 2022).
15. Yanagi, S.; Sato, T.; Kangawa, K.; Nakazato, M. The Homeostatic Force of Ghrelin. *Cell Metab.* **2018**, *27*, 786–804. [CrossRef] [PubMed]
16. Kim, J.Y. Optimal Diet Strategies for Weight Loss and Weight Loss Maintenance. *J. Obes. Metab. Syndr.* **2021**, *30*, 20–31. [CrossRef] [PubMed]
17. Valente, M.; Garbo, R.; Filippi, F.; Antonutti, A.; Ceccarini, V.; Tereshko, Y.; Di Lorenzo, C.; Gigli, G.L. Migraine Prevention through Ketogenic Diet: More than Body Mass Composition Changes. *J. Clin. Med.* **2022**, *11*, 4946. [CrossRef]
18. Huynh, P.; Calabrese, P. Pathophysiological Abnormalities in Migraine Ameliorated by Ketosis: A Proof-of-Concept Review. *J. Integr. Neurosci.* **2022**, *21*, 167. [CrossRef] [PubMed]
19. Bongiovanni, D.; Benedetto, C.; Corvisieri, S.; Del Favero, C.; Orlandi, F.; Allais, G.; Sinigaglia, S.; Fadda, M. Effectiveness of ketogenic diet in treatment of patients with refractory chronic migraine. *Neurol. Sci.* **2021**, *42*, 3865–3870. [CrossRef]

20. Paoli, A.; Mancin, L.; Giacona, M.C.; Bianco, A.; Caprio, M. Effects of a ketogenic diet in overweight women with polycystic ovary syndrome. *J. Transl. Med.* **2020**, *18*, 104. [CrossRef]
21. Weber, D.D.; Aminzadeh-Gohari, S.; Tulipan, J.; Catalano, L.; Feichtinger, R.G.; Kofler, B. Ketogenic diet in the treatment of cancer—Where do we stand? *Mol. Metab.* **2020**, *33*, 102–121. [CrossRef] [PubMed]
22. Liu, H.; Yang, Y.; Wang, Y.; Tang, H.; Zhang, F.; Zhang, Y.; Zhao, Y. Ketogenic diet for treatment of intractable epilepsy in adults: A meta-analysis of observational studies. *Epilepsia Open* **2018**, *3*, 9–17. [CrossRef]
23. Sreenivas, S. Keto Diet for Beginners—Nourish by WebMD. Available online: https://www.webmd.com/diet/keto-diet-for-beginners (accessed on 4 July 2022).
24. Casadei, K.; Kiel, J. Anthropometric Measurement. In *StatPearls*; StatPearls Publishing: Treasure Island, FL, USA, 2022. Available online: https://www.ncbi.nlm.nih.gov/books/NBK537315/ (accessed on 4 July 2022).
25. Sattler, K.M.; Deane, F.P.; Tapsell, L.; Kelly, P.J. Gender differences in the relationship of weight-based stigmatisation with motivation to exercise and physical activity in overweight individuals. *Health Psychol. Open* **2018**, *5*, 2055102918759691. [CrossRef] [PubMed]
26. Arnotti, K.; Bamber, M. Fruit and Vegetable Consumption in Overweight or Obese Individuals: A Meta-Analysis. *West J. Nurs. Res.* **2020**, *42*, 306–314. [CrossRef] [PubMed]
27. U.S. Department of Agriculture, Agricultural Research Service. USDA Food and Nutrient Database for Dietary Studies 2011–2012. Food Surveys Research Group Home Page. 2014. Available online: https://www.ars.usda.gov/northeast-area/beltsville-md-bhnrc/beltsville-human-nutrition-research-center/food-surveys-research-group/docs/fndds-download-databases/ (accessed on 4 July 2022).
28. Chang, T.; Ravi, N.; Plegue, M.A.; Sonneville, K.R.; Davis, M.M. Inadequate Hydration, BMI, and Obesity Among US Adults: NHANES 2009–2012. *Ann. Fam. Med.* **2016**, *14*, 320–324. [CrossRef] [PubMed]
29. Suliman, S.; van den Heuvel, L.L.; Kilian, S.; Bröcker, E.; Asmal, L.; Emsley, R.; Seedat, S. Cognitive insight is associated with perceived body weight in overweight and obese adults. *BMC Pub. Health* **2021**, *21*, 534. [CrossRef]
30. Giavarina, D. Understanding Bland Altman analysis. *Biochem. Med.* **2015**, *25*, 141–151. [CrossRef]
31. Wing, R.R.; Phelan, S. Long-term weight loss maintenance. *Am. J. Clin. Nutr* **2005**, *82*, 222–225. [CrossRef]
32. Paoli, A.; Bosco, G.; Camporesi, E.M.; Mangar, D. Ketosis, ketogenic diet and food intake control: A complex relationship. *Front. Psychol.* **2015**, *6*, 27. [CrossRef] [PubMed]
33. Koch, H.; Weber, Y.G. The glucose transporter type 1 (Glut1) syndromes. *Epilepsy Behav.* **2019**, *91*, 90–93. [CrossRef] [PubMed]
34. Poff, A.M.; Ari, C.; Seyfried, T.N.; D'Agostino, D.P. The Ketogenic Diet and Hyperbaric Oxygen Therapy Prolong Survival in Mice with Systemic Metastatic Cancer. *PLoS ONE* **2013**, *8*, e65522. [CrossRef]
35. Okada, C.; Imano, H.; Muraki, I.; Yamada, K.; Iso, H. The Association of Having a Late Dinner or Bedtime Snack and Skipping Breakfast with Overweight in Japanese Women. *J. Obes.* **2019**, *2019*, 2439571. [CrossRef]
36. Dowis, K.; Banga, S. The Potential Health Benefits of the Ketogenic Diet: A Narrative Review. *Nutrients* **2021**, *13*, 1654. [CrossRef]
37. Kosić, M.; Benković, M.; Jurina, T.; Valinger, D.; Gajdoš Kljusurić, J.; Tušek, A.J. Analysis of Hepatic Lipid Metabolism Model: Simulation and Non-Stationary Global Sensitivity Analysis. *Nutrients* **2022**, *14*, 4992. [CrossRef] [PubMed]
38. Markovik, G.; Knights, V.; Nikolovska Nedelkovska, D.; Damjanovski, D. Statistical analysis of results in patients applying the sustainable diet indicators. *J. Hyg. Eng. Des.* **2020**, *30*, 35–39.
39. Markovikj, G.; Knights, V. Model of optimization of the sustainable diet indicators. *J. Hyg. Eng. Des.* **2022**, *39*, 169–175.
40. Thomas, D.M.; Scioletti, M.; Heymsfield, S.B. Predictive Mathematical Models of Weight Loss. *Curr. Diab. Rep.* **2019**, *19*, 93. [CrossRef] [PubMed]
41. Pasiakos, S.M.; Cao, J.J.; Margolis, L.M.; Sauter, E.R.; Whigham, L.D.; McClung, J.P.; Rood, J.C.; Carbone, J.W.; Combs, G.F., Jr.; Young, A.J. Effects of high-protein diets on fat-free mass and muscle protein synthesis following weight loss: A randomized controlled trial. *FASEB J.* **2013**, *27*, 3837–3847. [CrossRef]

Review

The Impact Once-Weekly Semaglutide 2.4 mg Will Have on Clinical Practice: A Focus on the STEP Trials

Khaled Alabduljabbar [1,2], Werd Al-Najim [2] and Carel W. le Roux [2,*]

[1] Department of Family Medicine and Polyclinics, King Faisal Specialist Hospital and Research Centre, Riyadh 11211, Saudi Arabia; khalabduljabbar@kfshrc.edu.sa
[2] Diabetes Complications Research Centre, Conway Institute, University College Dublin, D04 V1W8 Dublin, Ireland; werd.al-najim@ucd.ie
* Correspondence: carel.leroux@ucd.ie

Abstract: Obesity is a complex and chronic disease that raises the risk of various complications. Substantial reduction in body weight improves these risk factors. Lifestyle changes, including physical activity, reduced caloric ingestion, and behavioral therapy, have been the principal pillars in the management of obesity. In recent years, pharmacologic interventions have improved remarkably. The Semaglutide Treatment Effect in People with Obesity (STEP) program is a collection of phase-III trials geared toward exploring the utility of once-weekly 2.4 mg semaglutide administered subcutaneously as a pharmacologic agent for patients with obesity. All the STEP studies included diet and exercise interventions but at different intensities. This review paper aims to explore the impact of the behavioral programs on the effect of semaglutide 2.4 mg on weight loss. The results of the STEP trials supported the efficacy of high-dose, once-weekly 2.4 mg semaglutide on body weight reduction among patients with obesity with/without diabetes mellitus. Semaglutide was associated with more gastrointestinal-related side effects compared to placebo but was generally safe and well tolerated. In all the STEP studies, despite the varying intestines of the behavioral programs, weight loss was very similar. For the first time, there may be a suggestion that these behavioral programs might not increase weight reduction beyond the effect of semaglutide. Nevertheless, the importance of nutritional support during substantial weight loss with pharmacotherapy needs to be re-evaluated.

Keywords: semaglutide; obesity; STEP program; weight loss; weight management; clinical trial; GLP-1

Citation: Alabduljabbar, K.; Al-Najim, W.; le Roux, C.W. The Impact Once-Weekly Semaglutide 2.4 mg Will Have on Clinical Practice: A Focus on the STEP Trials. *Nutrients* 2022, 14, 2217. https://doi.org/10.3390/nu14112217

Academic Editors: Javier Gomez-Ambrosi and Sara Baldassano

Received: 19 April 2022
Accepted: 24 May 2022
Published: 26 May 2022

Publisher's Note: MDPI stays neutral with regard to jurisdictional claims in published maps and institutional affiliations.

1. Introduction

Obesity is a complex and chronic disease and has a wide array of complications, including hypertension, hypercholesteremia, type 2 diabetes, cardiovascular disease, and some cancers [1–6]. Lifestyle interventions, comprising physical activity, reduced caloric ingestion, and behavioral therapy, have been the principal pillars in the management of obesity supported by pharmacotherapy and bariatric surgery [7–9]. However, weight loss maintenance has remained challenging [10]. Pharmacotherapy is usually used for individuals with a body mass index (BMI) ≥ 30 kg/m^2 or ≥ 27 kg/m^2 with ≥ 1 coexisting obesity complication [7–9], but the cost, efficacy, and tolerability curbs its utilization [11].

Only a few obesity medications have received approval by the US Food and Drug Administration (FDA), namely naltrexone-bupropion combination, phentermine-topiramate combination, orlistat, setmelanotide, liraglutide, and semaglutide [12–14]. These medications, except phentermine-topiramate, are also approved by the European Medicines Agency (EMA) to be used in Europe [15,16]. The mechanism of actions and the approval status for these medications are presented in Table 1.

Table 1. Mechanism of action and approval status of main obesity medications [15,16].

Medication	Mechanism of Action	Year of FDA Approval	Year of EMA Approval
Naltrexone-bupropion	Reduces energy consumption via potential synergistic effects on pro-opiomelanocortin neurons.	2014	2015
Phentermine-topiramate	Phentermine is an amphetamine-like appetite suppressant working through inhibition of noradrenaline reuptake in the hypothalamus, while topiramate is an anticonvulsant, which has some weight-loss effects, but its mechanism of action is not fully understood.	2012	Not approved
Orlistat	Decreases fat absorption by inhibiting the gastric and pancreatic lipases.	1999	1998
Setmelanotide	Melanocortin 4 (MC4) receptor agonist, works by restoring impaired MC4 receptor pathway activity caused by genetic deficits.	2020	2021
Liraglutide	Glucagon-like peptide-1 (GLP-1) receptor agonist that reduces hunger. Additionally, increases satiety.	2014	2015
Semaglutide	Glucagon-like peptide-1 (GLP-1) receptor agonist that reduces hunger. Additionally, increases satiety.	2021	2022

Semaglutide belongs to the family of glucagon-like peptide-1 analogs. Mechanistically, semaglutide is an incretin, which blocks glucagon release, postpones gastric clearing, reduces energy intake, stimulates satiety, and reduces hunger and appetite via peripheral and central nervous system actions [17]. Semaglutide was initially approved for the management of type 2 diabetes mellitus [18]. The observation that the GLP-1 analogs reduce body weight prompted the exploration of this class of medications as drugs to treat obesity [19–21].

The Semaglutide Treatment Effect in People with Obesity (STEP) program is a collection of 15 multi-institutional, phase-III, randomized, double-blind, placebo-controlled trials geared toward the authorization of once-weekly 2.4 mg semaglutide administered subcutaneously as an obesity medication. Each trial was designed to investigate the efficacy and safety of 2.4 mg semaglutide in people with overweight or obesity, taking in consideration patients' ethnicities, certain comorbidities, different age groups, or the parallel control arm interventions. Six of the program trials (STEP 1–4, 6, and 8) were published; the STEP 5 trial has been completed but not yet published, and the remining trials, including STEP 7, have not been completed yet. Herein, we document a narrative review focused on the clinical summary of the STEP trials, highlight limitations, and outline future directions, with a specific focus on the potential future role of lifestyle changes in obesity management involving such effective medications.

2. The STEP 1 Trial

The STEP 1 trial (ClinicalTrials.gov identifier: NCT03548935) included adults with obesity or overweight (BMI $\geq$ 27 kg/m^2) with at least one obesity complication [22]. Major exclusion criteria included diabetes mellitus, HbA1c $\geq$ 6.5%, or the use of anti-obesity medications in the past 12 weeks. In a 2:1 ratio, the trial randomized 1961 adults to either semaglutide or placebo. Semaglutide was administered in a dose-escalated fashion: the initial once-weekly dose of 0.25 mg was sustained for four weeks; the dose was then titrated to 0.5 mg, 1 mg, 1.7 mg, and 2.4 mg every four weeks. The 2.4 mg once-weekly dose was then maintained for 54 weeks. Overall, the duration of the study was 75 weeks; the treatment (semaglutide or placebo) lasted for 68 weeks, trailed by a follow-up interval of 7 weeks with no medication. The protocol included an unsupervised lifestyle intervention administered to all participants, consisting of a daily 500 kcal deficit diet and weekly 150 min of physical activity. The average age and BMI of the participants were 46 years and 37.9 kg/m^2, respectively. The majority of the participants were females (74.1%) and of White ethnicity (75.1%). Less than half of the participants had prediabetes (43.7%).

Semaglutide with lifestyle intervention resulted in more weight loss over 68 weeks compared to placebo with lifestyle intervention (mean difference (MD) = –12.4%, 95% confidence interval (CI): -13.4, -11.5). Moreover, the proportions of participants treated with semaglutide achieving $\geq$5%, $\geq$10%, and $\geq$15% weight loss at week 68 were 86.4%, 69.1%, and 50.5%, respectively. In addition, the semaglutide arm had substantial improvements in various anthropometric (BMI and waist circumference), inflammatory (C-reactive protein), blood pressure (diastolic and systolic), glycemic (HbA1c and fasting plasma glucose), and lipid (total cholesterol, triglycerides, and low-density lipoprotein cholesterol) parameters in contrast to the placebo arm. Semaglutide also substantially improved physical function scores compared to the placebo assessed by the 36-item Short Form Health Survey (SF-36) and the Impact of Weight on Quality of Life–Lite Clinical Trials Version (IWQOL-Lite-CT) questionnaire.

The rate of any reported side effect was higher with semaglutide contrasted with placebo (89.7% vs. 86.4%, respectively). The number of reported serious side effects was greater in the semaglutide arm compared to the placebo arm. The rate of drug termination was also greater in the semaglutide arm (7.0% vs. 3.1%), mostly due to gastrointestinal-related symptoms (4.5% vs. 0.8%). Gallbladder-related symptoms occurred in 2.6% of patients in the semaglutide arm and 1.2% in the placebo arm. The most commonly documented side effects in $\geq$10% of the semaglutide vs. placebo patients were nausea (44.2% vs. 17.4%), diarrhea (31.5% vs. 15.9%), vomiting (24.8% vs. 6.6%), constipation (23.4% vs. 9.5%), and nasopharyngitis (21.5% vs. 20.3%). The rates of hypoglycemia, acute pancreatitis, and injection site reactions were infrequent in the participants who received semaglutide (0.6%, 0.2%, and 5%, respectively).

In summary, among patients with BMI $\geq$ 27 kg/m^2, the STEP 1 trial concluded that once-weekly 2.4 mg semaglutide plus usual lifestyle adjustment was more beneficial than lifestyle interventions alone in reducing body weight and other cardiometabolic risk factors.

3. The STEP 2 Trial

The STEP 2 trial (ClinicalTrials.gov identifier: NCT03552757) included adults with BMI $\geq$ 27 kg/m^2 and HbA1c ranging from 7% to 10%; all participants were diagnosed with type 2 diabetes mellitus $\geq$6 months prior to study screening [23]. In a 1:1:1 ratio, the trial randomized 1210 participants to 2.4 mg semaglutide, 1.0 mg semaglutide, or placebo. All the participants in this study had the same lifestyle intervention as the STEP 1 trial. Semaglutide was administered in a dose-escalated fashion until reaching the targeted maintenance doses. Overall, the study duration was 75 weeks; the treatment (semaglutide or placebo) lasted for 68 weeks, trailed by a follow-up period of 7 weeks with no medication. Participants' average age and BMI were 55 years and 35.7 kg/m^2, respectively. The average HbA1c and interval of type 2 diabetes mellitus were 8.1% and 8 years. Slightly more than half of the participants were females (50.9%) and of White ethnicity (62.1%).

The 2.4 mg semaglutide with lifestyle intervention reduced body weight more than placebo and lifestyle intervention during the 68 weeks (MD = –6.2%, 95% CI: −7.3, −5.2). Moreover, the proportions of participants who had ≥5%, ≥10%, and ≥15% weight loss at week 68 were 68.8%, 45.6%, and 25.8%, respectively. Semaglutide also improved systolic blood pressure, HbA1c, waist circumference, and physical function scores. In addition, the analysis of exploratory secondary endpoints revealed beneficial reductions in lipid (triglycerides, very-low-density lipoprotein cholesterol, and free fatty acids), glycemic (HbA1c, fasting serum insulin, and fasting plasma glucose), inflammatory (C-reactive protein), and blood pressure (diastolic) profiles in support of the semaglutide 2.4 mg arm compared to the placebo arm.

The rate of any reported side effect was greater in the 2.4 mg semaglutide arm contrasted with the placebo arm (87.6% vs. 76.9%). Moreover, the number of reported serious side effects was comparable between both treatment arms. Additionally, the rate of drug termination was higher in the semaglutide 2.4 mg arm (6.2% vs. 3.5%), mostly secondary to gastrointestinal-related symptoms (4.2% vs. 1.0%). Gallbladder-related symptoms occurred in only 0.2% and 0.7% of the semaglutide 2.4 mg and placebo arms, respectively. In contrast with the placebo arm, the most commonly documented side effects in ≥10% of the semaglutide 2.4 mg patients included nausea (33.7% vs. 9.2%), diarrhea (21.3% vs. 11.9%), vomiting (21.8% vs. 2.7%), constipation (17.4% vs. 5.5%), and nasopharyngitis (16.9% vs. 14.7%). The rates of hypoglycemia, acute pancreatitis, and injection site reactions were infrequent in the semaglutide 2.4 mg arm (5.7%, 0.2%, and 3.0%, respectively).

In summary, among adults with type 2 diabetes mellitus and BMI $\geq 27 \text{ kg/m}^2$, the STEP 2 trial concluded that once-weekly 2.4 mg semaglutide plus lifestyle modification was better than lifestyle modification alone for weight loss and other cardiometabolic risk factors.

4. The STEP 3 Trial

The STEP 3 trial (ClinicalTrials.gov identifier: NCT03611582) included adults with the same inclusion and exclusion criteria as the STEP 1 trial [22,24]. In a 1:1 ratio, the trial randomized 611 participants to either semaglutide combined with very intensive behavior therapy or placebo with very intensive behavior therapy. The intensive behavior therapy was the major difference between STEP 3 compared to STEP 1 and STEP 2 trials [22,23]. It comprised a low-calorie diet during the opening 8 weeks, in addition to concentrated behavioral therapy sessions and physical exercise during the 68 weeks.

The participants were provided with a low-calorie diet (1000–1200 kcal/day) served as meal replacements for the first 8 weeks. Then, they were gradually transferred to hypocaloric diet (1200–1800 kcal/day) of conventional food for the remainder of the trial. After eight weeks, the calorie intake was calculated based on randomization body weight unless the participant's BMI reached $\leq 22.5 \text{ kg/m}^2$. The recommended caloric intake was then re-calculated with no energy deficit until the end of the trial.

Physical activity was prescribed from randomization and was tailored to achieve a goal of 100 min of physical activity/week. Participants were counseled to incorporate moderate-intensity activities within the exercises and were requested to increase their weekly physical activity target by 25 min every four weeks to reach 200 min/week. Furthermore, a total of 30 counseling sessions of intensive behavioral therapy were provided over the 68 weeks, covering various topics related to dietary changes, physical activities, and behavioral strategies to ensure the appropriate implementation and compliance with the intervention.

Similar to the STEP 1 and STEP 2 trials [22,23], semaglutide was administered in a dose-escalated fashion until reaching the targeted maintenance dose. The duration of the study was 75 weeks; the treatment (semaglutide with very intensive behavior therapy or placebo with very intensive behavior therapy) lasted for 68 weeks, trailed by a follow-up period of 7 weeks with no medication. The average age and BMI of participants were 46 years and 38 kg/m^2, respectively. The majority of the research participants were females (81.0%) and of White ethnicity (76.1%).

Semaglutide with very intensive behavior therapy resulted in a more significant weight loss from baseline to week 68 compared to placebo with very intensive behavior therapy (MD = –10.3%, 95% CI: −12.0, −8.6). Moreover, the proportions of research participants who achieved ≥5%, ≥10%, and ≥15% body weight loss at week 68 were 86.6%, 75.3%, and 55.8%, respectively. The placebo with intensive behavior therapy was also effective, albeit less so than semaglutide with intensive behavior, in causing ≥5%, ≥10%, and ≥15% weight loss at 68 weeks in 46.7%, 27.0%, and 13.2% of patients, respectively. Semaglutide had significant improvements in systolic blood pressure and waist circumference but was not different in the physical functioning scores compared to the placebo arm. Similar to the previous STEP trials, the analysis of exploratory secondary endpoints revealed beneficial reductions in various lipid, glycemic, inflammatory, and blood pressure (diastolic) profiles in support of the semaglutide arm compared to the placebo arm.

The rate of ≥1 reported side effect was comparable between the semaglutide and placebo arms (95.8% vs. 96.1%). Moreover, the number of reported serious side effects was greater in the semaglutide arm contrasted with the placebo arm (9.1% vs. 2.9%). Additionally, the rate of drug termination was higher with semaglutide (5.9% vs. 2.9%), mostly secondary to gastrointestinal-related symptoms (3.4% vs. 0.0%). Gallbladder-related symptoms took place in only 4.9% and 1.5% of the semaglutide and placebo arms, respectively. In contrast with the placebo arm, the most common side effects in ≥10% of the semaglutide arm included nausea (58.2% vs. 22.1%), constipation (36.9% vs. 24.5%), diarrhoea (36.1% vs. 22.1%), vomiting (27.3% vs. 10.8%), and nasopharyngitis (22.1% vs. 24.0%). The rates of hypoglycemia, acute pancreatitis, and injection site reactions were infrequent among individuals who received the semaglutide therapy (0.5%, 0%, and 5.4%, respectively).

In summary, the STEP 3 trial concluded that semaglutide plus intensive behavior therapy, including an initial low-calorie intake and rigorous behavioral therapy, culminated in clinically meaningful improvements in body weight and other cardiometabolic risk factors compared with the placebo treatment.

5. The STEP 4 Trial

The STEP 4 trial (ClinicalTrials.gov identifier: NCT03548987) [25] included adults with the same inclusion and exclusion criteria as the STEP 1 and STEP 3 trials [22,24]. For all research participants (*n* = 902), semaglutide was administered in a dose-escalated fashion for a 20-week run-in period (16 weeks of dose intensification starting with 0.25 mg until reaching 2.4 mg, trailed by 4 weeks of maintenance dose 2.4 mg). Only patients who were able to tolerate semaglutide 2.4 mg were included in the randomization period, thus excluding those who could not achieve the top dose of the medication. Overall, the duration of the study was 75 weeks; the run-in period lasted for 20 weeks, trailed by a randomization in 2:1 ratio (*n* = 803) to either 2.4 mg semaglutide or placebo and followed by a follow-up period of 7 weeks with no medication. The lifestyle intervention was a 500 kcal deficit diet and 150 min of exercise per week, similar to STEP 1, but less intensive than STEP 3 [22,24]. The average age and BMI of participants were 46 years and 38.4 kg/m^2, respectively. The majority of the research participants were females (79.0%) and of White ethnicity (83.7%).

After the 20-week run-in interval, the average weight loss was 10.6%, and several improvements were witnessed in blood pressure (systolic and diastolic), HbA1c, lipid parameters, and waist circumference. Between week 20 to week 68, participants randomized to ongoing semaglutide continued to lose weight as opposed to those randomized to the placebo arm who gained weight during the same period (MD = –14.8%, 95% CI: −16.0, −13.5). Moreover, the continued semaglutide arm achieved significant decreases in systolic blood pressure (MD = –3.9 mmHg, 95% CI: −5.8, −2.0) and waist circumference (MD = −9.7 cm, 95% CI: −10.9, −8.5). Furthermore, the physical function scores were significantly better with continued semaglutide. Similar, to the earlier STEP trials [22–24], the analysis of exploratory secondary endpoints revealed beneficial reductions in various lipid and glycemic profiles in support of the continued semaglutide arm contrasted with the switched placebo arm.

The rate of any adverse event was greater in the continued semaglutide arm than the switched placebo arm (81.3% vs. 75.0%). The number of serious side effects was higher in the continued semaglutide arm contrasted with the switched placebo arm (7.7% vs. 5.6%). However, the rate of drug termination was comparable between both arms (2.4% vs. 2.2%). Gastrointestinal and gallbladder-related symptoms took place in 41.9% and 2.8% of the continued semaglutide arm and in 26.1% and 3.7% of the switched placebo arm. In contrast with the switched placebo arm, the most commonly documented side effects in ≥5% of the continued semaglutide arm included diarrhoea (14.4% vs. 7.1%), nausea (14.0% vs. 4.9%), constipation (11.6% vs. 6.3%), nasopharyngitis (10.8% vs. 14.6%), and vomiting (10.3% vs. 3.0%). The rates of hypoglycemia, acute pancreatitis, and injection site reactions were infrequent in the continued semaglutide arm (0.6%, 0%, and 2.6%, respectively).

In summary, the STEP 4 trial concluded that once-weekly continued 2.4 mg semaglutide after a 20-week run-in interval plus standard lifestyle modifications led to sustained body weight loss over the next 48 weeks contrasted with individuals who switched to placebo who started regaining weight.

6. The STEP 5 Trial

The STEP 5 trial (ClinicalTrials.gov identifier: NCT03693430) included adults with the same inclusion and exclusion criteria as the STEP 1, STEP 3, and STEP 4 trials [22,24–26]. In a 1:1 ratio, the trial randomized 304 participants to either semaglutide with standard lifestyle modifications of a 500 kcal deficit diet and 150 min of exercise per week or placebo with standard lifestyle modifications of a 500 kcal deficit diet and 150 min of exercise per week. Semaglutide was administered in a dose-escalated fashion until reaching the targeted maintenance dose of 2.4 mg (end of week 20), which was continued until week 104. The study continued for 111 weeks; the treatment (semaglutide or placebo) lasted for 104 weeks, trailed by a follow-up period of 7 weeks with no medication. The average age and BMI of participants were 47 years and 38.5 kg/m^2, respectively. The majority of the research participants were females (78.0%) and of White ethnicity (93.1%).

Semaglutide with standard lifestyle modifications achieved more weight loss from baseline to week 104 contrasted with placebo with standard lifestyle modifications (MD = −12.6%, 95% CI: −15.3, −9.8). Moreover, the proportions of research participants with ≥5%, ≥10%, ≥15%, and ≥20% weight loss at week 104 with semaglutide were 77.1%, 61.8%, 52.1%, and 36.1%, respectively. In addition, the semaglutide arm had significant improvements in various cardiovascular risk factors, such as systolic blood pressure (MD = −4.2 mmHg, 95% CI: −7.3, −1.0), diastolic blood pressure (MD = −3.7 mmHg, 95% CI: −6.1, −2.1), C-reactive protein (MD = −53.1%, 95% CI: −63.2, −40.0), and waist circumference (MD = −9.2 cm, 95% CI: −12.2, −6.2). The semaglutide arm also had significant improvements in various metabolic risk factors, such as HbA1c (MD = −0.33%, 95% CI: −0.41, −0.25), fasting plasma glucose (MD = −9.2 mg/dL, 95% CI: −12.0, −6.5), fasting serum insulin (MD = −27.4%, 95% CI: −39.3, −13.3), and triglycerides (MD = −22.0%, 95% CI: −29.8, −13.2).

The rate of any adverse event was greater after semaglutide compared to placebo (96.1% vs. 89.5%). The number of reported serious side effects was unexpectedly lower in the semaglutide arm contrasted with the placebo arm (7.9% vs. 11.8%). The rate of drug termination was similar in the semaglutide and placebo arms (5.9% vs. 4.6%). Gastrointestinal and gallbladder-related symptoms took place in 82.2% and 2.6% of the semaglutide arm. In contrast, gastrointestinal and gallbladder-related symptoms took place in 53.9% and 1.3% of the placebo arm. The rates of hypoglycemia, acute pancreatitis, and injection site reactions were rare in the semaglutide arm (2.6%, 0%, and 6.6%, respectively).

In summary, the STEP 5 trial concluded that once-weekly semaglutide dose (plus lifestyle modifications) led to sustained body after two years of treatment, improved cardiovascular and metabolic risk factors, and depicted satisfactory safety profile compared with placebo.

7. The STEP 6 Trial

The STEP 6 trial (ClinicalTrials.gov identifier: NCT03811574) included east Asian adults, with or without type 2 diabetes, who reported a failed weight loss dietary attempt and had a BMI $\geq$ 27 kg/m^2 with two or more weight-related medical problems or a BMI $\geq$ 35 kg/m^2 with at least one weight-related medical problem [27]. The main exclusion criteria were previous or planned anti-obesity treatment or surgery and bodyweight changes of 5 kg or more in the past 3 months before screening. The Asian ethnicity was the major difference between STEP 6 compared to the previous STEP trials (the majority were White) [22–26]. In a 4:1:2:1 ratio, the trial randomized 401 participants to either semaglutide 2.4 mg or placebo, or semaglutide 1.7 mg or placebo. The dose was administered in an escalated fashion until reaching the targeted doses. All the participants were advised to follow the standard lifestyle modifications similar to STEP 5 trial. Overall, the duration of the study was 75 weeks; the treatment lasted for 68 weeks, trailed by a follow-up interval of 7 weeks with no medication. The average age and BMI of participants were 51 years and 31.9 kg/m^2, respectively. All the research participants were Asian (100%), and the majority were males (63%).

The semaglutide with lifestyle intervention in both doses (2.4 mg and 1.7 mg) reduced body weight more than placebo with lifestyle intervention during the 68 weeks (MD = -11.06%, 95% CI: -12.88, -9.24 and MD = -7.52%, 95% CI: -9.62, -5.43, respectively). Moreover, the proportions of participants who had $\geq$5%, $\geq$10%, and $\geq$15% weight loss at week 68 with semaglutide 2.4 mg were 83%, 61%, and 41%, respectively. The treatment arm also showed significant reductions in waist circumference, systolic blood pressure, and HbA1c. In addition, the analysis of exploratory secondary endpoints revealed favorable reductions among semaglutide groups in BMI, fasting plasma glucose, C-reactive protein, and plasminogen activator inhibitor-1, lipid profile (except for high-density lipoprotein cholesterol). An improvement in the physical function score was noted in the semaglutide 2.4 mg group. From baseline to week 68, greater reductions in abdominal visceral fat area were observed in the semaglutide 2.4 mg (-40%) and 1.7 mg (-22.2%) groups than the placebo group (-6.9%).

The rate of reported adverse events was 86% in the semaglutide 2.4 mg group, 82% in the semaglutide 1.7 mg group, and 79% in the placebo group. Unexpectedly, the percentage of serious adverse events was lower in the semaglutide 2.4 mg arm (5%) contrasted with semaglutide 1.7 mg and placebo arms (7% each). The rate of drug termination was higher in semaglutide groups (3%) compared to placebo (1%). Gallbladder-related symptoms took place in only 1% in all groups. Gastrointestinal-related symptoms, which were mostly mild to moderate, were more common in semaglutide 1.7 mg group (64%) than semaglutide 2.4 mg group (59%) or placebo group (30%). The rates of hypoglycemia and acute pancreatitis were 0% in all arms. Injection site reactions were reported in only four participants in the semaglutide 2.4 mg arm.

In summary, among east Asian patients with BMI $\geq$ 27 kg/m^2, with or without type 2 diabetes, the STEP 6 trial concluded that once-weekly 2.4 mg semaglutide plus lifestyle adjustment led to significant reductions in body weight, abdominal visceral fat, and other cardiometabolic risk factors compared with placebo in this population.

8. The STEP 8 Trial

The STEP 8 trial (ClinicalTrials.gov identifier: NCT04074161) was an open label with treatment arms and double-blinded against matched placebo arms [28]. It included adults with the same inclusion and exclusion criteria as the STEP 1 trial. In a 3:1:3:1 ratio, the trial randomized 338 participants to either once-weekly semaglutide (dose-escalation to 2.4 mg over 16 weeks), or matching placebo, or once-daily liraglutide (dose escalation to 3.0 mg over 4 weeks), or matching placebo. Both semaglutide and liraglutide are long-acting

GLP-1 analogs. As a result of the substitution of amino acids that prevents the degeneration of dipeptidyl peptidase 4 and addition of C18 fatty acids, semaglutide has a half-life of 165 h, whereas liraglutide's half-life is about 13 h [26]. All the participants in this trial had the same lifestyle intervention as the STEP 1 trial. The study continued for 75 weeks; the treatments lasted for 68 weeks, trailed by a 7-week follow-up period with no medications. The mean age and BMI of participants were 49 years and 37.5 kg/m^2, respectively. The majority of the research participants were females (78.4%) and of White ethnicity (73.7%).

Semaglutide with lifestyle modifications resulted in a more significant weight loss from baseline to week 68 compared to liraglutide with lifestyle modifications (MD = -9.4%, 95% CI: -12.0, -6.8). Furthermore, the proportions of the semaglutide patients who achieved $\geq 5\%$, $\geq 10\%$, $\geq 15\%$, and $\geq 20\%$ body weight loss at week 68 were 87.2%, 70.9%, 55.6%, and 38.5%, respectively. Liraglutide was also effective, albeit less than semaglutide, in causing $\geq 5\%$, $\geq 10\%$, $\geq 15\%$, and $\geq 20\%$ weight loss at 68 weeks in 58.1%, 25.6%, 12%, and 6% of participants, respectively. At week 68, reductions in BMI, waist circumference, blood pressure, HbA1c, fasting plasma glucose, triglyceride, total cholesterol, very-low-density lipoprotein cholesterol, and C-reactive protein levels were significantly greater with semaglutide compared to liraglutide.

The rate of any reported adverse events was 95.2% with semaglutide, 96.1% with liraglutide, and 95.3% with placebo. The number of reported serious side effects was higher with liraglutide (11%) compared to semaglutide (7.9%) or placebo (7.1%). Moreover, drug termination was more common in the liraglutide arm (12.6%) vs. semaglutide (3.2%) and placebo (3.5%). Gastrointestinal- and gallbladder-related symptoms were reported in 84.1% and 0.8% with semaglutide, 82.7% and 3.1% with liraglutide, and 55.3% and 1.2% with placebo. Hypoglycemia and acute pancreatitis were reported only with the liraglutide group (0.8% both). The injection site reactions were observed with liraglutide (11%) and placebo (5.9%) but not with semaglutide (0%).

In summary, the STEP 8 trial concluded that once-weekly semaglutide with lifestyle modifications was significantly superior to once-daily liraglutide with lifestyle modifications in body weight reduction and other cardiometabolic risk factors improvement.

9. Discussion

The STEP program demonstrated that once-weekly semaglutide with various intensity of lifestyle modifications was superior to placebo or once-daily liraglutide with lifestyle modifications in body weight reduction and other cardiometabolic risk factors improvement. The main secondary efficacy endpoints are summarized in Table 2. The STEP 2 trial included individuals with type 2 diabetes mellitus and obesity [23]. In the STEP 6 trial, only 25% of the patients had diabetes [27]. Conversely, the STEP 1, 3, 4, 5, and 8 trials did not include patients with type 2 diabetes mellitus [22,24–26,28], which may explain the superior weight loss in STEP 1, 3–6, and 8. The purpose of the STEP trials was for semaglutide 2.4 mg to gain regulatory approval and, as such, the two primary efficacy outcomes were percentage weight loss and the proportion of individuals achieving $\geq 5\%$ weight loss at the endpoint.

In a recent meta-analysis of randomized controlled trials comparing the efficacy of different obesity medications, it showed that the percentage of bodyweight reduction from baseline with phentermine-topiramate was 7.97%, naltrexone-bupropion was 4.11%, orlistat was 3.16%, and liraglutide was 4.68%. Phentermine-topiramate and naltrexone-bupropion combinations were associated with the most adverse events. Their findings suggested that semaglutide might be the most effective among all the different obesity medications [15].

Table 2. Change in the main efficacy endpoints from baseline to end of treatment for STEP 1–6 and 8 [22–28].

Parameter	STEP 1		STEP 2			STEP 3		STEP 4 [b]		STEP 5		STEP 6			STEP 8	
	Semaglutide 2.4 mg	Placebo	Semaglutide 2.4 mg	Semaglutide 1 mg	Placebo	Semaglutide 2.4 mg	Placebo	Semaglutide 2.4 mg	Placebo	Semaglutide 2.4 mg	Placebo	Semaglutide 2.4 mg	Semaglutide 1.7 mg	Placebo	Semaglutide 2.4 mg	Liraglutide 3 mg
Body weight change (%)	−14.9	−2.4	−9.6	−7.0	−3.4	−16	−5.7	−7.9	6.9	−15.2	−2.6	−13.2	−9.6	−2.1	−15.8	−6.4
Participants with ≥5% weight loss (%)	86.4	31.5	68.8	57.1	28.5	86.6	47.6	88.7	47.6	77.1	34.4	83	72	21	87.2	58.1
Participants with ≥10% weight loss (%)	69.1	12.0	45.6	28.7	8.2	75.3	75.3	79.0	20.4	61.8	13.3	61	42	5	70.9	25.6
Participants with ≥15% weight loss (%)	50.5	4.9	25.8	13.7	3.2	55.8	13.2	63.7	9.2	52.1	7.0	41	24	3	55.6	12.0
WC (cm)	−13.54	−4.13	−9.4	−6.7	−4.5	−14.6	−6.3	−6.4	3.3	−14.4	−5.2	−11.1	−7.7	−1.8	−13.2	−6.6
BMI (kg/m^2)	−5.54	−0.92	−3.5	−2.5	−1.3	−6.0	−2.2	−2.6	2.2	NA	NA	−4.3	−3.1	−0.6	NA	NA
CRP [a] (mg/dL)	0.47	0.85	0.51	0.58	0.83	−59.6	−22.9	NA	NA	−56.7	−7.8	0.39	0.64	0.92	−52.6	−24.5
SBP (mmHg)	−6.16	−1.06	−3.9	−2.9	−0.5	−6.3	−1.6	0.5	4.4	−5.7	−1.6	−11.0	−12.0	−5.0	−5.7	−2.9
DBP (mmHg)	−2.83	−0.42	−1.6	−0.6	−0.9	−3.0	−0.8	0.3	0.9	−4.5	−0.8	−5	−5	−3	−5.0	−0.5
HbA1c (%)	−0.45	−0.15	−1.6	−1.5	−0.4	−0.51	−0.27	−0.1	0.1	−0.43	−0.10	−1.0	−0.9	0.0	−0.2	−0.1
FPG (mg/dL)	−8.35	−0.48	−2.1	−1.8	−0.1	−6.73	−0.65	−0.8	6.7	−7.6	1.6	−19.3	−18.3	1.7	−8.3	−4.3
TC [a]	0.97	1.00	0.99	0.98	0.99	−3.8	2.1	5	11	NA	NA	0.91	0.93	1.00	−7.1	−0.1
LDL [a]	0.97	1.01	1.00	0.99	1.00	−4.7	2.6	1	8	NA	NA	0.86	0.90	0.95	−6.5	0.9
TG [a]	0.78	0.93	0.78	0.83	0.91	−22.5	−6.5	−6	15	−19.0	3.7	0.79	0.78	1.05	−20.7	−11.0

[a] The data were presented as ratios of end of treatment to baseline for STEP 1, 2, and 6 and change % for STEP 3–5 and 8; [b] Change from week 20 (run-in period) to end of treatment. Abbreviations: WC, waist circumference; BMI, body mass index; CRP, C-reactive protein; SBP, systolic blood pressure; DBP, diastolic blood pressure; HbA1c, hemoglobin A1c; FPG, fasting plasma glucose; TC, total cholesterol; LDL, low-density lipoprotein; TG, triglyceride; NA, not available.

The STEP 4 trial (a withdrawal study) appraised the cessation of semaglutide therapy after a 20-week run-in interval. STEP 4 showed that individuals who continued semaglutide therapy had sustained a significant body weight loss contrasted with those who switched to placebo who started regaining weight [25]. Six of the STEP trials employed a semaglutide dose of 2.4 mg, except for the STEP 2 and STEP 6 trials, which also had an arm using semaglutide 1.0 mg and 1.7 mg, respectively. STEP 2 and 6 demonstrated that the higher semaglutide dose resulted in more body weight loss but also had fewer adverse effects, reflecting a dose–response effect. The STEP 5 trial had the longest duration of all STEP trials (104 weeks on medication) and explored the long-term effect of 2.4 mg semaglutide contrasted with placebo on body weight and various cardiometabolic risk factors over a two-year period. In the STEP 8 trial, the reduction in body weight was significantly greater with weekly semaglutide injection when compared to daily liraglutide injection, accompanied by significant improvements in various cardiometabolic risk factors. The analysis of exploratory secondary endpoints in all the STEP trials revealed beneficial effects on blood pressure, glycemic, lipid, inflammatory, and anthropometric parameters.

Overall, semaglutide had a good safety profile without any new safety signals not previously detected in other GLP-1 analogs. The rate of serious adverse events and proportion of side effects leading to drug termination was generally similar to other GLP-1 analogs. The tolerability profile and main adverse events for STEP 1–6 and 8 are presented in Table 3. The vast majority of the drug-associated adverse events were mild gastrointestinal-related symptoms. The rates of hypoglycemia, acute pancreatitis, gallbladder-related symptoms, and injection site reactions were low in the semaglutide groups and often comparable with the placebo groups. These will, however, need to be monitored in post-market-surveillance schemes.

The STEP trials have several strengths, including the scientifically robust methodologies, as reflected by the phase-III, large-sized, multicentric, double-blind, and placebo-controlled study designs. Limitations include the unintentional biased gender and ethnicity, as the vast majority of the recruited research participants were White females. Patients were recruited from routine clinical services where the usual demographic is reflected in the trials with a preponderance of females. These sociodemographic factors could have introduced a bias in the pooled outcomes. However, in the STEP 6 trial, all participants were east Asian, and the majority were males, and the outcomes were almost similar to the other STEP trials. Another limitation includes the short-term follow-up interval of roughly 68 weeks. This limitation was partly addressed in the STEP 5 trial, which provided a much longer follow-up of two years. However, obesity is a chronic disease and will require chronic treatment.

All STEP trials included a lifestyle intervention. However, only the STEP 3 trial incorporated very intensive lifestyle modifications, which included a low-calorie intake during the opening 8 weeks and then an additional 30 weeks of intensive behavioral therapy sessions with registered dieticians [24]. As a consequence, the patients in the placebo arm lost almost double the amount of weight recorded in the placebo arms of STEP 1, 2, 5, and 6 [22,23,26,27]. The placebo arm in STEP 3 lost 5.7% of weight, while the placebo arm in STEP 1 lost 2.4%, STEP 2 lost 3.4%, STEP 5 lost 2.6%, and STEP 6 lost 2.1%. Weight loss in the placebo arm of STEP 4 was 5%, but these patients were busy regaining weight after being treated with semaglutide for 20 weeks before being switched to only receiving standard lifestyle modifications. However, the approach of short-term drug treatment followed by standard lifestyle modifications in STEP 4 appeared as effective as very intensive lifestyle modification with placebo treatment. It was striking that the total weight loss achieved at the end of the treatment period in the semaglutide arms for STEP 1 was 14.9%, STEP 3 was 16%, STEP 4 was 17.4%, STEP 5 was 15.2%, STEP 6 was 13.2%, and STEP 8 was 15.8%. Only STEP 3 had very intensive lifestyle modifications, and it was expected that, similar to SCALE intensive behavior therapy, the addition of the intensive lifestyle modifications to semaglutide 2.4 mg would have added significantly more weight loss than when semaglutide 2.4 mg was combined with standard lifestyle modifications [29,30]. This raises the question of whether semaglutide 2.4 mg requires any lifestyle modification to be effective.

Table 3. Tolerability profile and main adverse events for STEP 1–6 and 8 [22–28].

Parameter, n (%)	STEP 1		STEP 2			STEP 3		STEP 4 [a]		STEP 5		STEP 6			STEP 8	
	Semaglutide 2.4 mg	Placebo	Semaglutide 2.4 mg	Semaglutide 1 mg	Placebo	Semaglutide 2.4 mg	Placebo	Semaglutide 2.4 mg	Placebo	Semaglutide 2.4 mg	Placebo	Semaglutide 2.4 mg	Semaglutide 1.7 mg	Placebo	Semaglutide 2.4 mg	Liraglutide 3 mg
Any adverse event	1171 (89.7)	566 (86.4)	353 (87·6)	329 (81·8)	309 (76·9)	390 (95.8)	196 (96.1)	435 (81.3)	201 (75.0)	NA (96.1)	NA (89.5)	171 (86)	82 (82)	80 (79)	120 (95.2)	122 (96.1)
Serious adverse event	128 (9.8)	42 (6.4)	40 (9.9)	31 (7.70)	37 (9.2)	37 (9.1)	6 (2.9)	41 (7.7)	15 (5.6)	NA (7.9)	NA (11.8)	10 (5)	7 (7)	7 (7)	10 (7.9)	14 (11.0)
Adverse events leading to trial product discontinuation	92 (7.0)	20 (3.1)	25 (6.2)	20 (5.0)	14 (3.5)	24 (5.9)	24 (5.9)	13 (2.4)	6 (2.2)	NA (5.9)	NA (4.6)	5 (3)	3 (3)	0	4 (3.2)	16 (12.6)
Gastrointestinal disorders	969 (74.2)	314 (47.9)	256 (63.5)	231 (57.5)	138 (34.3)	337 (82.8)	129 (63.2)	224 (41.9)	70 (26.1)	125 (82.2)	82 (53.9)	118 (59)	64 (64)	30 (30)	106 (84.1)	105 (82.7)
Gallbladder-related disorders	34 (2.6)	8 (1.2)	1 (0.2)	4 (1.0)	3 (0.7)	20 (4.9)	3 (1.5)	15 (2.8)	10 (3.7)	4 (2.6)	2 (1.3)	2 (1)	1 (1)	1 (1)	1 (0.8)	4 (3.1)
Hypoglycemia	8 (0.6)	5 (0.8)	23 (5.7)	22 (5.5)	12 (3.0)	2 (0.5)	0	3 (0.6)	3 (1.1)	4 (2.6)	0	0	0	0	0	1 (0.8)
Acute pancreatitis	3 (0.2)	0	1 (0.2)	0	1 (0.2)	0	0	0	0	0	0	0	0	0	0	1 (0.8)
Injection site reactions	65 (5.0)	44 (6.7)	12 (3.0)	6 (1.5)	10 (2.5)	22 (5.4)	12 (5.9)	14 (2.6)	6 (2.2)	10 (6.6)	15 (9.9)	4 (2)	0	0	0	14 (11.0)
Diarrhoea	412 (31.5)	104 (15.9)	86 (21.3)	89 (22.1)	48 (11.9)	147 (36.1)	45 (22.1)	77 (14.4)	19 (7.1)	NA	NA	32 (16)	22 (22)	6 (6)	35 (27.8)	23 (18.1)
Constipation	306 (23.4)	62 (9.5)	70 (17.4)	51 (12.7)	22 (5.5)	150 (36.9)	50 (24.5)	62 (11.6)	17 (6.3)	NA	NA	52 (26)	19 (19)	3 (3)	49 (38.9)	40 (31.5)
Nausea	577 (44.2)	114 (17.4)	136 (33.7)	129 (32.1)	37 (9.2)	237 (58.2)	45 (22.1)	75 (14.0)	13 (4.9)	NA	NA	35 (18)	18 (18)	4 (4)	77 (61.1)	75 (59.1)
Vomiting	324 (24.8)	43 (6.6)	88 (21.8)	54 (13.4)	11 (2.7)	111 (27.3)	22 (10.8)	55 (10.3)	8 (3.0)	NA	NA	19 (9)	10 (10)	2 (2)	32 (25.4)	32 (25.4)
Nasopharyngitis	281 (21.5)	133 (20.3)	68 (16.9)	47 (11.7)	59 (14.7)	90 (22.1)	90 (22.1)	58 (10.8)	39 (14.6)	NA	NA	53 (27)	24 (24)	18 (18)	10 (7.9)	11 (8.7)

[a] Data from week 20 (run-in period) to end of treatment.

Other successful obesity treatments, such as bariatric surgery, do not appear to require lifestyle modifications to provide any more weight loss within the first year after surgery [31,32]. This may be due to the subject's obesity being effectively treated with surgery. The changes in food intake behavior after successful obesity treatment may not be amplified by giving the patient a more stringent diet or exercise regimen. If this is true, then the cost of providing effective obesity care will significantly reduce in the first year because it is often the requirement for lifestyle modifications, which makes it difficult for practitioners to prescribe obesity treatments. This does not mean that lifestyle modifications may not be helpful in the longer term because, as the STEP trials have shown, there are also non-responders and partial responders to semaglutide. The addition of lifestyle modification may provide additional weight loss to those who only partially respond to the medication and thus result in substantial additional health gain; for example, in STEP 3 trial, 75.3% of patients achieved >10% weight loss compared to 69.1% in STEP 1.

Another major unexplored consequence of >15% weight loss with semaglutide may be the inevitable lean muscle mass loss. This is also evident after liraglutide and bariatric surgery [32–35]. The challenge is that patients who consume so few calories because of the effective medication cannot consume enough protein in their daily intake to stop them from becoming catabolic and losing muscle. Exercising these patients further may only result in them becoming more catabolic and losing even more muscle mass [36]. Thus, nutritional therapies once patients are in a steep negative energy balance may have to focus on optimizing protein intake to prevent muscle mass loss. This may further improve the functional gains made by patients if they can achieve 15% weight loss and maintain most of the lean muscle mass [31,37]. These hypotheses, however, require further testing, as our suggestions are purely speculation based on the similarities between trials, which used intensive or less intensive lifestyle changes.

10. Conclusions

In summary, the results of the STEP trials supported the efficacy of high-dose, once-weekly 2.4 mg semaglutide on body weight reduction among individuals with obesity. While semaglutide resulted in more gastrointestinal-related side effects, the medication appeared generally safe and well tolerated. The drug may be so effective that the role of nutritional therapy may have to be redefined, and a shift away from using nutritional therapy to achieve more weight loss to rather using nutritional therapy to achieve more health gain may be required.

Author Contributions: Conceptualization, C.W.l.R. and K.A.; Methodology, K.A., W.A.-N., and C.W.l.R.; Resources, K.A.; Writing—Original Draft Preparation, K.A.; Writing—Review and Editing, W.A.-N. and C.W.l.R.; Supervision, C.W.l.R.; All authors have read and agreed to the published version of the manuscript.

Funding: This research received no external funding.

Institutional Review Board Statement: Not applicable.

Informed Consent Statement: Not applicable.

Data Availability Statement: Not applicable.

Conflicts of Interest: C.W.l.R. reports grants from the Irish Research Council, Science Foundation Ireland, Anabio, and the Health Research Board. He serves on the advisory boards of Novo Nordisk, Herbalife, GI Dynamics, Eli Lilly, Johnson & Johnson, Sanofi Aventis, AstraZeneca, Janssen, Bristol-Myers Squibb, Glia, and Boehringer Ingelheim. C.W.l.R. is a member of the Irish Society for Nutrition and Metabolism outside the area of work commented on here. He has been the chief medical officer and director of the Medical Device Division of Keyron since January 2011. Both of these are unremunerated positions. C.W.l.R. was a previous investor in Keyron, which develops endoscopically implantable medical devices intended to mimic the surgical procedures of sleeve gastrectomy and gastric bypass. He continues to provide scientific advice to Keyron for no remuneration. The other authors declare no conflict of interest.

References

1. Blüher, M. Obesity: Global epidemiology and pathogenesis. *Nat. Rev. Endocrinol.* **2019**, *15*, 288–298. [CrossRef]
2. Neeland, I.J.; Poirier, P.; Després, J.P. Cardiovascular and Metabolic Heterogeneity of Obesity: Clinical Challenges and Implications for Management. *Circulation* **2018**, *137*, 1391–1406. [CrossRef] [PubMed]
3. Fruh, S.M. Obesity: Risk factors, complications, and strategies for sustainable long-term weight management. *J. Am. Assoc. Nurse Pract.* **2017**, *29*, S3–S14. [CrossRef]
4. Parker, E.D.; Folsom, A.R. Intentional weight loss and incidence of obesity-related cancers: The Iowa Women's Health Study. *Int. J. Obes. Relat. Metab. Disord.* **2003**, *27*, 1447–1452. [CrossRef] [PubMed]
5. Sharma, V.; Coleman, S.; Nixon, J.; Sharples, L.; Hamilton-Shield, J.; Rutter, H.; Bryant, M. A systematic review and meta-analysis estimating the population prevalence of comorbidities in children and adolescents aged 5 to 18 years. *Obes. Rev.* **2019**, *20*, 1341–1349. [CrossRef] [PubMed]
6. Guh, D.P.; Zhang, W.; Bansback, N.; Amarsi, Z.; Birmingham, C.L.; Anis, A.H. The incidence of co-morbidities related to obesity and overweight: A systematic review and meta-analysis. *BMC Public Health* **2009**, *9*, 88. [CrossRef] [PubMed]
7. Garvey, W.T.; Mechanick, J.I.; Brett, E.M.; Garber, A.J.; Hurley, D.L.; Jastreboff, A.M.; Nadolsky, K.; Pessah-Pollack, R.; Plodkowski, R. American association of clinical endocrinologists and American college of endocrinology comprehensive clinical practice guidelines for medical care of patients with obesity. *Endocr. Pract.* **2016**, *22* (Suppl. 3), 1–203. [CrossRef] [PubMed]
8. Yumuk, V.; Tsigos, C.; Fried, M.; Schindler, K.; Busetto, L.; Micic, D.; Toplak, H. European Guidelines for Obesity Management in Adults. *Obes. Facts* **2015**, *8*, 402–424. [CrossRef]
9. Wharton, S.; Lau, D.C.W.; Vallis, M.; Sharma, A.M.; Biertho, L.; Campbell-Scherer, D.; Adamo, K.; Alberga, A.; Bell, R.; Boulé, N.; et al. Obesity in adults: A clinical practice guideline. *Cmaj* **2020**, *192*, E875–E891. [CrossRef]
10. Sumithran, P.; Prendergast, L.A.; Delbridge, E.; Purcell, K.; Shulkes, A.; Kriketos, A.; Proietto, J. Long-term persistence of hormonal adaptations to weight loss. *N. Engl. J. Med.* **2011**, *365*, 1597–1604. [CrossRef]
11. Bessesen, D.H.; Van Gaal, L.F. Progress and challenges in anti-obesity pharmacotherapy. *Lancet Diabetes Endocrinol.* **2018**, *6*, 237–248. [CrossRef]
12. Singh, G.; Krauthamer, M.; Bjalme-Evans, M. Wegovy (semaglutide): A new weight loss drug for chronic weight management. *J. Investig. Med.* **2021**, *70*, 5–13. [CrossRef] [PubMed]
13. Khera, R.; Murad, M.H.; Chandar, A.K.; Dulai, P.S.; Wang, Z.; Prokop, L.J.; Loomba, R.; Camilleri, M.; Singh, S. Association of Pharmacological Treatments for Obesity With Weight Loss and Adverse Events: A Systematic Review and Meta-analysis. *JAMA* **2016**, *315*, 2424–2434. [CrossRef] [PubMed]
14. Markham, A. Setmelanotide: First Approval. *Drugs* **2021**, *81*, 397–403. [CrossRef]
15. Shi, Q.; Wang, Y.; Hao, Q.; Vandvik, P.O.; Guyatt, G.; Li, J.; Chen, Z.; Xu, S.; Shen, Y.; Ge, L.; et al. Pharmacotherapy for adults with overweight and obesity: A systematic review and network meta-analysis of randomised controlled trials. *Lancet* **2022**, *399*, 259–269. [CrossRef]
16. Burcelin, R.; Gourdy, P. Harnessing glucagon-like peptide-1 receptor agonists for the pharmacological treatment of overweight and obesity. *Obes. Rev.* **2017**, *18*, 86–98. [CrossRef]
17. Isaacs, D.; Prasad-Reddy, L.; Srivastava, S.B. Role of glucagon-like peptide 1 receptor agonists in management of obesity. *Am. J. Health Syst. Pharm.* **2016**, *73*, 1493–1507. [CrossRef]
18. Avgerinos, I.; Michailidis, T.; Liakos, A.; Karagiannis, T.; Matthews, D.R.; Tsapas, A.; Bekiari, E. Oral semaglutide for type 2 diabetes: A systematic review and meta-analysis. *Diabetes Obes. Metab.* **2020**, *22*, 335–345. [CrossRef]
19. Kinzig, K.P.; D'Alessio, D.A.; Seeley, R.J. The diverse roles of specific GLP-1 receptors in the control of food intake and the response to visceral illness. *J. Neurosci.* **2002**, *22*, 10470–10476. [CrossRef]
20. Schick, R.R.; Zimmermann, J.P.; vorm Walde, T.; Schusdziarra, V. Peptides that regulate food intake: Glucagon-like peptide 1-(7–36) amide acts at lateral and medial hypothalamic sites to suppress feeding in rats. *Am. J. Physiol. Regul. Integr. Comp. Physiol.* **2003**, *284*, R1427–R1435. [CrossRef]
21. Alruwaili, H.; Dehestani, B.; le Roux, C.W. Clinical Impact of Liraglutide as a Treatment of Obesity. *Clin. Pharmacol.* **2021**, *13*, 53–60. [CrossRef]
22. Wilding, J.P.H.; Batterham, R.L.; Calanna, S.; Davies, M.; Van Gaal, L.F.; Lingvay, I.; McGowan, B.M.; Rosenstock, J.; Tran, M.T.D.; Wadden, T.A.; et al. Once-Weekly Semaglutide in Adults with Overweight or Obesity. *N. Engl. J. Med.* **2021**, *384*, 989. [CrossRef]
23. Davies, M.; Færch, L.; Jeppesen, O.K.; Pakseresht, A.; Pedersen, S.D.; Perreault, L.; Rosenstock, J.; Shimomura, I.; Viljoen, A.; Wadden, T.A.; et al. Semaglutide 2·4 mg once a week in adults with overweight or obesity, and type 2 diabetes (STEP 2): A randomised, double-blind, double-dummy, placebo-controlled, phase 3 trial. *Lancet* **2021**, *397*, 971–984. [CrossRef]
24. Wadden, T.A.; Bailey, T.S.; Billings, L.K.; Davies, M.; Frias, J.P.; Koroleva, A.; Lingvay, I.; O'Neil, P.M.; Rubino, D.M.; Skovgaard, D.; et al. Effect of Subcutaneous Semaglutide vs Placebo as an Adjunct to Intensive Behavioral Therapy on Body Weight in Adults With Overweight or Obesity: The STEP 3 Randomized Clinical Trial. *JAMA* **2021**, *325*, 1403–1413. [CrossRef]
25. Rubino, D.; Abrahamsson, N.; Davies, M.; Hesse, D.; Greenway, F.L.; Jensen, C.; Lingvay, I.; Mosenzon, O.; Rosenstock, J.; Rubio, M.A.; et al. Effect of Continued Weekly Subcutaneous Semaglutide vs. Placebo on Weight Loss Maintenance in Adults with Overweight or Obesity: The STEP 4 Randomized Clinical Trial. *JAMA* **2021**, *325*, 1414–1425. [CrossRef]

26. Garvey, W.T.; Batterham, R.L.; Bhatta, M.; Buscemi, S.; Christensen, L.N.; Frias, J.P.; Jódar, E.; Kandler, K.; Rigas, G.; Wadden, T.A.; et al. Two year Effect of Semaglutide 2.4 mg vs. Placebo in Adults with Overweight or Obesity: STEP 5. In Proceedings of the 39th Annual Meeting of The Obesity Society (TOS), ObesityWeek, 1–5 November 2021.
27. Kadowaki, T.; Isendahl, J.; Khalid, U.; Lee, S.Y.; Nishida, T.; Ogawa, W.; Tobe, K.; Yamauchi, T.; Lim, S. Semaglutide once a week in adults with overweight or obesity, with or without type 2 diabetes in an east Asian population (STEP 6): A randomised, double-blind, double-dummy, placebo-controlled, phase 3a trial. *Lancet Diabetes Endocrinol.* **2022**, *10*, 193–206. [CrossRef]
28. Rubino, D.M.; Greenway, F.L.; Khalid, U.; O'Neil, P.M.; Rosenstock, J.; Sørrig, R.; Wadden, T.A.; Wizert, A.; Garvey, W.T.; Investigators, S. Effect of Weekly Subcutaneous Semaglutide vs. Daily Liraglutide on Body Weight in Adults with Overweight or Obesity without Diabetes: The STEP 8 Randomized Clinical Trial. *JAMA* **2022**, *327*, 138–150. [CrossRef]
29. Wadden, T.A.; Tronieri, J.S.; Sugimoto, D.; Lund, M.T.; Auerbach, P.; Jensen, C.; Rubino, D. Liraglutide 3.0 mg and Intensive Behavioral Therapy (IBT) for Obesity in Primary Care: The SCALE IBT Randomized Controlled Trial. *Obesity* **2020**, *28*, 529–536. [CrossRef]
30. Pi-Sunyer, X.; Astrup, A.; Fujioka, K.; Greenway, F.; Halpern, A.; Krempf, M.; Lau, D.C.W.; le Roux, C.W.; Violante Ortiz, R.; Jensen, C.B.; et al. A Randomized, Controlled Trial of 3.0 mg of Liraglutide in Weight Management. *N. Engl. J. Med.* **2015**, *373*, 11–22. [CrossRef]
31. Tabesh, M.R.; Maleklou, F.; Ejtehadi, F.; Alizadeh, Z. Nutrition, Physical Activity, and Prescription of Supplements in Pre- and Post-bariatric Surgery Patients: A Practical Guideline. *Obes. Surg.* **2019**, *29*, 3385–3400. [CrossRef]
32. Nuijten, M.A.H.; Eijsvogels, T.M.H.; Monpellier, V.M.; Janssen, I.M.C.; Hazebroek, E.J.; Hopman, M.T.E. The magnitude and progress of lean body mass, fat-free mass, and skeletal muscle mass loss following bariatric surgery: A systematic review and meta-analysis. *Obes. Rev.* **2022**, *23*, e13370. [CrossRef] [PubMed]
33. Colquitt, J.L.; Pickett, K.; Loveman, E.; Frampton, G.K. Surgery for weight loss in adults. *Cochrane Database Syst. Rev.* **2014**, *8*, CD003641. [CrossRef] [PubMed]
34. Astrup, A.; Carraro, R.; Finer, N.; Harper, A.; Kunesova, M.; Lean, M.E.J.; Niskanen, L.; Rasmussen, M.F.; Rissanen, A.; Rössner, S.; et al. Safety, tolerability and sustained weight loss over 2 years with the once-daily human GLP-1 analog, liraglutide. *Int. J. Obes.* **2012**, *36*, 843–854. [CrossRef] [PubMed]
35. Park, J.S.; Kwon, J.; Choi, H.J.; Lee, C. Clinical effectiveness of liraglutide on weight loss in South Koreans: First real-world retrospective data on Saxenda in Asia. *Medicine* **2021**, *100*, e23780. [CrossRef]
36. King, W.C.; Bond, D.S. The importance of preoperative and postoperative physical activity counseling in bariatric surgery. *Exerc. Sport Sci. Rev.* **2013**, *41*, 26–35. [CrossRef]
37. Sherf Dagan, S.; Goldenshluger, A.; Globus, I.; Schweiger, C.; Kessler, Y.; Kowen Sandbank, G.; Ben-Porat, T.; Sinai, T. Nutritional Recommendations for Adult Bariatric Surgery Patients: Clinical Practice. *Adv. Nutr.* **2017**, *8*, 382–394. [CrossRef]

 *nutrients*

Article

Does Bypass of the Proximal Small Intestine Impact Food Intake, Preference, and Taste Function in Humans? An Experimental Medicine Study Using the Duodenal-Jejunal Bypass Liner

Madhawi M. Aldhwayan [1], Werd Al-Najim [2,*], Aruchuna Ruban [3], Michael Alan Glaysher [4], Brett Johnson [5], Navpreet Chhina [6], Georgios K. Dimitriadis [7], Christina Gabriele Prechtl [8], Nicholas A. Johnson [8], James Patrick Byrne [4], Anthony Peter Goldstone [6], Julian P. Teare [3], Carel W. Le Roux [2,9] and Alexander Dimitri Miras [5,9]

[1] Community Health Sciences Department, College of Applied Medical Sciences, King Saud University, Riyadh 11451, Saudi Arabia; maldhwayan@ksu.edu.sa

[2] Diabetes Complications Research Centre, Conway Institute, School of Medicine, University College Dublin, D04 V1W8 Dublin, Ireland; carel.leroux@ucd.ie

[3] Department of Surgery and Cancer, Imperial College London, London SW7 2BX, UK; aruchuna.ruban@nhs.net (A.R.); j.teare@imperial.ac.uk (J.P.T.)

[4] Department of General Surgery, University Hospital Southampton NHS Foundation Trust, Southampton SO17 1BJ, UK; michaelglaysher@me.com (M.A.G.); james.byrne@uhs.nhs.uk (J.P.B.)

[5] Department of Metabolism, Digestion and Reproduction, Imperial College London, London SW7 2BX, UK; brett.johnson@nhs.net (B.J.); a.miras@nhs.net (A.D.M.)

[6] PsychoNeuroEndocrinology Research Group, Centre for Neuropsychopharmacology, Division of Psychiatry, Department of Brain Sciences, Imperial College London, Hammersmith Hospital, London SW7 2BX, UK; navchhina@gmail.com (N.C.); tony.goldstone@imperial.ac.uk (A.P.G.)

[7] Department of Endocrinology, King's College Hospital NHS Trust, London SW7 2BX, UK; georgios.dimitriadis@kcl.ac.uk

[8] Imperial Clinical Trials Unit, Department of Public Health, Imperial College London, London SW7 2BX, UK; c.prechtl@imperial.ac.uk (C.G.P.); nicholas.johnson@imperial.ac.uk (N.A.J.)

[9] Centre for Diabetes, Ulster University, BT52 1SA Coleraine, Ireland

[*] Correspondence: werd.al-najim@ucd.ie

Citation: Aldhwayan, M.M.; Al-Najim, W.; Ruban, A.; Glaysher, M.A.; Johnson, B.; Chhina, N.; Dimitriadis, G.K.; Prechtl, C.G.; Johnson, N.A.; Byrne, J.P.; et al. Does Bypass of the Proximal Small Intestine Impact Food Intake, Preference, and Taste Function in Humans? An Experimental Medicine Study Using the Duodenal-Jejunal Bypass Liner. *Nutrients* **2022**, *14*, 2141. https://doi.org/10.3390/nu14102141

Academic Editor: Javier Gómez-Ambrosi

Received: 29 April 2022
Accepted: 16 May 2022
Published: 20 May 2022

Publisher's Note: MDPI stays neutral with regard to jurisdictional claims in published maps and institutional affiliations.

Abstract: The duodenal-jejunal bypass liner (Endobarrier) is an endoscopic treatment for obesity and type 2 diabetes mellitus (T2DM). It creates exclusion of the proximal small intestine similar to that after Roux-en-Y Gastric Bypass (RYGB) surgery. The objective of this study was to employ a reductionist approach to determine whether bypass of the proximal intestine is the component conferring the effects of RYGB on food intake and sweet taste preference using the Endobarrier as a research tool. A nested mechanistic study within a large randomised controlled trial compared the impact of lifestyle modification with vs. without Endobarrier insertion in patients with obesity and T2DM. Forty-seven participants were randomised and assessed at several timepoints using direct and indirect assessments of food intake, food preference and taste function. Patients within the Endobarrier group lost numerically more weight compared to the control group. Using food diaries, our results demonstrated similar reductions of food intake in both groups. There were no significant differences in food preference and sensory, appetitive reward, or consummatory reward domain of sweet taste function between groups or changes within groups. In conclusion, the superior weight loss seen in patients with obesity and T2DM who underwent the Endobarrier insertion was not due to a reduction in energy intake or change in food preferences.

Keywords: Endobarrier; obesity; food preferences; eating behaviour; taste function

---

1. Introduction

The impressive weight loss observed after RYGB surgery is caused predominantly through a reduction in appetite and hence food intake [1,2]. However, a subgroup of patients also change other aspects of their eating behaviour, including food preference [1]. This shift away from energy-dense sweet and/or fatty foods to less energy-dense options is thought to be an additional mechanism underlying weight loss [3,4]. The gap in our current knowledge is which component of the RYGB gut manipulations underlies the observed changes in dietary behaviour.

The manipulations with RYGB include the formation of a small gastric pouch which is anastomosed to the proximal jejunum, bypass of the stomach and proximal small intestine through which the biliopancreatic secretions still flow and mix with food at the jejuno-jejunal anastomosis and throughout the common channel.

Animal models of the duodenal-jejunal bypass operation have contributed to our understanding of the role of proximal intestinal bypass on eating behaviour. Mice that underwent duodenal-jejunal bypass (DJB) surgery exhibited lower sugar intake in a sweet-seeking task compared to sham-operated mice [5]. The mechanism was thought to involve disrupted gut-brain signalling in the DJB mice, in which duodenal glucose infusions caused a higher release of dopamine than jejunal glucose infusions in the dorsal striatum of sham mice. This effect was significantly diminished in DJB mice [5]. This observation leads to the hypothesis that bypass of the proximal small intestine might be the component of the RYGB manipulations responsible for the reduction in the preference for sweet/fatty foods after surgery.

We adopted a reductionist approach and used the duodenal-jejunal bypass liner (Endobarrier device, GI Dynamics, Lexington, MA, USA) as a research tool to enable us to address our hypothesis in humans. The Endobarrier is a 60 cm fluoropolymer sheath that is inserted endoscopically, anchored at the duodenal bulb and lines 60 cm of the proximal small intestine. We previously demonstrated in the largest RCT in the field that the Endobarrier causes superior weight loss to lifestyle modification in people with obesity and T2DM [6].

The aim of this experimental medicine study was to determine the impact of the Endobarrier device on food intake, food preferences and taste function in humans.

2. Materials and Methods

2.1. Patients and Study Design

This was a nested mechanistic study within a larger randomised controlled trial comparing the impact of lifestyle modification with vs. without Endobarrier insertion in patients with obesity and T2DM [6]. The study took place in two academic centres, investigational sites—Imperial College London and University of Southampton. Patients were recruited and followed up in the NIHR Imperial and Southampton Clinical Research Facilities. A complete description of the trial protocol was previously published [7]. In brief, the trial was conducted over 2 years (1 year of treatment and 1 year follow up), 160 participants were randomized at a 1:1 ratio to one of the two study arms. For this nested study, data were collected at 5 time points (mechanistic visits): at baseline (2 weeks before intervention), 10 days, 6 months, 12 months, and 24 months post-intervention (Figure 1).

The Endobarrier is an impermeable fluoropolymer sleeve inserted endoscopically through the duodenum and into the jejunum. The sleeve is open at both ends allowing for chyme passage from the stomach into the lower jejunum, bypassing nutrient absorption along its length by creating a barrier between the partially digested food and the absorptive surface of the small intestine [8]. Implanting the device takes an average of 45 min, and the implant is performed under general anaesthetic. The device barbs are anchored to the duodenal bulb 5–10 mm away from the pylorus. The sleeve then extends for 60 cm through the duodenum by peristalsis movement. Device explant is also done under general anaesthesia, taking, on average, 30 min to perform. The participant is usually discharged to home the same day following recovery from the anaesthetic.

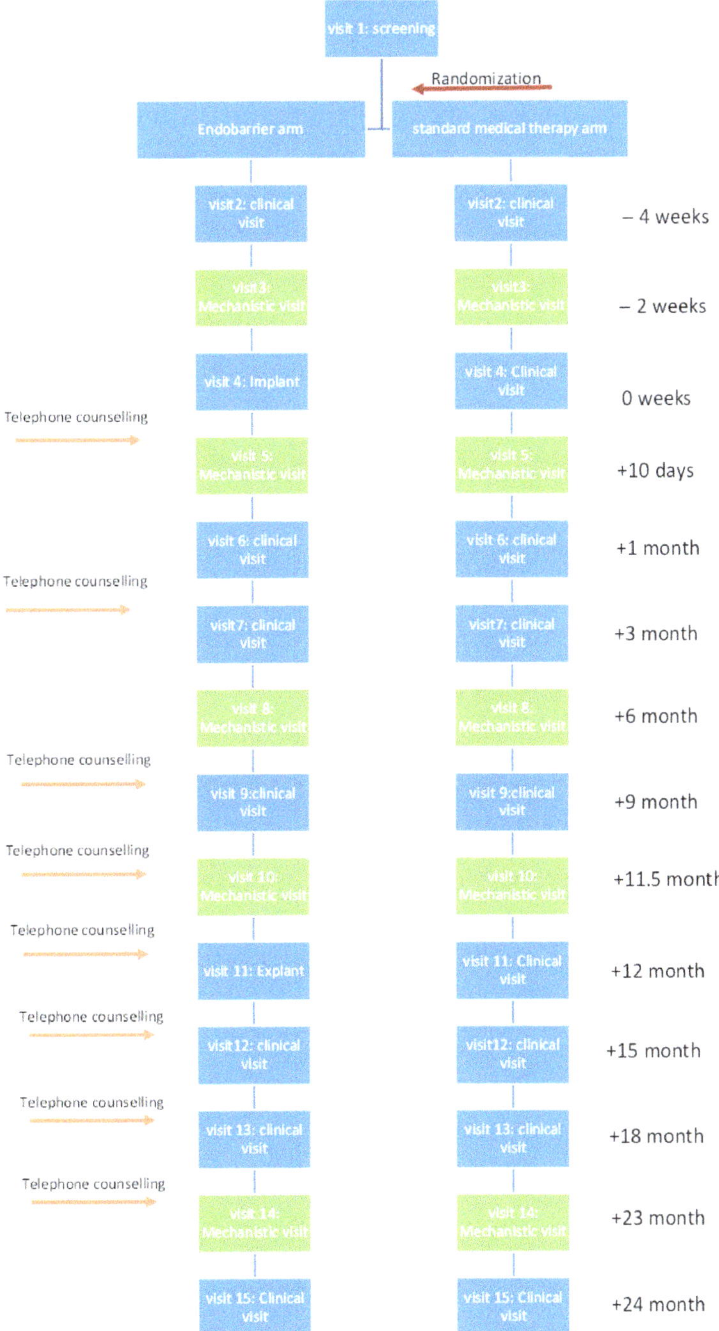

Figure 1. Trial design.

2.2. Dietary and Physical Activity Counselling

All participants' dietary history and current dietary behaviour were assessed at baseline. A qualified dietitian counselled participants regarding their diet and physical activity. The dietary counselling programme was intended to provide each participant with lifestyle and behavioural modification information and impart good eating practices. Guidelines for daily total requirements were between 1200 and 1500 kilocalories for women and between 1500 and 1800 kilocalories for men. Participants were advised to eat regularly every day (five times/day), to control their portion sizes, to increase their intake of low glycaemic index (GI) and high-protein foods, and to reduce their intake of foods high in fat, sugar, and alcohol.

Participants in both groups were advised to include more physical activity in their daily routine, like walking more every day and climbing the stairs instead of taking the lift or escalators. They were also asked to start with short periods of low-intensity exercise and increase the intensity and duration slowly. Their goal was to include 150 min/week of moderate-intensity and 75 min/week of vigorous-intensity aerobic activity and muscle-strengthening activities more than two days a week.

2.3. Liquid Diet

All participants followed a liquid diet during the seven days before and 13 days (±3 days) after the DJBL insertion, or the fourth clinical visit for the control group. The liquid diet was based on a liquid meal replacement—Fortisip compact® meal replacements (Nutricia Ltd., Trowbridge, UK): four bottles (125 mL each, energy: 300 kcal; carbohydrates: 49%; fat: 35%; protein: 16%) for women and five bottles for men daily. Allowed in addition to this were: milk, flavoured milk, water, low-sugar squashes, vegetable juices, tea or coffee without sugar, unsweetened puree fruit juice, or clear soups. After the liquid diet, participants in both groups were advised to follow a low-calorie diet.

2.4. Anthropometric Measurements

Weight was measured at all visits, in bare feet, and wearing light clothes. Height without shoes was recorded at the baseline visit. Body mass index (BMI) was calculated. Percentage of body composition (fat mass, fat-free mass in kg and %) were obtained using a bio-electrical impedance analysis machine MC-780MA (TANITA Corporation, Japan).

2.5. Food Intake and Macronutrient Selection

Participants were asked to complete a weighed food diary for 3 days, 2 weekdays and 1 weekend at baseline (2 weeks before intervention), 10 days, 6 months, 12 months, and 24 months post-intervention. Information from the diaries was entered and analysed using Dietplan7 software (Forestfield Software Ltd. West Sussex, UK) to obtain total daily caloric intake and percentage contribution from carbohydrates, protein, and fat.

2.6. Assessment of Taste Function

2.6.1. Sensory Domain of Sweet Taste

The detection threshold for sweet taste was measured using the method of constant stimuli [9]. In brief, 112 polystyrene cups were presented in 8 blocks; each block consists of 14 cups, including 7 concentrations of sucrose and 7 water stimuli randomly organised. An amount of 15 mL of the sucrose solution and water was presented in each cup. Participants were asked to taste the solution, swirl it around properly and expel it without swallowing. Then they were asked to describe the quality of the solution they were testing, if it was sweet or water. After each stimulus participants were asked to rinse their mouth with water (the same water used to prepare the solutions) before tasting the next stimulus. Participant answers were recorded on a template scoring sheet. The detection test was performed on the morning of the study day after an overnight fast. All solutions were prepared using water (Caledonian Still Natural Mineral Water, Sainsbury's Supermarkets Ltd., London, UK) and sucrose (Sigma-Aldrich, Dorset, UK) and presented at room temperature. Seven sucrose concentrations were used (2.1, 6.25, 12.5, 25, 50, 100, and 300 mM). All participants performed the above-described test at baseline (2 weeks before intervention), at 10 days, and at 6 months. Participants were asked to come to the research facility at 8 am after over-night fast. Each visit had a different random arrangement of the cups to minimise learning and familiarisation.

The data collected from the sucrose detection test allows for the derivation of a psychometric function, which is a mathematical equation that plots the performance of participants against the physical aspect (concentration) of the stimulus. The performance was measured as a percentage of correct responses (responses where the participant was able to detect the stimulus correctly).

A 'hit' was defined as when the participant correctly reported that the stimulus was different from water when sucrose was presented. A 'false alarm' (FA) was defined when the participant incorrectly reported that the stimulus was different from water when water was presented. The hit rate for a given sucrose concentration was adjusted for the false alarm rate to derive a 'corrected hit rate' using the following equation:

$$Corrected\ hit\ rate = \frac{P(hit) - P(FA)}{1.0 - P(FA)}$$

where $P(hit)$ = the proportion of sucrose trials (cups) of a given concentration that were hit, and $P(FA)$ = the proportion of water trials that were false alarms. Thus, when the uncorrected hit rate is equal to the false alarm rate, the corrected hit rate = 0.

Concentration–response curves were fitted to the corrected hit rate values for each participant for the three tested occasions (2 weeks pre, 10-days and 6 months post-intervention) to derive a family of individual psychometric functions using the following logistic equation:

$$f(x) = \frac{a}{1 + 10^{((\log 10(x) - c)*b)}}$$

where $\log 10(x) = \log 10$ concentration, a = the upper asymptote of performance (maximum performance = 1), b = slope, and c = the log10 concentration at $1/2$ a performance (i.e., EC50, defined as half-maximal effective concentration). We defined the c parameter as the threshold because it represents the inflexion point of the psychometric function and thus optimally represents horizontal shifts in the sensitivity.

Only c-values of the individual curve fits for the participants who had fits that accounted for at least 85% of the variance were compared. C-values were calculated using Mystat® (Systat® 12) software (Cranes Software International Ltd., Palo Alto, CA, USA). The shifts in the c parameters between groups and within groups were assessed.

2.6.2. Appetitive Reward Domain of Taste Function

The appetitive reward value of sweet/fat taste was measured using the validated method of the progressive ratio task [10]. In brief, this is a computer task in which participants were seated in front of a screen with a plate of 20 chocolate candies (M&M® crispy candies, Mars UK Limited, Slough, UK), each one containing approximately 4 kcal (energy contribution: 43.7% sugars, 44.1% fat). They were asked to click on the mouse button continuously until they received a message on the screen, allowing them to consume their reward (one M&M's only). The required number of clicks increased progressively after each reward (candy). The first ratio was ten clicks with a geometric increase of two (i.e., 10, 20, 40, 80, etc.) for every ratio afterwards. Participants were allowed to terminate the task at any point by pressing the spacebar button on the keyboard. This test was carried out on two occasions, two weeks pre- and six months post-intervention. Testing occurred 3 h after consuming a standardised meal of 250 mL of Fortisip Compacts vanilla flavour, (Energy: 600 kcal, carbohydrates: 74.2 g, fat: 23.2 g, protein: 24 g). The total number of clicks and clicks in the last completed ratio (breakpoint) were recorded. In addition, the number of consumed and remaining candies were calculated from the plate after the termination of the task to cross-check and validate the participants followed the instructions. Comparisons between groups and within groups were assessed.

2.6.3. Consummatory Reward Domain of Taste Function

The consummatory reward value of sweet taste was measured using a validated methodology [9]. In brief, 30 polystyrene cups were presented in 3 blocks; each block consists of 10 cups, comprising 5 cups of the 5 different sucrose concentrations and 5 cups of water for rinsing after sweet solutions. Odd number cups contained the sucrose solutions, and even number cups contained the rinsing water. An amount of 15 mL of the sucrose solution and water was presented in each cup. All solutions were stored at 4 °C and presented cold for testing. Solutions were prepared using still water (Caledonian Still Natural Mineral Water, Sainsbury's Supermarkets Ltd., London, UK: pH 7.4, calcium 27 mg/L, chloride 6.4 mg/L, bicarbonate 103 mg/L, magnesium 6.9 mg/L, sulphate 10.6 mg/L, sodium 6.6 mg/L). Sucrose was from (Sigma-Aldrich, Dorset, UK), five different concentrations of sucrose were used (0, 50, 100, 200, 400 mM). Two different visual analogue scales were used to assess the liking of the sweet drinks as follows:

The Hedonic General Labeled Magnitude Scale: This visual analogue scale was used to rate the pleasantness of the sweetness of the solution relative to any liking feeling they had ever experienced. This is a vertical scale with the middle anchor representing the ideal rating ('Neutral') with a value of zero (0), and measurements of the most positive ('Strongest liking of any kind') representing the highest value of +100, and most negative rating ('Strongest disliking of any kind') with the least value of −100 located at the lowest end of the scale.

The 'Just About Right' scale: This visual analogue scale was used to compare the sweetness of the solution as compared to the ideal sweetness of the participant's preferred soft drink. This was a vertical visual analogue scale, having a middle point where the ideal rating was situated *('Just right: My ideal sweetness in a drink')* which corresponded to the value of zero (0), while the upper end of the scale measured the most positive *('Far too sweet: I would never drink it)* corresponded to a value of +100, and the most negative rating *('Far too little sweetness: I would never drink it')* which corresponded to a value of −100, and this was at the lower end of the scale.

All participants performed the above-described test on three occasions: 2 weeks pre-intervention, 10 days, and 6 months post-intervention. Participants completed this test after the sensory domain task and still in the fasting state. Each visit had a different random arrangement of the cups to minimise learning and familiarisation.

2.7. Statistical Analyses

The mixed model analysis was used to investigate the treatment effect on the variables of interest over time, allowing us to perform both between-groups and within-group comparisons. The model included fixed effects for the visit (time of assessment), group (DJBL or control) and their corresponding interaction (group×visit), as well as a random intercept effect for each patient. The model was adapted to include a third level where appropriate (for example, sucrose concentrations).

All participants who attended baseline and at least one visit were included in the analysis. Analysis results are presented in the form of Type-III test results of fixed effects (*p*-values) and their subsequent estimates (mean ± SD). Any parameter that produced a significant result ($p < 0.05$) in the analysis was considered for post-hoc testing of least-square means to investigate any potential effect in more detail. The Pearson test was used for linear regressions. Statistical analysis was completed using IBM statistics SPSS 24, and graphs were generated using GraphPad Prism version 8.

The trial was approved by the Fulham Research Ethics Committee on 10 July 2014 (reference 14/LO/0871) and conducted in accordance with the Declaration of Helsinki.

3. Results

3.1. Baseline Characteristics

Forty-seven participants took part in this study, 27 in the Endobarrier group and 20 in the control group, 55% of the participants were male (Table 1). Within the control group, 1 participant withdrew from the study at visit 5 (10 days after intervention), 4 participants withdrew at visit 8 (6 months after intervention). Within the Endobarrier group, 2 participants withdrew from the study at visit 5 (10 days after intervention), 5 participants withdrew at visit 8 (6 months after intervention).

Table 1. Baseline characteristics of participants.

	Control ($n = 20$)	Endobarrier ($n = 27$)
Age (years)	54 ± 6	52 ± 8
Sex (M/F)	8/12	18/9
Weight (kg)	101.3 ± 14.4	109.4 ± 18.9
BMI (kg/m^2)	36 ± 4	36 ± 5
Bio-impedance body fat (%)	42 ± 8	39 ± 7
HbA1c (mmol/mol)	70 ± 12	76 ± 11
Diabetes duration (years)	7 (1–25)	8 (2–19)
HOMA-IR	5.43 ± 3.6	5.36 ± 1.8

Data are presented as mean $\pm$ SD or median (range).

3.2. Body Weight

There was a significant reduction in total body weight within each group at 10 days, 6 and 12 months compared to baseline ($p < 0.001$) but no significant differences between groups. At 12 months the Endobarrier group lost $11 \pm 5\%$ total body weight vs. $8 \pm 8\%$ in the control group, while at 24 months, the Endobarrier group lost $4 \pm 5\%$ vs. $7 \pm 7\%$ in the control group (($p < 0.02$, Figure 2).

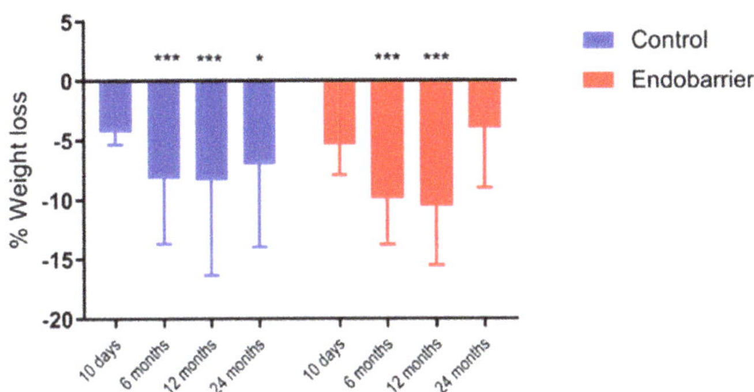

Figure 2. Percentage weight loss throughout the study. * $p < 0.05$, *** $p < 0.001$ compared to baseline within the same group. Data given as mean $\pm$ SD.

3.3. Food (Energy) Intake and Macronutrient Selection

Total daily caloric intake from the three-day food diary was significantly reduced within both groups at all time points compared to baseline, but there were no significant differences between the groups (Table 2).

Within the Endobarrier group, there was a significant increase in the % contribution from carbohydrates at 10 days, a significant increase in the % contribution from protein at 12 months, and a significant decrease in the % contribution from fat at 10 days. Within the control group, there was a significant increase in the % contribution from carbohydrates at

10 days, a significant decrease in the % contribution from protein at 10 days and an increase at 6 months. However, there were no significant differences between groups (Table 3).

Table 2. Total daily caloric intake (kcal).

		Group						Mixed Model Analysis	
		Control				Endobarrier		Effect	*p*-Value
	n				*n*				
Baseline	17	1740	±	285	24	1911	±	506	
10 Days	17	1194	±	203 ***	22	1097	±	407 ***	Group	0.51
6 months	14	1443	±	321 *	16	1575	±	410 **	Time	<0.001
12 months	13	1504	±	470 *	13	1423	±	647 ***	Group × Time	0.25
24 months	14	1525	±	494	12	1788	±	761	

Results presented as mean ± SD. * $p < 0.05$ ** $p < 0.01$ *** $p < 0.001$ compared to baseline within the same group.

Table 3. Percentage contribution of macronutrient to daily energy intake.

		Group						Mixed Model Analysis	
		Control				Endobarrier		Effect	*p*-Value
		Carbohydrates (% of Total Caloric Intake)							
	n								
Baseline	17	40	±	8	24	40	±	7	
10 Days	17	47	±	2 **	22	46	±	6 **	Group	0.83
6 months	14	39	±	9	16	41	±	7	Time	<0.001
12 months	13	41	±	9	13	37	±	8	Group × Time	0.50
24 months	14	40	±	7	12	42	±	7	
		Protein (% of Total Caloric Intake)							
Baseline	17	19	±	4	24	19	±	5	
10 Days	17	16	±	1 *	22	19	±	6	Group	0.89
6 months	14	24	±	5 **	16	21	±	6	Time	<0.001
12 months	13	21	±	4	13	22	±	7 *	Group × Time	0.05
24 months	14	22	±	7	12	19	±	5	
		Fat (% of Total Caloric Intake)							
Baseline	17	38	±	7	24	38	±	6	
10 Days	17	37	±	2	22	35	±	4 *	Group	0.90
6 months	14	36	±	10	16	36	±	7	Time	0.50
12 months	13	36	±	9	13	38	±	7	Group × Time	0.60
24 months	14	37	±	8	12	36	±	7	

Results presented as mean ± SD. * $p < 0.05$ compared to baseline within the same group. ** $p < 0.01$ compared to baseline within the same group.

3.4. Sensory Domain of Sweet Taste

There was no significant change in the curves of mean corrected hit rate both within and between groups at baseline, 10 days, and 6 months post intervention (Figure 3).

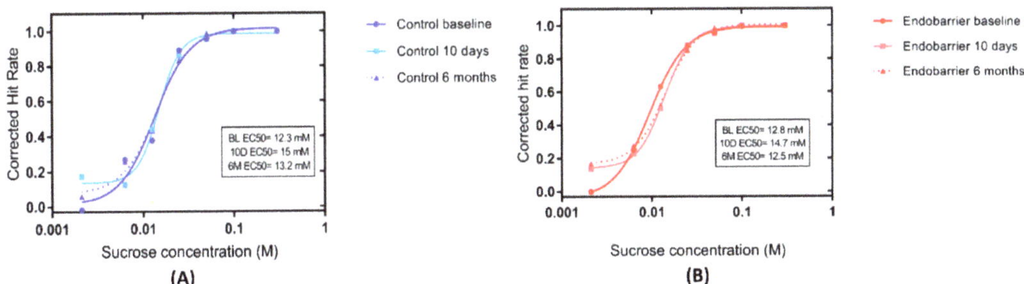

Figure 3. Sweet taste detection. Curves of the mean corrected hit rate over time for (**A**) controls (blue) $n = 16$ and (**B**) Endobarrier (red) $n = 25$ groups as a function of sucrose concentration. The EC50 was derived from the c-parameter in the curve fit and represented the concentration at which the corrected hit rate reaches 50% of the maximum asymptote.

3.5. Appetitive Reward Value of Sweet Taste

There was no significant change in the breakpoint either within or between groups ($p = 0.12$ for group $\times$ time interaction) (Figure 4).

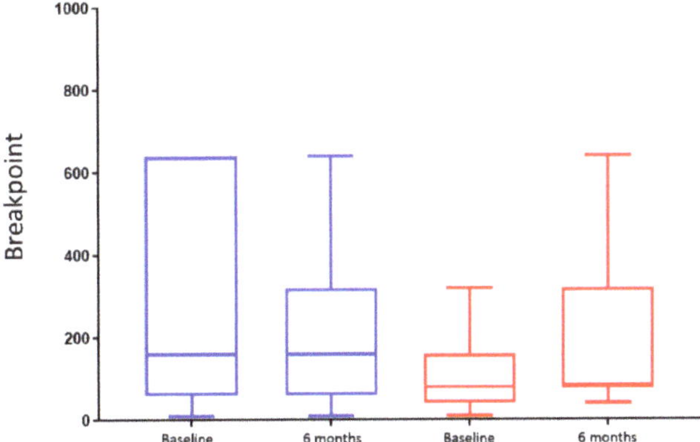

Figure 4. Breakpoint at the progressive ratio task. Box plot of the breakpoint for chocolate candies in control (blue) $n = 9$ and Endobarrier (red) $n = 11$ groups. The lower and upper boundaries of the box represent 25th and 75th percentiles, respectively. Lower and upper whiskers represent 10th and 90th percentiles, respectively. The line in the middle of the box represents the median.

3.6. Consummatory Reward Value of Sweet Taste

There was no significant change in the consummatory reward value of sweet taste both within and between groups using Just About Right scale (Figure 5) and Hedonic General Labeled Magnitude Scale (Figure 6) at baseline, 10 days, and 6 months post intervention.

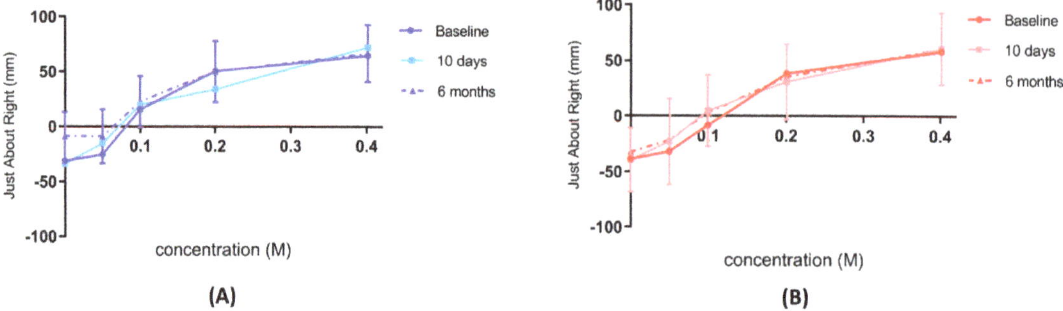

Figure 5. Just About Right scale ratings of sweet taste. Consummatory reward value of sweet taste assessed by Just About Right scale for (**A**) controls *n* = 19 (blue) and (**B**) Endobarrier *n* = 24 (red groups). Data are presented as the mean rating at each concentration ± SD.

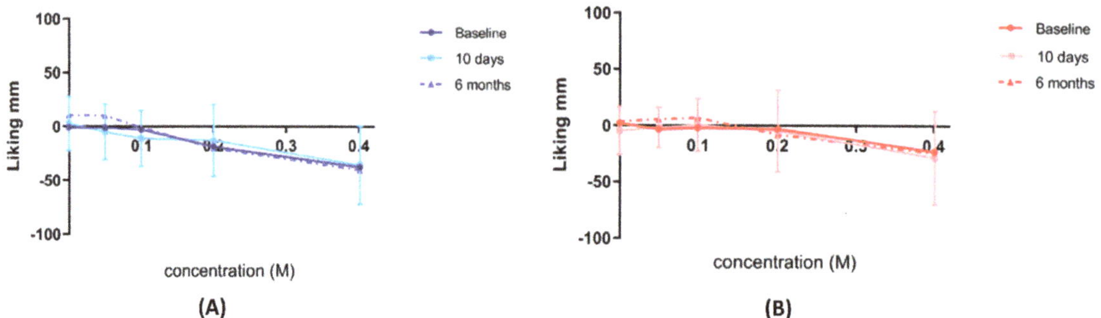

Figure 6. Hedonic general Labeled Magnitude Scale ratings of sweet taste. Consummatory reward value of sweet taste assessed by the Hedonic general Labeled Magnitude Scale for (**A**) control *n* = 19 (blue) and (**B**) Endobarrier *n* = 24 (red) groups. Data are presented as the mean rating at each concentration ± SD.

4. Discussion

This is, to our knowledge, the first experimental medicine study to assess the mechanisms of action of the Endobarrier device on weight loss as a nested study within an RCT. Patients in the Endobarrier group lost numerically more weight than the control group. We assessed several measures of dietary behaviour and identified significant changes on specific aspects within groups but no significant differences between groups.

To date, there has been limited literature on the effect of Endobarrier on food intake. Using food diaries, our results demonstrated similarly reduced food intake within both the Endobarrier and the control groups. Similarly, a recent case series study of patients with obesity and T2DM demonstrated reduced food intake at 36 weeks after the Endobarrier implant using a semi-quantitative Food Frequency Questionnaire [11]. In contrast, a prospective observational study of two groups, a group of patients with obesity and normal glucose-tolerance, and another group of matched metformin-treated patients T2DM who underwent Endobarrier implant, demonstrated lower food intake only at one week [12]. This was followed by a return to baseline food intake at explantation (26 weeks), despite ongoing weight loss. This was the only human study so far to use an ad libitum meal to assess food intake. Of note, this study also did not include a control group for comparison [12]. Among animal models, a study comparing food intake between diet-induced obese rats after endoluminal sleeve insertion and sham-operated controls showed reduced food intake in the sleeve group compared to no change in the control group at eight weeks [13].

Alternative mechanisms of weight loss after Endobarrier have been proposed including increased energy expenditure in both human and animal models. We cannot exclude that the numerically superior weight loss observed among our Endobarrier group might be attributed to an increase in energy expenditure. Rohde et al. reported an increase in resting energy expenditure using indirect calorimetry in patients with obesity but not among patients with T2DM after Endobarrier implant [12]. Similarly, Munoz et al. in their animal model, demonstrated an increase of 13% in total and 9% in resting energy expenditure among Endoluminal sleeve treated rats compared to shams [13].

Nutrient malabsorption has been proposed as a possible mechanism of weight loss after the Endobarrier, due to the bypass of 60 cm of the small intestine. However, when fat malabsorption was measured using ^{13}C mixed triglyceride breath test in patients with obesity and T2DM, there was no evidence of reduced intraluminal lipolytic activity suggesting that fat malabsorption does not take place [14]. Similarly, no evidence of food malabsorption was found in rats treated with an endoluminal sleeve, as measured by the difference in calories consumed and excreted in the stool using direct calorimetry [13].

Another plausible mechanism that could explain the weight loss in the Endobarrier group is gut inflammation. The insertion of a foreign body in the intestine could have triggered a low-grade inflammatory state. Gut inflammation can cause weight loss due to several mechanisms including increased resting energy expenditure, and the action of proinflammatory cytokines [15]. Against this hypothesis is the fact that we measured plasma concentrations of C-reactive protein in the main clinical RCT [6] and were not found to be elevated after intervention. We would also have expected gut inflammation to decrease appetite and thus total daily energy intake, but we did not observe this in our study. In this report we do not present appetite ratings or gut hormone measurements. Whilst enhanced post-prandial concentrations of plasma GLP-1 and PYY were reported in some studies [16–18], the magnitude of the increase was modest and the findings inconsistent [16,19].

Weight regain after Endobarrier explant is reported in several studies. Interestingly, in the studies that had a control group, the Endobarrier group had the most weight regain compared to the control group [20]. This was in line with our findings, where the Endobarrier group had around 7% weight regain compared to 1% in the control group. Similarly, Villarasa et al., in their recent prospective trial concluded a total percentage weight loss of about 15% at the time of explantation (48 weeks) followed by weight regain during the next year, maintaining only 7% of the total weight loss; there was no control group in this study [17].This rebound demonstrates that the Endobarrier works only when it is in situ and does not have any long term learning effects on eating behaviour. One explanation for the magnitude of the rebound might be attributed to the absence of abdominal discomfort that patients commonly report, resulting in increased meal size, caloric intake and subsequent weight regain.

The role of the duodenum in food preferences and reward has been investigated in animal models of the DJB procedure, which like the Endobarrier, involves bypass of the proximal small intestine [5]. In line with our results, Qu et al. recently demonstrated that sweet preference was not different between DJB mice and sham-operated mice in a two-bottle sweet preference test [21]. Reduced preference appeared only after prolonged exposure to the sweet solutions indicating a learning effect [21]. Similarly, in their animal model, Zhang et al. demonstrated that DJB mice preferred the flavours of intragastric infusions of metabolised glucose compared to the flavours of non-metabolised glucose [22]. Nevertheless, the surgical duodenal bypass did not affect the ability of mice to differentiate (prefer) between the flavours of metabolised versus non-metabolised glucose solutions [22]. The same study also showed that reward circuits in the brain responded to intra-portal mesenteric infusions of the metabolised glucose only, suggesting a post-absorptive role for glucose preference and reward.

The absence of changes in food preference and taste function are reminiscent of some of the studies in humans and animals undergoing RYGB [23,24]. In one of the most comprehensive studies in the literature, food preferences did not change in a group of patients undergoing RYGB, but the subgroup of patients who experiences changes in food preferences lost more weight [2]. This finding suggests that changes in food preferences do not take place in everyone but in those that do, they contribute to weight loss as an additional mechanism.

The strengths of this study include its randomised design, two trial sites, length of follow-up, multidisciplinary team involved in patients care, and delivery of an intensive medical intervention throughout the study period. In addition, we used complementary measures of eating behaviour, including assessment of food intake, taste detection thresholds, appetitive and consummatory reward value of sweet taste with various sweet concentrations. Despite the length of the study, the same two dietitians carried out dietary analyses throughout the study period to reduce variability. In addition, participants in both groups received the exact behavioural and dietary modification instructions from a single dietitian throughout the study.

The major limitation in our study was its unblinded design. In addition, in the smaller mechanistic sub-set of participants having these dietary assessments, the Endobarrier insertion resulted in numerically superior weight loss, which was not as pronounced as in the main RCT [6]. There are also inherent limitations to using verbal/written reports, especially in a trial that is not double-blinded. The problem of under-reporting of food intake among patients with and without obesity is common when using indirect measures of food intake [25]. It would have been preferable to measure these aspects of eating behaviour using a buffet meal or a 24-h residential stay. The study days were long and included numerous tasks which might have contributed to participant fatigue, which could have been avoided if the tasks had been performed on separate days. Only sweet taste assessments were made and not fat or combined sweet/fat (other than the progressive ratio task). Assessments were generally performed in the fasted state and may have been different in the post-prandial state, except for the progressive ratio task, which was assessed post prandially Furthermore, sample sizes declined over time due to drop-out during the trial. Finally, we did not measure energy expenditure or calorie malabsorption as alternative mechanisms causing weight loss after the Endobarrier.

5. Conclusions

In conclusion, this experimental medicine study demonstrated that reduction of self-reported energy intake, changes in food preferences, and sweet taste were not the mechanisms underlying the weight loss observed after Endobarrier insertion in people with obesity and T2DM.

Author Contributions: Conceptualization, W.A.-N., A.R., M.A.G., C.G.P., J.P.B., A.P.G., J.P.T., C.W.L.R. and A.D.M.; Data curation, M.M.A., W.A.-N., N.A.J. and J.P.B.; Formal analysis, M.M.A., W.A.-N., N.A.J., A.P.G., J.P.T., C.W.L.R. and A.D.M.; Funding acquisition, C.G.P., A.P.G. and J.P.T.; Investigation, M.M.A., W.A.-N., A.R., M.A.G., N.C., G.K.D., C.G.P., J.P.B., A.P.G., J.P.T., C.W.L.R. and A.D.M.; Methodology, W.A.-N., N.C., C.G.P., N.A.J., J.P.B., A.P.G., J.P.T., C.W.L.R. and A.D.M.; Project administration, M.M.A., W.A.-N., A.R., M.A.G., B.J., C.G.P., A.P.G., J.P.T., C.W.L.R. and A.D.M.; Resources, C.G.P., A.P.G. and J.P.T.; Software, C.G.P. and A.P.G.; Supervision, C.G.P., J.P.B., A.P.G., J.P.T., C.W.L.R. and A.D.M.; Visualization, M.M.A.; Writing—original draft, M.M.A., A.P.G., J.P.T., C.W.L.R. and A.D.M.; Writing—review and editing, M.M.A., W.A.-N., A.R., M.A.G., B.J., N.C., G.K.D., C.G.P., N.A.J., J.P.B., A.P.G., J.P.T., C.W.L.R. and A.D.M. All authors have read and agreed to the published version of the manuscript.

Funding: This study was funded by the Efficacy and Mechanism Evaluation Programme, a Medical Research Council and National Institute for Health Research (NIHR) partnership reference 12/10/04. C.W.R is funded by the Irish Research Council (IRCLA/2017/234) and The Health Research Board (USIRL-2016-2). A.D.M. has been supported from grants from the JP Moulton Charitable Foundation, National Institute of Health Research, Imperial College Healthcare Charity and Novo Nordisk; N.C. has been supported by grants from Wellcome Trust. The Division of Diabetes, Endocrinology and Metabolism is funded by grants from the UK Medical Research Council (MRC), Biotechnology and

Biological Sciences Research Council, and the NIHR; an Integrative Mammalian Biology Capacity Building Award; and an FP7- HEALTH-2009–241592 EuroCHIP grant. It is also supported by the NIHR Biomedical Research Center Funding Scheme. M.A. was funded by the Department of Community Health Sciences, College of Applied Medical Sciences, King Saud University, Riyadh, Saudi Arabia.

Institutional Review Board Statement: The trial was approved by the Fulham Research Ethics Committee on 10 July 2014 (reference 14/LO/0871) and conducted in accordance with the Declaration of Helsinki.

Informed Consent Statement: Informed consent was obtained from all subjects involved in the study.

Data Availability Statement: The data presented in this study are available on request from the corresponding author. The data are not publicly available due to privacy.

Acknowledgments: The devices and nutritional supplements were kindly provided free of charge by GI Dynamics and Nutricia Advanced Medical Nutrition respectively. Infrastructure support was provided by the NIHR Imperial Biomedical Research Center, NIHR Imperial and Clinical Research Facility, London, NIHR Southampton Biomedical Research Center UK and NIHR Southampton clinical research facility, Southampton, UK.

Conflicts of Interest: A.R. received travel fees support from GI Dynamics. A.D.M. has received honoraria for presentations and advisory board contribution by Novo Nordisk, Boehringer Ingelheim, AstraZeneca, Johnson & Johnson and research grant funding from Fractyl. A.P.G. reports funding supported by UK Medical Research Council and Wellcome Trust, outside of the submitted work, was on a Data Safety Monitoring Board for Novo Nordisk, and has received honoraria for presentations and advisory board contribution by Janssen, Pfizer, Novo Nordisk, Zafgen, Soleno Therapeutics Inc, and Millendo Theapeutics Inc, and Merck. C.W.R. is a member of scientific advisory board for Herbalife, GI Dynamics, NovoNordisk, Keyron, Sanofi, has provided ad hoc consulting for Ethicon and Fractyl, occasional speaking engagement for MSD, Boehringer Ingelheim and Lilly. J.P.T. received travel fees support from GI Dynamics. W.A. has received honoraria for presentations and educational grants from Novo Nordisk. The rest of the authors report no conflicts of interest. "The funders had no role in the design of the study; in the collection, analyses, or interpretation of data; in the writing of the manuscript, or in the decision to publish the results".

References

1. Kapoor, N.; al Najim, W.; Menezes, C.; Price, R.K.; O'Boyle, C.; Bodnar, Z.; Spector, A.C.; Docherty, N.G.; le Roux, C.W. A Comparison of Total Food Intake at a Personalised Buffet in People with Obesity, before and 24 Months after Roux-en-Y-Gastric Bypass Surgery. *Nutrients* **2021**, *13*, 3873. [CrossRef] [PubMed]
2. Nielsen, M.S.; Rasmussen, S.; Christensen, B.J.; Ritz, C.; le Roux, C.W.; Schmidt, J.B.; Sjödin, A. Bariatric Surgery Does Not Affect Food Preferences, but Individual Changes in Food Preferences May Predict Weight Loss. *Obesity* **2018**, *26*, 1879–1887. [CrossRef] [PubMed]
3. Miller, G.D.; Norris, A.; Fernandez, A. Changes in Nutrients and Food Groups Intake Following Laparoscopic Roux-en-Y Gastric Bypass (RYGB). *Obes. Surg.* **2014**, *24*, 1926–1932. [CrossRef] [PubMed]
4. Ullrich, J.; Ernst, B.; Wilms, B.; Thurnheer, M.; Schultes, B. Roux-en Y Gastric Bypass Surgery Reduces Hedonic Hunger and Improves Dietary Habits in Severely Obese Subjects. *Obes. Surg.* **2012**, *23*, 50–55. [CrossRef]
5. Han, W.; Tellez, L.A.; Niu, J.; Medina, S.; Ferreira, T.; Zhang, X.; Su, J.; Tong, J.; Schwartz, G.J.; Pol, A.V.D.; et al. Striatal Dopamine Links Gastrointestinal Rerouting to Altered Sweet Appetite. *Cell Metab.* **2015**, *23*, 103–112. [CrossRef]
6. Ruban, A.; Miras, A.D.; Glaysher, M.A.; Goldstone, A.P.; Prechtl, C.G.; Johnson, N.; Chhina, N.; Al-Najim, W.; Aldhwayan, M.; Klimowska-Nassar, N.; et al. Duodenal-Jejunal Bypass Liner for the management of Type 2 Diabetes Mellitus and Obesity: A Multicenter Randomized Controlled Trial. *Ann. Surg.* **2021**, *275*, 440. [CrossRef]
7. Glaysher, M.A.; Mohanaruban, A.; Prechtl, C.G.; Goldstone, A.P.; Miras, A.; Lord, J.; Chhina, N.; Falaschetti, E.; Johnson, N.A.; Al-Najim, W.; et al. A randomised controlled trial of a duodenal-jejunal bypass sleeve device (EndoBarrier) compared with standard medical therapy for the management of obese subjects with type 2 diabetes mellitus. *BMJ Open* **2017**, *7*, e018598. [CrossRef]
8. Ruban, A.; Ashrafian, H.; Teare, J.P. The EndoBarrier: Duodenal-Jejunal Bypass Liner for Diabetes and Weight Loss. *Gastroenterol. Res. Pract.* **2018**, *2018*, 7823182. [CrossRef]
9. Bueter, M.; Miras, A.D.; Chichger, H.; Fenske, W.; Ghatei, M.A.; Bloom, S.R.; Unwin, R.J.; Lutz, T.; Spector, A.C.; le Roux, C.W. Alterations of sucrose preference after Roux-en-Y gastric bypass. *Physiol. Behav.* **2011**, *104*, 709–721. [CrossRef]

10. Miras, A.D.; Jackson, R.N.; Jackson, S.N.; Goldstone, A.P.; Olbers, T.; Hackenberg, T.; Spector, A.C.; le Roux, C.W. Gastric bypass surgery for obesity decreases the reward value of a sweet-fat stimulus as assessed in a progressive ratio task. *Am. J. Clin. Nutr.* **2012**, *96*, 467–473. [CrossRef]

11. Obermayer, A.; Tripolt, N.; Aziz, F.; Högenauer, C.; Aberer, F.; Schreiber, F.; Eherer, A.; Sourij, C.; Stadlbauer, V.; Svehlikova, E.; et al. EndoBarrier™ Implantation Rapidly Improves Insulin Sensitivity in Obese Individuals with Type 2 Diabetes Mellitus. *Biomolecules* **2021**, *11*, 574. [CrossRef] [PubMed]

12. Rohde, U.; Bsc, C.A.F.; Vilmann, P.; Langholz, E.; Friis, S.U.; Krakauer, M.; Rehfeld, J.F.; Holst, J.J.; Vilsbøll, T.; Knop, F.K. The impact of EndoBarrier gastrointestinal liner in obese patients with normal glucose tolerance and in patients with type 2 diabetes. *Diabetes Obes. Metab.* **2016**, *19*, 189–199. [CrossRef] [PubMed]

13. Muñoz, R.; Carmody, J.S.; Stylopoulos, N.; Davis, P.; Kaplan, L.M. Isolated duodenal exclusion increases energy expenditure and improves glucose homeostasis in diet-induced obese rats. *Am. J. Physiol. Integr. Comp. Physiol.* **2012**, *303*, R985–R993. [CrossRef] [PubMed]

14. McMaster, J.J.; Rich, G.G.; Shanahan, E.R.; Do, A.T.; Fletcher, L.M.; Kutyla, M.J.; Tallis, C.; Jones, M.P.; Talley, N.J.; Macdonald, G.A.; et al. Induction of Meal-related Symptoms as a Novel Mechanism of Action of the Duodenal-Jejunal Bypass Sleeve. *J. Clin. Gastroenterol.* **2020**, *54*, 528–535. [CrossRef] [PubMed]

15. Elsherif, Y.; Alexakis, C.; Mendall, M. *Determinants of Weight Loss Prior to Diagnosis in Inflammatory Bowel Disease: A Retrospective Observational Study*; Gastroenterology Research and Practice: London, UK, 2014; p. 762191.

16. Jirapinyo, P.; Haas, A.V.; Thompson, C.C. Effect of the Duodenal-Jejunal Bypass Liner on Glycemic Control in Patients With Type 2 Diabetes With Obesity: A Meta-analysis With Secondary Analysis on Weight Loss and Hormonal Changes. *Diabetes Care* **2018**, *41*, 1106–1115. [CrossRef] [PubMed]

17. Vilarrasa, N.; de Gordejuela, A.G.R.; Casajoana, A.; Duran, X.; Toro, S.; Espinet, E.; Galvao, M.; Vendrell, J.; López-Urdiales, R.; Pérez, M.; et al. Endobarrier(R) in Grade I Obese Patients with Long-Standing Type 2 Diabetes: Role of Gastrointestinal Hormones in Glucose Metabolism. *Obes. Surg.* **2017**, *27*, 569–577. [CrossRef]

18. de Jonge, C.; Rensen, S.S.; Verdam, F.J.; Vincent, R.P.; Bloom, S.R.; Buurman, W.A.; le Roux, C.W.; Schaper, N.C.; Bouvy, N.D.; Greve, J.W.M. Endoscopic Duodenal–Jejunal Bypass Liner Rapidly Improves Type 2 Diabetes. *Obes. Surg.* **2013**, *23*, 1354–1360. [CrossRef]

19. Kaválková, P.; Mraz, M.; Trachta, P.; Klouckova, J.; Cinkajzlová, A.; Lacinová, Z.; Haluzíková, D.; Beneš, M.; Vlasáková, Z.; Burda, V.; et al. Endocrine effects of duodenal–jejunal exclusion in obese patients with type 2 diabetes mellitus. *J. Endocrinol.* **2016**, *231*, 11–22. [CrossRef]

20. Koehestanie, P.; de Jonge, C.; Berends, F.; Janssen, I.M.; Bouvy, N.D.; Greve, J.W.M. The Effect of the Endoscopic Duodenal-Jejunal Bypass Liner on Obesity and Type 2 Diabetes Mellitus, a Multicenter Randomized Controlled Trial. *Ann. Surg.* **2014**, *260*, 984–992. [CrossRef]

21. Qu, T.; Han, W.; Niu, J.; Tong, J.; de Araujo, I.E. On the roles of the Duodenum and the Vagus nerve in learned nutrient preferences. *Appetite* **2019**, *139*, 145–151. [CrossRef]

22. Zhang, L.; Han, W.; Lin, C.; Li, F.; De Araujo, I.E. Sugar Metabolism Regulates Flavor Preferences and Portal Glucose Sensing. *Front. Integr. Neurosci.* **2018**, *12*, 57. [CrossRef] [PubMed]

23. Moizé, V.; Andreu, A.; Flores, L.; Torres, F.; Ibarzabal, A.; Delgado, S.; Lacy, A.; Rodriguez, L.; Vidal, J. Long-Term Dietary Intake and Nutritional Deficiencies following Sleeve Gastrectomy or Roux-En-Y Gastric Bypass in a Mediterranean Population. *J. Acad. Nutr. Diet.* **2013**, *113*, 400–410. [CrossRef] [PubMed]

24. Vinolas, H.; Barnetche, T.; Ferrandi, G.; Monsaingeon-Henry, M.; Pupier, E.; Collet, D.; Gronnier, C.; Gatta-Cherifi, B. Oral Hydration, Food Intake, and Nutritional Status Before and After Bariatric Surgery. *Obes. Surg.* **2019**, *29*, 2896–2903. [CrossRef] [PubMed]

25. Hill, R.; Davies, P.S.W. The validity of self-reported energy intake as determined using the doubly labelled water technique. *Br. J. Nutr.* **2001**, *85*, 415–430. [CrossRef] [PubMed]

Article

Are People with Obesity Attracted to Multidisciplinary Telemedicine Approach for Weight Management?

Luisa Gilardini [1,*], Raffaella Cancello [1], Luca Cavaggioni [1], Amalia Bruno [1], Margherita Novelli [1], Sara P. Mambrini [2,3], Gianluca Castelnuovo [4,5] and Simona Bertoli [1,3]

1   Obesity Unit—Laboratory of Nutrition and Obesity Research, Department of Endocrine and Metabolic Diseases, IRCCS Istituto Auxologico Italiano, Via Ariosto 13, 20145 Milan, Italy; r.cancello@auxologico.it (R.C.); cavaggioni.luca@gmail.com (L.C.); am.bruno@auxologico.it (A.B.); m.novelli@auxologico.it (M.N.); simona.bertoli@unimi.it (S.B.)
2   Laboratory of Metabolic Research, S. Giuseppe Hospital, Istituto Auxologico Italiano, IRCCS, 28824 Piancavallo, Italy; s.mambrini@auxologico.it
3   International Center for the Assessment of Nutritional Status (ICANS), Department of Food, Environmental and Nutritional Sciences (DeFENS), University of Milan, 20133 Milan, Italy
4   Psychology Research Laboratory, Istituto Auxologico Italiano IRCCS, 28824 Piancavallo, Italy; gianluca.castelnuovo@auxologico.it
5   Department of Psychology, Catholic University of Milan, 20123 Milan, Italy
*   Correspondence: l.gilardini@auxologico.it; Tel.: +39-026-1911-2561; Fax: +39-026-1911-2541

Abstract: The forced isolation due to the COVID-19 pandemic interrupted the lifestyle intervention programs for people with obesity. This study aimed to assess: (1) the behaviors of subjects with obesity towards medical care during the pandemic and (2) their interest in following a remotely delivered multidisciplinary program for weight loss. An online self-made survey addressed to subjects with obesity was linked to the official website of our institute. Four hundred and six subjects completed the questionnaire (90% females, 50.2 ± 11.6 years). Forty-six percent of the subjects cancelled any scheduled clinical assessments during the pandemic, 53% of whom had chronic disease. Half of the subjects were prone to following a remotely delivered lifestyle intervention, especially with a well-known health professional. About 45% of the respondents were favorable towards participating in remote psychological support and nutritional intervention, while 60% would practice physical activity with online tools. Male subjects and the elderly were more reluctant than those female and younger, especially for online psychological support. Our survey showed an interest on the part of the subjects with obesity to join a multidisciplinary weight loss intervention remotely delivered. Male subjects and the elderly seem less attracted to this intervention, and this result highlights that, even with telemedicine, the approach to weight management should be tailored.

Keywords: obesity; lifestyle intervention; telemedicine; COVID-19 pandemic

Citation: Gilardini, L.; Cancello, R.; Cavaggioni, L.; Bruno, A.; Novelli, M.; Mambrini, S.P.; Castelnuovo, G.; Bertoli, S. Are People with Obesity Attracted to Multidisciplinary Telemedicine Approach for Weight Management? *Nutrients* **2022**, *14*, 1579. https://doi.org/10.3390/nu14081579

Academic Editor: Javier Gómez-Ambrosi

Received: 11 March 2022
Accepted: 8 April 2022
Published: 11 April 2022

Publisher's Note: MDPI stays neutral with regard to jurisdictional claims in published maps and institutional affiliations.

1. Introduction

Coronavirus disease 2019 (COVID-19) is a severe acute respiratory syndrome caused by SARS-CoV-2 that was first discovered in Wuhan, China in December 2019 and rapidly spread to the rest of the world [1]. The infection is highly transmissible, and the number of those infected with COVID-19 has now reached more than 364 million patients and over 5,500,000 deaths.

To contrast and contain the pandemic, at the beginning of March 2020, the Italian Government adopted restriction measurements consisting of a temporary closure of all nonessential activities, strengthening the measures aimed to increase personal hygiene, symptom monitoring, early diagnosis, and patient isolation [2]. The lockdown was repeated in November 2020 and March 2021. All these restrictions required individuals to stay at home, leading to modifications in lifestyles and daily life habits, especially for those in frail categories, such as subjects with obesity. In this context, people are prone in buying large quantities of ultra-processed, unhealthy food to cope with fear, boredom, or anxiety evoked by the worldwide pandemic [3,4]. Moreover, in this difficult situation, individuals with eating disorders may be at a high risk of relapsing or of a worsening of the severity of their disorder [5,6]. Combined with a decrease in the levels of physical activity registered [7], the impaired nutritional habits could lead to weight gain. An Italian study in a small cohort of individuals with obesity showed a significant weight gain 1 month after the beginning of the lockdown period [8]. The increased risk of weight gain during lockdown and the evidence that any degree of obesity has been associated with poor prognosis in patients with SARS-CoV-2 infection [9] have pointed out the importance of ongoing support for obese subjects to manage the disease during the pandemic. Indeed, in Italy, all non-urgent medical visits, including clinical practices planned for obesity management, were deferred to ensure social distancing and reduce the virus spread. Since the resumption of clinical activities after the lockdowns, the fear of contracting the infection in healthcare places, such as hospitals and clinics, has grown in people, and there is still a significant reduction in access to clinical care programs, including metabolic rehabilitation for obesity. The global COVID-19 pandemic has led to a revolution in many fields, promoting alternative strategies with the use of technology. In addition to this, the medical sector has also promoted the usefulness of telehealth and telemedicine, but benefits and barriers in using technology should be considered when dealing with patients [10]. It is worth noticing that telemedicine may represent a novel, effective option in obesity management. In fact, it has previously been demonstrated that video visits with physicians and dieticians can be effective in driving weight loss compared to the standard care [11,12]. IRCCS Istituto Auxologico Italiano (https://www.auxologico.it/, accessed on 10 March 2022) is a specialized national center for obesity care. In Italy, the prevalence of obesity was estimated as 10.9% in 2019, and it is higher in men (11.7%) than in women (10.3%), with increasing prevalence from the north to the south of the country [13]. In our institute, we conduct a 3-month multidisciplinary program aimed at weight loss in obesity-suffering subjects with the involvement of dieticians, physicians, psychologists, and exercise physiologists. The intervention includes individual interviews with the health professionals, nutritional/psychological group sessions, and a one-hour session of moderate intensity physical activity under the supervision of a physical trainer [14]. At the end of the rehabilitation, patients were given an appointment for the regular three months of follow-up visits. During the pandemic, we were forced to stop this rehabilitation program, and we wondered what the barriers were for the patients with obesity to perform the intervention with telemedicine. We conducted a self-reported online survey among the newsletter readers of Istituto Auxologico Italiano from October 2020 to March 2021 to observe the perception of telemedicine of patients with obesity.

The purpose of this survey was to investigate (1) the behaviors of obese subjects towards medical care during the pandemic and (2) their opinion on the possibility of following a remotely delivered program for weight loss, including psychological, nutritional, and physical activity interventions.

2. Materials and Methods

The present study is a cross-sectional design carried out using an online self-made questionnaire (from October 2020 to March 2021) adopting a Google online survey platform (Google LLC, Mountain View, CA, USA). A link to the electronic survey is present on the official website of the IRCCS Istituto Auxologico Italiano (www.auxologico.it, accessed on 10 March 2022) and was shared via the local institute newsletter. Registration to receive the newsletter is open to everyone, but usually, the subscribers are patients who use the services of our institute for healthcare. All participants were requested to provide informed consent through an appropriate checkbox in the survey regarding research purposes. Participants' answers were anonymous, in accordance with Google's privacy policy (https://policies.google.com/privacy?hl=it, accessed on 20 February 2022). Each participant was identified by a progressive anonymous number. The self-made survey included a questionnaire composed by 34 questions broken down into three sections:

(1) personal anonymous data (age, gender, zip code, education, current work, presence of chronic diseases, availability of an IT tool, weight, and height); (2) disease management during the pandemic (we asked if the patient missed scheduled control visits for any health issue, if they contacted the doctor, and by what means); and (3) obesity management during the pandemic and opinion on the use of telemedicine (whether the patient is prone to following a nutritional, psychological, and physical activity program remotely, what is the best modality, and what are the critical issues).

The text of the questionnaire is linked as an annex (Supplementary Materials S1). The inclusion criteria were body mass index values $\geq$ 30 kg/m^2 and ages $\geq$ 18 years. Participants with lower BMI values were excluded by statistical analysis. The participants were asked multiple- or single-choice questions or questions whose answers required the interviewee to enter numeric data. For example: Enter your weight.

The study was conducted in full agreement with the national and international regulations and the Declaration of Helsinki (2000). Participants independently completed an anonymous online questionnaire, explicitly agreeing to participate in the survey. Participants' personal information were made anonymous to maintain and protect confidentiality. The anonymous nature of the web survey did not allow us to trace in any way sensitive personal data. Therefore, the present web survey study did not require approval by ethics committee. Once completed, each questionnaire was transmitted to the Google platform, and the final database was downloaded as a Microsoft Excel sheet.

Statistical Analysis

Continuous variables were expressed as the mean $\pm$ standard deviation (SD) and categorical data as frequencies and proportions. Differences between groups were calculated using the Student's *t*-test for independent samples and analysis of variance for the comparison of multiple groups. Frequencies were compared using a χ2 test. All analyses were performed using SPSS version 26.0 (SPSS Inc., Chicago, IL, USA). A *p*-value < 0.05 was considered statistically significant.

3. Results

In total, 465 subjects completed the entire questionnaire. Fifty-nine subjects were only overweight and were excluded. Table 1 showed the characteristics of 406 obese subjects. The subjects were prevalently female, and the age group with the highest frequency (59.4 percent) was 41–60 years old. The mean BMI was 38.0 $\pm$ 6.1 kg/m^2. Men had more severe degrees of obesity compared to women (40.1 $\pm$ 5.8 vs. 37.8 $\pm$ 5.8 kg/m^2, p < 0.05), had fewer previous diet attempts (more than three diets: 62.5% vs. 83.7%, p < 0.001), and tended to have more obesity-related chronic diseases (52.4% vs. 68.3%, p = 0.05) than women. There were no differences in age and education levels between the sexes. The percentage of subjects with obesity-related chronic diseases increased with age (34% in subjects aged <40 years (y), 56% in subjects aged 41–60 y, and 71% in subjects aged >60 y, p < 0.0001) but not with the degree of obesity. The degree of obesity, gender, and education level were similar across the age groups. Only four subjects (all females) had no electronic tools, three of whom were >65 years. A sedentary behavior was present in 41%. The most declared activity by physically active subjects was "walking".

Table 1. Characteristics of 406 subjects with obesity who answered the questionnaire.

Participant Characteristics	
Age, years	50.2 ± 11.6
Female, %	90
Educational level	
Primary school, %	5.4
Secondary school, %	57
University, %	37.5
BMI, kg/m^2	38.0 ± 6.1
Class of BMI	
Class I (BMI 30–34.9)%	36.3
Class II (BMI 35–39.9)%	29.8
Class III (BMI ≥ 40)%	32.3
Subjects with obesity related chronic diseases, %	54.1
Subjects who practise physical activity, %	59
Subjects with at least one diet attempt, %	96
Subjects with at least one electronic tool, %	99

Data are expressed as mean ± SD or percentage (%).

3.1. Disease Management during Pandemic

The answers given by the participants to the questions about healthcare during pandemic and their opinions on remote visits are summarized in Table 2. Patients who cancelled or postponed a scheduled medical examination (46%) were similar in age, sex, and degree of obesity compared to those who did not but had more frequent chronic diseases (53.2% vs. 37.8%, $p < 0.005$). Of the subjects who cancelled a visit, fifty-three percent contacted the medical doctor in another way, especially by phone, mail, and the WhatsApp (WhatsApp Inc. 2020, Manlo Park, CA, USA). These subjects had more chronic diseases (57.7% vs. 45.7%, $p < 0.05$) than those who did not contact the doctor. Only 24% of subjects believed that the doctor could understand their state of health well through a video consultation, but the percentage rose to 62% if the subject had already met the therapist in a face-to-face visit.

Table 2. Answers given by the participants to questions about medical care during the pandemic with relative percentage (%).

Have you cancelled or postponed any scheduled clinical assessments during the pandemic?
No, 53.7%
Yes, 46.3%
Due to the impossibility of a visit during this period, have you contacted your doctor for the management of your complications in any other way?
No, never, 47.9%
Yes, by WhatsApp or phone message, 12.7%
Yes, by email, 17.1%
Yes, by telephone, 20.7%
Yes, by video consulting, 1.6%
Do you think that, during this period, a remote medical video consultation could help you have less health risks?
No, 12.8%
Yes, 59.0%
I don't know, 28.1%
Do you think your doctor can understand your health through a video consultation?
No, 12.8%
I don't know, 24.5%
Yes, 24.4%
Yes, but only if he has already met me during a face-to-face visit, 37.7%

3.2. Obesity Management and Telemedicine

Fifty-five percent of subjects were open to following a remotely delivered lifestyle intervention, 23% only if the health professional was known, 15.4% were undecided, and 6.6% refused. There was no difference in age, sex, degree of obesity, and having a chronic disease in the four groups (Figure 1). Most patients believed that the cost of the online lifestyle intervention should be provided by the national health system. Forty-five percent of the respondents to the survey felt favorably about participating in remote psychological regular support. Men were more reluctant than women (37.5% of males refused this type of therapy vs. 18% of females, $p < 0.005$). Subjects older than 60 years and with chronic diseases tended to be less disposed toward remote psychological therapy than younger subjects and those without chronic diseases (Figure 2). Most people would prefer a video consultation with the psychologist once a week. Another preferred modality is an online interview as needed. The minority would like to have two talks a week. An online nutritional intervention would have been accepted by 46% of the subjects. Males and the elderly were more opposed or indecisive towards this intervention than females and those younger (Figure 3). According to most, nutritional therapy should take place by teleconsulting with a dietician once a week. They also welcomed the sending of video conferencing/written materials concerning diet/nutrition/health. Group activities online once a week was the least welcome option. About 60% of subjects would practice physical activity with online tools supervised by a trainer, while 14% are not interested. The subjects who were more predisposed toward the online program were females and those younger (Figure 4). Concerning physical activity, most of the samples in the study believed that the best way to practice physical activity at a distance was real-time online group lessons, once a week, followed by the reception of training tables and one-on-one meetings via the video platform with an exercise physiologist. The deterring factors of the possibility of doing physical activity online were in order of frequency: laziness, no deterrent, the lack of an appropriate space at home, the difficulty of using electronic tools, and the fear of getting hurt.

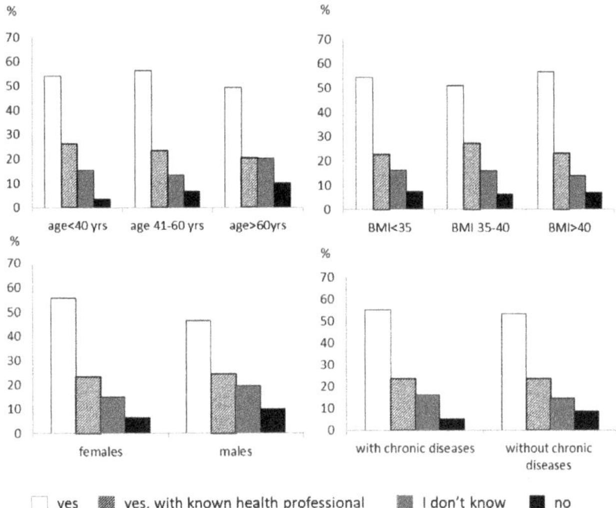

Figure 1. The figure shows the answer to the question: "Would you be like to start an online multidisciplinary intervention for weight management?". The responders are divided by age, sex, degree of obesity, and the presence of chronic diseases.

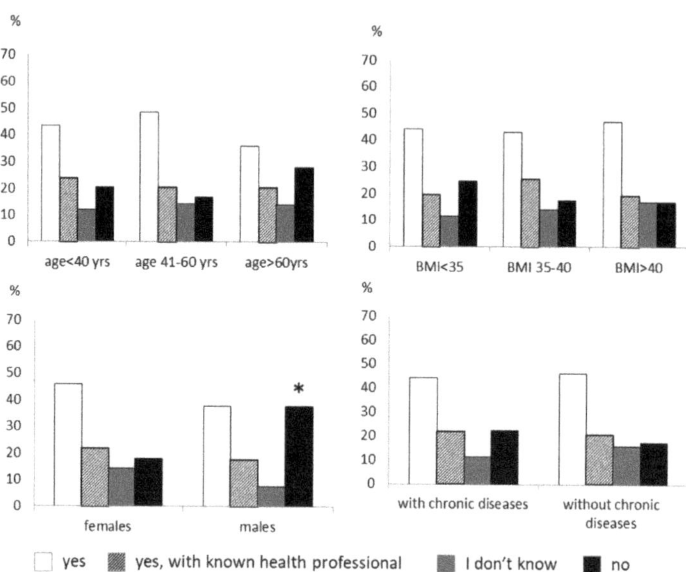

Figure 2. The figure shows the answer to the question: "Would you undergo a remotely delivered psychological intervention?". The responders are divided by age, sex, degree of obesity, and the presence of chronic diseases. * $p < 0.005$ vs. females.

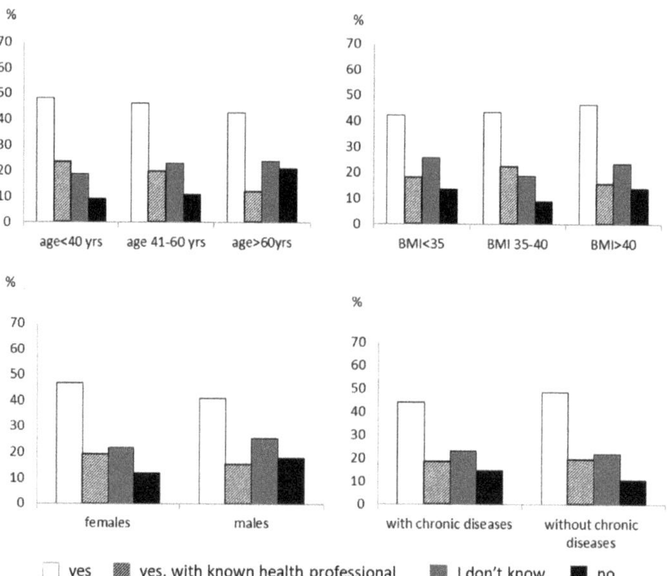

Figure 3. The figure shows the answer to the question: "Would you will join an online nutritional intervention?". The responders are divided by age, sex, degree of obesity, and the presence of chronic diseases.

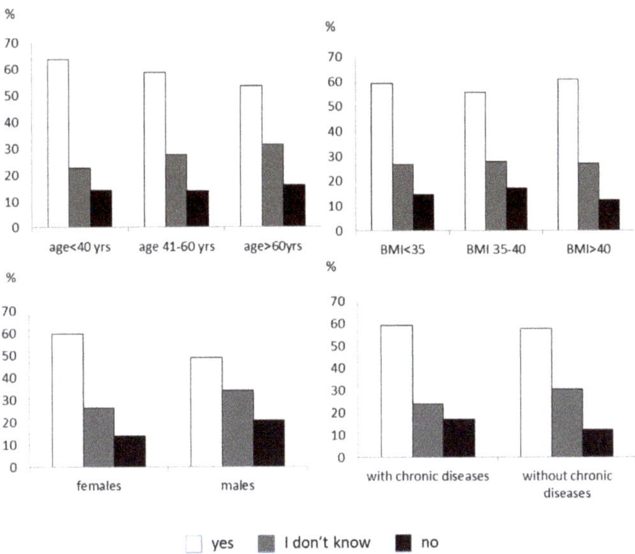

Figure 4. The figure shows the answer to the question: "Would you like to practice physical activity online supervised by an exercise physiologist?". The responders are divided by age, sex, degree of obesity, and the presence of chronic diseases.

4. Discussion

Our study is the first investigation detecting habits and behaviors of obese subjects towards medical care during the pandemic and providing the opinions of subjects with obesity of a remotely delivered program for weight loss, including psychological, nutritional, and physical activity interventions through telemedicine. The COVID-19 pandemic has had not only deleterious effects on infected people but also on the non-COVID-19 patients who have not been able to receive the same level of assistance as before. In Italy, from 15 March 2020, outpatient visits were limited to no deferrable ones, while other appointments were postponed or cancelled to conserve resources and reduce the risk of viral transmission. Our survey highlighted that half of the individuals with obesity responders missed all scheduled medical examinations, especially those with chronic diseases. In according with this finding, a questionnaire coordinated by the Italian National Institute of Health and aimed at people over 65 years revealed that 44% of 1200 subjects interviewed were resigned during the pandemic to missing at least one medical examination (or diagnostic test) that they would need [15]. Bonora et al. showed that the number of visits for diabetic subjects performed during the lockdown period was 47.7% lower than in the same month of the previous 2 years and that the reduction of visits was significantly greater for aged type 2 diabetes patients with heavier complication burdens and complex pharmacotherapies than for younger ones with less complicated diabetes [16]. A study conducted in North Carolina (USA) reported that 53% of outpatient cardiology encounters were cancelled in 2020, and individuals who utilized telehealth tended to be younger, with fewer comorbidities, than those cancelled or referred care [17]. In our survey, patients contacted the doctor, especially via email and WhatsApp, indicating that these media are replacing telephone calls. Although about 60% of the survey participants thought that online consulting might decrease the risk of being infected, only 24% believed that the doctor could perceive their health conditions without an in-person visit. The percentage increased if the subject met the health professional during a previous face-to-face visit. Telemedicine could be a huge opportunity for obesity management during the COVID-19 pandemic, but acceptance by subjects with obesity could be a critical issue. Obesity is a disease in which physician–

patient communication is fundamental, and the relationship between the patient and the health professional is important for the success of the intervention [18].

For this reason, it was important to investigate how remotely delivered interventions could be perceived in these patients and if there were any phenotypes of patients who were more reluctant toward telemedicine. Our study demonstrated that more than half of the subjects with obesity were willing to participate in an online multidisciplinary lifestyle intervention, but familiarity with the therapist was a conditional factor in the greater acceptance of the therapy. When we analyzed the perceptions of the three interventions separately (psychological, nutritional, and physical activity), we found that the subjects were more reluctant toward online psychological and nutritional therapy than a physical activity program. Men and the elderly tended to be less interested to an online intervention than women and those younger. In particular, about 40% of men with obesity refused online psychological therapy. It is known that men remain less likely than women to access psychotherapy or participate in lifestyle modification programs, including weight loss intervention [19,20]. In our cohort, men had a more severe degree of obesity and associated chronic diseases than women, suggesting that men seek help for their condition (in this case, were attracted by a survey on a weight loss program) later than women and when their health is already compromised with other complications. Thus, appealing and innovative approaches that improve their weight status are needed for men. In a study investigating men's experiences and perspectives regarding social support after bariatric surgery, male patients reported feeling alone and isolated during the weight loss support groups consisting primarily of women, and they preferred online social support [21]. Other studies confirmed a positive and good response by men to telemedicine [22,23]. Despite this, in our study, we found some reluctance by men to participate in an online weight loss program, especially as regards psychological intervention. Since, as already mentioned above, physician–patient communication is fundamental in obesity management, a face-to-face consultant with the healthcare professional is necessary to explain the modalities and benefits of online lifestyle interventions and to decrease men's distrust of this type of approach.

Few studies have examined telemedicine weight loss interventions for older people [24]. Telemedicine may have several advantages, eliminating mobility impediments and the risk of COVID-19 infection. Indeed, older obese subjects have the major risk of severe disease and/or death from COVID-19. In our study, about 20% of subjects were >60 years old, and they had more chronic diseases than those younger. They were less attracted to telemedicine, probably because they felt fragile and needed an in-person consultation with the healthcare professional, including to monitor in their presence their health status. Furthermore, although most of them had an electronic tool and the educational level was no different compared to young subjects, it is possible that they had an inability to manipulate the technology, as well as a cognitive impairment issue.

It is necessary to promote remotely delivered weight loss interventions tailored for older people that overcome technological and cognitive barriers. The responders chose as the best mode for a nutritional intervention an online visit with a dietician, while real-time group lessons were with an exercise physiologist to practice physical activity at home. The online consultation was probably preferred, because it allows to create interactions between users through facial expressions and voice tone, confirming how important the relationship is with a therapist in the nutritional intervention in obesity management. The online physical activity group lessons have an advantage in that the exercise physiologist can give a visual demonstration of the exercises and that group members can support each other. As a matter of fact, telemedicine may represent a novel, effective option when treating obese patients, but the benefits and barriers in using technology should be carefully considered [10].

Our findings suggest that we could use a "hybrid" model for the management of weight loss during the pandemic. From a practical viewpoint, we could speculate on using a face-to-face consultation for the initial evaluation in order to create a strong patient–

physician relationship, and the subsequent visits could be performed remotely, interspersed with traditional visits, especially with frail subjects (i.e., men and elderly). Finally, it is plausible to use telemedicine in providing physical activity group lessons.

The limitations of this survey are also worthy of discussion: (1) It was a self-reported survey, and the study group may not be representative of all the population with obesity, because the survey was addressed and diffused through the newsletter of our institute. In fact, although the newsletter was open to everyone, some subjects may have been unintentionally excluded—in particular, those who did not have an electronic tool or the needed knowledge and skills to carry out the online survey. (2) The gender disproportion confirms the lower interest of men in weight loss issues than women. (3) Information on a previous SARS-CoV-2 infection was not investigated in this survey. This point may have played a role in patients' decisions to accept or refuse a remote management approach. (4) The questionnaire did not investigate the knowledge of the subjects on how to use the technological aspect of telemedicine. Although almost all the interviewees stated that they have a technological tool (smartphone, computer, or tablet), this does not mean that they were able to use it. This is especially true for the elderly and for those with lower education levels.

5. Conclusions

In conclusion, our survey showed an interest on the part of subjects with obesity to join a multidisciplinary weight loss intervention remotely delivered. Even if with caution, given a possible bias during the recruitment, we noticed that men and the elderly were more reluctant than women and those younger to participate in an online nutritional and psychological intervention. This result highlights once again that, even with telemedicine, the approach to weight management should be tailored. As our survey revealed an interest in telemedicine on the part of people with obesity, we can imagine using this approach also in the post-COVID-19 period. In fact, the implementation of telemedicine in obesity care could minimize patient travel time and missed work, expanding the possibility of treatment to a greater number of subjects with obesity in order to sustain higher adherence to lifestyle changes.

Supplementary Materials: The following supporting information can be downloaded at https://www.mdpi.com/article/10.3390/nu14081579/s1: Supplementary Material S1: The questionnaire.

Author Contributions: Conceptualization, L.G., S.B., R.C., L.C. and G.C.; methodology, S.B., R.C., L.C., M.N. and A.B.; data curation, L.G., R.C, A.B. and S.P.M.; writing—original draft preparation, L.G., S.B., R.C. and L.C.; and writing—review and editing and supervision, L.G., S.B., R.C., L.C. and G.C. All authors have read and agreed to the published version of the manuscript.

Funding: This research was funded by Italian Ministry of Health.

Institutional Review Board Statement: The study was conducted in full agreement with the national and international regulations and the Declaration of Helsinki. The anonymous nature of the web survey did not allow to trace, in any way, sensitive personal data. Therefore, the present web survey study did not require approval by an ethics committee.

Informed Consent Statement: All participants were requested to provide informed consent through an appropriate checkbox in the survey regarding research purposes.

Data Availability Statement: The datasets used during the current study are available from the corresponding author upon reasonable request.

Acknowledgments: The authors acknowledge Daniela Barzaghi and Greta Giuliani (IRCCS Istituto Auxologico Italiano Marketing and Communication Office) for the technical support given.

Conflicts of Interest: The authors declare no conflict of interest.

References

1. Wang, C.; Horby, P.W.; Hayden, F.G.; Gao, G.F. A Novel Coronavirus Outbreak of Global Health Concern. *Lancet* **2020**, *395*, 470–473. [CrossRef]
2. Gatto, M.; Bertuzzo, E.; Mari, L.; Miccoli, S.; Carraro, L.; Casagrandi, R.; Rinaldo, A. Spread and Dynamics of the COVID-19 Epidemic in Italy: Effects of Emergency Containment Measures. *Proc. Natl. Acad. Sci. USA* **2020**, *117*, 10484–10491. [CrossRef] [PubMed]
3. Ammar, A.; Trabelsi, K.; Brach, M.; Chtourou, H.; Boukhris, O.; Masmoudi, L.; Bouaziz, B.; Bentlage, E.; How, D.; Ahmed, M.; et al. Effects of Home Confinement on Mental Health and Lifestyle Behaviours during the COVID-19 Outbreak: Insight from the ECLB-COVID19 Multicenter Study. *Biol. Sport* **2021**, *38*, 9–21. [CrossRef] [PubMed]
4. Zupo, R.; Castellana, F.; Sardone, R.; Sila, A.; Giagulli, V.A.; Triggiani, V.; Cincione, R.I.; Giannelli, G.; De Pergola, G. Preliminary Trajectories in Dietary Behaviors during the COVID-19 Pandemic: A Public Health Call to Action to Face Obesity. *Int. J. Environ. Res. Public Health* **2020**, *17*, 7073. [CrossRef]
5. Branley-Bell, D.; Talbot, C.V. Exploring the Impact of the COVID-19 Pandemic and UK Lockdown on Individuals with Experience of Eating Disorders. *J. Eat. Disord.* **2020**, *8*, 44. [CrossRef]
6. Giel, K.E.; Schurr, M.; Zipfel, S.; Junne, F.; Schag, K. Eating Behaviour and Symptom Trajectories in Patients with a History of Binge Eating Disorder during COVID-19 Pandemic. *Eur. Eat. Disord. Rev.* **2021**, *29*, 657–662. [CrossRef]
7. Deschasaux-Tanguy, M.; Druesne-Pecollo, N.; Esseddik, Y.; de Edelenyi, F.S.; Allès, B.; Andreeva, V.A.; Baudry, J.; Charreire, H.; Deschamps, V.; Egnell, M.; et al. Diet and Physical Activity during the Coronavirus Disease 2019 (COVID-19) Lockdown (March–May 2020): Results from the French NutriNet-Santé Cohort Study. *Am. J. Clin. Nutr.* **2021**, *113*, 924–938. [CrossRef]
8. Pellegrini, M.; Ponzo, V.; Rosato, R.; Scumaci, E.; Goitre, I.; Benso, A.; Belcastro, S.; Crespi, C.; De Michieli, F.; Ghigo, E.; et al. Changes in Weight and Nutritional Habits in Adults with Obesity during the "Lockdown" Period Caused by the COVID-19 Virus Emergency. *Nutrients* **2020**, *12*, 2016. [CrossRef]
9. Sanchis-Gomar, F.; Lavie, C.J.; Mehra, M.R.; Henry, B.M.; Lippi, G. Obesity and Outcomes in COVID-19: When an Epidemic and Pandemic Collide. *Mayo Clin. Proc.* **2020**, *95*, 1445–1453. [CrossRef]
10. Wijesooriya, N.R.; Mishra, V.; Brand, P.L.P.; Rubin, B.K. COVID-19 and Telehealth, Education, and Research Adaptations. *Paediatr. Respir. Rev.* **2020**, *35*, 38–42. [CrossRef]
11. Johnson, K.E.; Alencar, M.K.; Coakley, K.E.; Swift, D.L.; Cole, N.H.; Mermier, C.M.; Kravitz, L.; Amorim, F.T.; Gibson, A.L. Telemedicine-Based Health Coaching Is Effective for Inducing Weight Loss and Improving Metabolic Markers. *Telemed. E-Health* **2019**, *25*, 85–92. [CrossRef] [PubMed]
12. Alencar, M.K.; Johnson, K.; Mullur, R.; Gray, V.; Gutierrez, E.; Korosteleva, O. The Efficacy of a Telemedicine-Based Weight Loss Program with Video Conference Health Coaching Support. *J. Telemed. Telecare* **2019**, *25*, 151–157. [CrossRef] [PubMed]
13. Rapporto Osservasalute. 2020. Available online: https://www.Osservatoriosullasalute.It/Osservasalute/Rapporto-Osservasalute-2020 (accessed on 21 November 2021).
14. Gilardini, L.; Cancello, R.; Caffetto, K.; Cottafava, R.; Gironi, I.; Invitti, C. Nutrition Knowledge Is Associated with Greater Weight Loss in Obese Patients Following a Multidisciplinary Rehabilitation Program. *Minerva Endocrinol.* **2021**, *46*, 296–302. [CrossRef] [PubMed]
15. Epicentro, ISS. PASSI e PASSI d'argento e La Pandemia COVID-19. Primo Report Nazionale Del Modulo COVID. Available online: https://www.Epicentro.Iss.It/Coronavirus/Sars-Cov-2-Flussi-Dati-Confronto-Passi-Pda (accessed on 19 February 2022).
16. Bonora, B.M.; Morieri, M.L.; Avogaro, A.; Fadini, G.P. The Toll of Lockdown Against COVID-19 on Diabetes Outpatient Care: Analysis from an Outbreak Area in Northeast Italy. *Diabetes Care* **2021**, *44*, e18–e21. [CrossRef]
17. Wosik, J.; Clowse, M.E.B.; Overton, R.; Adagarla, B.; Economou-Zavlanos, N.; Cavalier, J.; Henao, R.; Piccini, J.P.; Thomas, L.; Pencina, M.J.; et al. Impact of the COVID-19 Pandemic on Patterns of Outpatient Cardiovascular Care. *Am. Heart J.* **2021**, *231*, 1–5. [CrossRef] [PubMed]
18. Albury, C.; Strain, W.D.; Brocq, S.L.; Logue, J.; Lloyd, C.; Tahrani, A. The Importance of Language in Engagement between Health-Care Professionals and People Living with Obesity: A Joint Consensus Statement. *Lancet Diabetes Endocrinol.* **2020**, *8*, 447–455. [CrossRef]
19. Pagoto, S.L.; Schneider, K.L.; Oleski, J.L.; Luciani, J.M.; Bodenlos, J.S.; Whited, M.C. Male Inclusion in Randomized Controlled Trials of Lifestyle Weight Loss Interventions. *Obesity* **2012**, *20*, 1234–1239. [CrossRef]
20. Robertson, C.; Archibald, D.; Avenell, A.; Douglas, F.; Hoddinott, P.; van Teijlingen, E.; Boyers, D.; Stewart, F.; Boachie, C.; Fioratou, E.; et al. Systematic Reviews of and Integrated Report on the Quantitative, Qualitative and Economic Evidence Base for the Management of Obesity in Men. *Health Technol. Assess.* **2014**, *18*, 1–424. [CrossRef]
21. Men's Experiences and Perspectives Regarding Social Support after Weight Loss Surgery. *J. Mens Health* **2016**, *12*, 25. [CrossRef]
22. Ventura Marra, M.; Lilly, C.; Nelson, K.; Woofter, D.; Malone, J. A Pilot Randomized Controlled Trial of a Telenutrition Weight Loss Intervention in Middle-Aged and Older Men with Multiple Risk Factors for Cardiovascular Disease. *Nutrients* **2019**, *11*, 229. [CrossRef]

23. Drew, R.J.; Morgan, P.J.; Kay-Lambkin, F.; Collins, C.E.; Callister, R.; Kelly, B.J.; Hansen, V.; Young, M.D. Men's Perceptions of a Gender-Tailored EHealth Program Targeting Physical and Mental Health: Qualitative Findings from the SHED-IT Recharge Trial. *Int. J. Environ. Res. Public. Health* **2021**, *18*, 12878. [CrossRef] [PubMed]
24. Batsis, J.A.; Pletcher, S.N.; Stahl, J.E. Telemedicine and Primary Care Obesity Management in Rural Areas—Innovative Approach for Older Adults? *BMC Geriatr.* **2017**, *17*, 6. [CrossRef] [PubMed]

Article

Personalized Metabolic Avatar: A Data Driven Model of Metabolism for Weight Variation Forecasting and Diet Plan Evaluation

Alessio Abeltino [1,2], Giada Bianchetti [1,2], Cassandra Serantoni [1,2], Cosimo Federico Ardito [3], Daniele Malta [3], Marco De Spirito [1,2] and Giuseppe Maulucci [1,2,*]

1. Neuroscience Department, Biophysics Section, Università Cattolica del Sacro Cuore, 00168 Rome, Italy
2. Fondazione Policlinico Universitario A. Gemelli IRCSS, 00168 Rome, Italy
3. RAN Innovation, Viale della Piramide Cestia, 00153 Rome, Italy
* Correspondence: giuseppe.maulucci@unicatt.it; Tel.: +39-06-30154265

Abstract: Development of predictive computational models of metabolism through mechanistic models is complex and resource demanding, and their personalization remains challenging. Data-driven models of human metabolism would constitute a reliable, fast, and continuously updating model for predictive analytics. Wearable devices, such as smart bands and impedance balances, allow the real time and remote monitoring of physiological parameters, providing for a flux of data carrying information on user metabolism. Here, we developed a data-driven model of end-user metabolism, the Personalized Metabolic Avatar (PMA), to estimate its personalized reactions to diets. PMA consists of a gated recurrent unit (GRU) deep learning model trained to forecast personalized weight variations according to macronutrient composition and daily energy balance. The model can perform simulations and evaluation of diet plans, allowing the definition of tailored goals for achieving ideal weight. This approach can provide the correct clues to empower citizens with scientific knowledge, augmenting their self-awareness with the aim to achieve long-lasting results in pursuing a healthy lifestyle.

Keywords: metabolism; deep learning; gated recurrent unit; wearables; forecasting; diet plans; digital nutrition

Citation: Abeltino, A.; Bianchetti, G.; Serantoni, C.; Ardito, C.F.; Malta, D.; De Spirito, M.; Maulucci, G. Personalized Metabolic Avatar: A Data Driven Model of Metabolism for Weight Variation Forecasting and Diet Plan Evaluation. *Nutrients* **2022**, *14*, 3520. https://doi.org/10.3390/nu14173520

Academic Editor: Javier Gómez-Ambrosi

Received: 26 July 2022
Accepted: 23 August 2022
Published: 26 August 2022

Publisher's Note: MDPI stays neutral with regard to jurisdictional claims in published maps and institutional affiliations.

1. Introduction

The global obesity epidemic has been spreading throughout most countries since the 1980s. Obesity contributes directly to incident cardiovascular risk factors, including dyslipidemia, type 2 diabetes, hypertension, and sleep disorders [1–3]. Obesity also leads to the development of cardiovascular diseases independently of other cardiovascular risk factors. More recent data highlight abdominal obesity, as determined by waist circumference, as a cardiovascular disease risk marker that is independent of body mass index [4,5]. Lifestyle modification and subsequent weight loss improve both metabolic syndrome and associated systemic inflammation and endothelial dysfunction, leading to a reduction of coronary artery disease, heart failure, and atrial fibrillation [6–9].

Quantifying lifestyle modifications to decrease cardiovascular risk is nowadays conceivable following the increased use of wearable devices, such as smartwatches, smart bands and impedance balances. These devices allow the real-time and remote monitoring of physiological parameters. As measurement and feedback systems become more refined and personalized, these devices can help people to change their lifestyles and improve wellbeing. Moreover, they have the potential to be linked into a wide range of lifestyle support services through community, public and private providers. An important improvement in managing the huge variety of wearable and portable devices comes from web-based applications. Several solutions exist on digital stores, but they mostly suffer from incomplete and not well-defined food databases and lack of personalization due to the scarce integration of the information flux: users must rely on different applications, furnishing partial and unrelated information about their metabolic state, where energy intake and expenditure are not directly related. To overcome this issue, we developed a digital web-based application (ArMOnIA) integrating dietary, anthropometric, and physical activity data [10]. Data flows from smart devices (smart band and impedance balance) and diet diaries are collected to build an accurate and personalized estimation of energy balance (accounting for individual body composition, age, and hydration state). We already showed, in a single-arm uncontrolled prospective study on self-monitored voluntary normal or overweight adults, that this application, by simply allowing the visualization of the energy balance in a dashboard, helps users to significantly decrease their average energy balance and consequently BMI in a period of 45 days [10]. The data streams provided by this platform can be analyzed relying on machine learning and artificial neural networks, with the aim to provide predictive and personalizable computational models of metabolism. In particular, the problem of the prediction of weight variations traditionally relies on estimations based on thermodynamic models depending on age, height, gender, and current weight [11]. However, diet predictors developed through these models have limited application because they assume weight stability and do not account for factors such as microbiome, variations in type and expression of genes linked to nutrition, and quality and quantity of physical activity. Some human genome-scale metabolic models (GEM), such as Recon3D [12], contain the human gene–protein reaction associations and can mechanistically predict metabolic fluxes. However, these complex models need long elaboration times or high-performance computing (HPC) and cannot be embedded in edge computing (EC) to improve scalability and performance. Moreover, the personalization of metabolic models remains challenging [13], as they require new methodological approaches to integrate molecular and physiological data. Data-driven models of human metabolism would constitute a reliable, fast, and continuously updating model for predictive analytics. These models could indeed offer crucial data for achieving the best weight forecasts and the creation of individualized diet and exercise plans. Differently from the well-established knowledge-driven models, data-driven models can account for all of the metabolic processes, from genetic predispositions to current microbiome composition, affecting weight changes. Relying on this information embedded in the model, they could provide for personalized weight forecasts and for the creation of individualized diet and exercise plans, with the aim to achieve long-lasting results in pursuing a healthy lifestyle. To this aim, here we developed a personalized model of end-user metabolism, the Personalized Metabolic Avatar (PMA), to estimate its reactions to diets. PMA consists of a gated recurrent unit (GRU) deep neural network allowing the prediction and simulation of personalized weight variations according to macronutrient composition [14–16] and daily energy balance [17] and allowing the generation of tailored diet plans. PMA may be adopted to gradually improve adoption of healthy habits in a person-specific fashion.

2. Materials and Methods

2.1. Study Population and Protocol

In this single-arm uncontrolled prospective study, a group of four adult volunteers (three normal and one overweight) recruited from our lab staff self-monitored daily their weight, diet and step count for more than 300 days using the ArMOnIA app, without predetermined objectives or intervention. Other assessment data were collected in-person via digital diaries. The four participants shared their personal data after signing an informed consent. The protocol is as follows:

- *Food diaries*: users must register daily the foods eaten during breakfast, lunch, dinner and snacks.
- *Physical activities (PA):* users must wear a smart band all day and all night, especially during physical activities where they have to specify the type of activity performed. These include: jogging, walking, swimming, working out, general sports, etc. When-

ever participants forget to track their own activities with the smart band, they must register them into the ArMOnIA app, where the calories burned from these activities are evaluated through the compendium [18]. This is also performed for other activities not monitored by the smart band, such as house cleaning, driving, etc.

- *Weight monitoring:* users have to weigh themselves barefoot every day after waking up using an impedentiometric balance.

2.2. Wearables and Devices

The following devices were chosen for tracking anthropometric and PA data:

- MiBand 6, a smart band (Xiaomi Inc.®, Beijing, China), for tracking PA and estimating calories burned during exercises (walking, running, etc.).
- Mi Body Composition Scale, an impedance balance (Xiaomi Inc.®, Beijing, China), for tracking anthropometric data such as: weight, resting metabolism, fat rate, muscle rate, bone mass.

These devices already had been used in three studies on PubMed, and 11 clinical trials had been performed using MiBand-1. Validation results in estimating RMR can be retrieved in a recent publication [19].

2.3. Data Collection, Storage and Retrieval through an Ad Hoc Developed Web App and Estimation of Personalized Energy Balance

A web application (www.apparmonia.com, accessed 21 July 2022) was developed in Python 3.8 with the libraries Django (https://www.djangoproject.com/, accessed 21 July 2022) and Django_plotly_dash (https://django-plotly-dash.readthedocs.io/en/latest/, accessed 21 July 2022) for data collection, storage, and visualization of energy balance through a dashboard [17].

The web application allows for data collection, storage, analysis and visualization. These are detailed below.

2.3.1. Data Collection

Data provided in-person through a digital diary: food and other activities not included in the smart band (home activities, music playing, driving, etc.).

Data from the smart band and impedance balance were retrieved through the ZEPP Life® app (Anhui Huami Information Technology Co., Ltd., Hefei, China).

2.3.2. Data Storage

Retrieved data underwent anonymization and are then stored into a NoSQL database (MongoDB®, New York, NY, USA, https://www.mongodb.com/, accessed 30 June2022).

2.3.3. Data Retrieval

The quantities retrieved from the database needed for the development of PMA were the following:

1. w is the weight acquired daily by the Mi Body Composition Scale.
2. m_C is the mass expressed in grams of total daily carbohydrate intake, m_L is the mass expressed in grams of total daily lipid intake, and m_P is the mass expressed in grams of total daily protein intake.
3. daily energy balance, EB, calculated according to the formula

$$EB = EI - TEE, \tag{1}$$

where EI is the daily energy intake, and TEE is the daily total energy expenditure.

EI is considered as the sum of all ingested calories as retrieved from the following databases: DIETABIT (www.dietabit.it, accessed 5 July 2022), CREA (www.crea.gov.it, accessed 5 July 2022), BDA (www.bda-ieo.it, accessed 5 July 2022), and OPENFOODFACTS (www.it.openfoodfacts.org, accessed 5 July 2022).

TEE is calculated according to the formula

$$TEE = RMR + TEA + TEF \tag{2}$$

where *TEA* is the thermic effect of activity, *RMR* is the resting metabolism ratio, both measured using the values provided by the ZEPP Life® app [19], and *TEF* is the thermic effect of food, referring to the energy expenditure related to food consumption [20] (i.e., digestion, absorption, assimilation, and storage), dependent on the amount and type of food consumed, which accounts for about 10% [21] of *TEE* and is estimated from food data through the following formula:

$$TEF = 0.095 \cdot (m_c \cdot 3.75) + 0.015 \cdot (m_L \cdot 9) + 0.25 \cdot (m_P \cdot 4) \tag{3}$$

2.4. Data Preprocessing

We considered energy balance and food composition as the main drivers of weight variations [10,22]. As already introduced in Section 2.3, the datasets used for the construction and testing of the model consisted of the following data:

- Weight: *w(t)* [kg]
- Energy balance: *EB(t)* [kcal]
- Daily carbohydrate intake: $m_c(t)$ [g]
- Daily protein intake: $m_p(t)$ [g]
- Daily lipid intake: $m_l(t)$ [g]
- Week cosine: $\cos(\frac{2}{7}\pi t)$
- Week sine: $\sin(\frac{2}{7}\pi t)$

In Figure 1A,B, sample *w(t)*, *EB(t)* and $m_c(t)$, $m_p(t)$, $m_l(t)$ time series are reported. The last two terms, *week cosine* and *week sine*, were introduced to account for seasonality that can affect diet and PA habits, as shown in previous studies [23]. So far, in Figure S1, we showed a violin plot of a representative user reporting the distribution of the energy balance through all days in a week. As we can see, there is a variation among days confirmed also by statistical tests (Section S1).

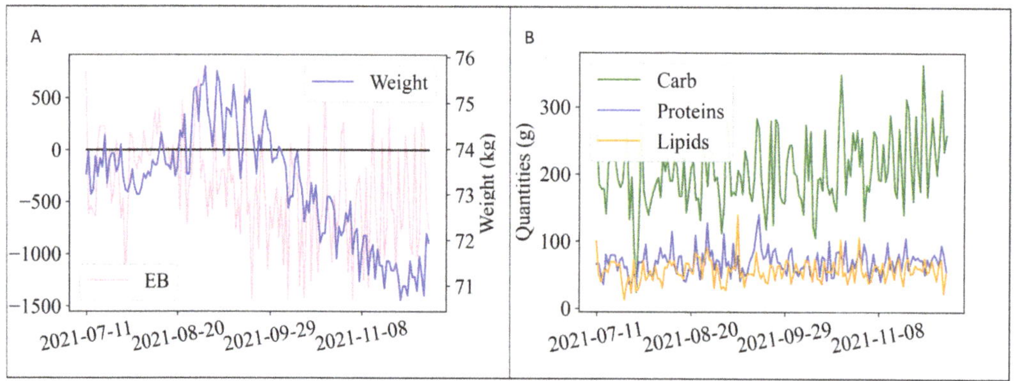

Figure 1. Time series describing user metabolism. (**A**) Representative time series for weight and EB. (**B**) Representative time series for food composition.

We then handled missing values (below 3% of the entire dataset) using the 'pad' method, taking values from the previous row. During imputing, test and train were separated to avoid crosstalk between the two sets.

EB(t) values can be affected by biases due to wrong insertion of food quantities, which are typically underestimated [24]. To account for these biases, we calculated for each time point the weekly variation of $EB_{week}(t)$ and the weekly weight variation of $\Delta w_{week}(t)$ and fitted with a linear regression model $EB_{week} = a \cdot \Delta w_{week} + b$ (Figure S2). b is the average bias in the estimation of the energy balance, which was subtracted from the estimated $EB(t)$.

2.5. PMA Development with RNN Network

PMA was shaped as output of a deep recurrent neural network, bridging the evolution of weight $w(t)$ and $m_c(t)$, $m_p(t)$, $m_l(t)$ (Section 2.4). Recurrent neural networks (RNN) are a very flexible class of neural networks, widely used to solve problems involving dependent data, such as time series. Therefore, this type of neural network best suited our application needs. Among the RNNs, we selected the mono-layer GRU (Section S3 [25]).

Data Preparation

Before deep learning can be used, time series forecasting problems must be re-framed as supervised learning problems. It is standard practice to use lagged observations (e.g., $t - 1$) as input variables to forecast the current time step (t). This is called *'multi-step forecasting'* [26]. Calling k the lagged observation, the supervised learning dataset is reframed as:

$$var_1(t - k) \ldots var_7(t - k) \ldots var_1(t - i) \ldots var_6(t - 1) \ldots var_7(t - 1) \ldots var_1(t)$$

where the overall time series are renamed with the string var_j, where j indicates the variable considered running from 1 to 7.

2.6. Model Selection

The first step in the development of the PMA is the definition of the architecture of the RNN used.

For this work, the most suitable architecture found for our application (Figure S3) was composed of the following layers:

- *Input layer*: weight and exogenous series such as EB and food composition (carbohydrate, protein and lipid content expressed in grams) at previous times with respect to the output (plus historical values from the time series target). This corresponds to the x_t of Equation (S4), defined as follows: $x_t = [EB(t - k), m_c(t - k), m_p(t - k), m_c(t - k), w(t - k), \ldots, EB(t - 1), m_c(t - 1), m_p(t - 1), m_c(t - 1)$, with k the lagged observation (specific for each user, as explained below).
- *Hidden layer*: a GRU neural network with the addition of a dropout layer (Section S4 [27,28]).
- *Output layer*: composed of one output, the weight $w(t + 1)$ at time $t + 1$.

After that, we needed to choose the best set of hyperparameters (HP) to allow the model to predict accurately for every dataset used.

HP are the number of neurons, the type of activation function, the batch size, the number of epochs, the dropout value and the lookback value k (how many time steps we look back for the forecasting of the time series target). HP tuning was carried out to find the possible best sets to build the model from a specific dataset and with a specific goal [29]. HP tuning consists of the scanning of macro-parameters for the reduction of a loss function. Typically, in time series forecasting, the tuning is carried out to reduce the root mean squared error (RMSE) of the test-train forecasting (see Equation (4) below). Nevertheless, for our study, we introduced several constraints in selecting HP to guarantee correct dynamics of weight variations. In order to do so, we performed a simulation for 7 days (described in Section 2.7), considering diet plans consisting of different EB values: $-1000, -500, 0, 500$ and 1000 kcal. HP that did not respect the following conditions were discarded:

- $w(t + 7) - w(t) > 0$ for $EB = 1000$ kcal
- $w(t + 7) - w(t) < 0$ for $EB = -1000$ kcal
- $w(t + 7)_{EB=1000} - w(t + 7)_{EB=-1000} < 10$ kg
- $w(t + 7) - w(t)$ has to be an increasing function of EB

After this preselection, the choice of the best set of parameters was then made through minimization of the RMSE of the test-train forecasting, evaluated according to the formula:

$$\text{RMSE} = \sqrt{\sum_{i=1}^{n} \frac{(\hat{y}_i - y_i)^2}{n}} \tag{4}$$

In the following, we report in detail the HP parameter scanning sets:

Number of neurons: The number of neurons in the hidden layer for the GRU neural network has to be adjusted to the solution complexity: the task with a more complex level to predict needs more neurons. To consider GRU with increasing complexity, the number of neurons was chosen from the following range: 50, 100, 150 and 200.

Activation function: The activation function of the GRU mono-layer is crucial to compute the input values into output values. We considered eight activation functions to tune: 'tanh', 'ReLU', 'sigmoid', 'softplus', 'softsign', 'selu', 'elu', 'exponential'. In Figure S4, we reported the activation functions 'tanh' and 'ReLU' as the most performant functions in our datasets.

Batch size: Batch size is the number of training data sub-samples for the input. The smaller batch size makes the learning process faster at the expense of the variance of validation dataset accuracy. To minimize the time of the learning process as much as possible, we set the range of this value with the following values: 8, 16, 32, 64, 128.

Number of epochs: The number of times a whole dataset is passed through the neural network model is called an epoch. One epoch means that the training dataset is passed forward and backward through the neural network once. The number of epochs must be tuned to gain the optimal result: too few epochs typically result in underfitting, while too many epochs lead to overfitting. Hence, we verified optimal agreement of the test loss and train loss through the plot of learning curves. Following this visual inspection (see Section 3.1 and Figure 2), the number of epochs available for tuning was limited to the set: 50, 100, 150, 200.

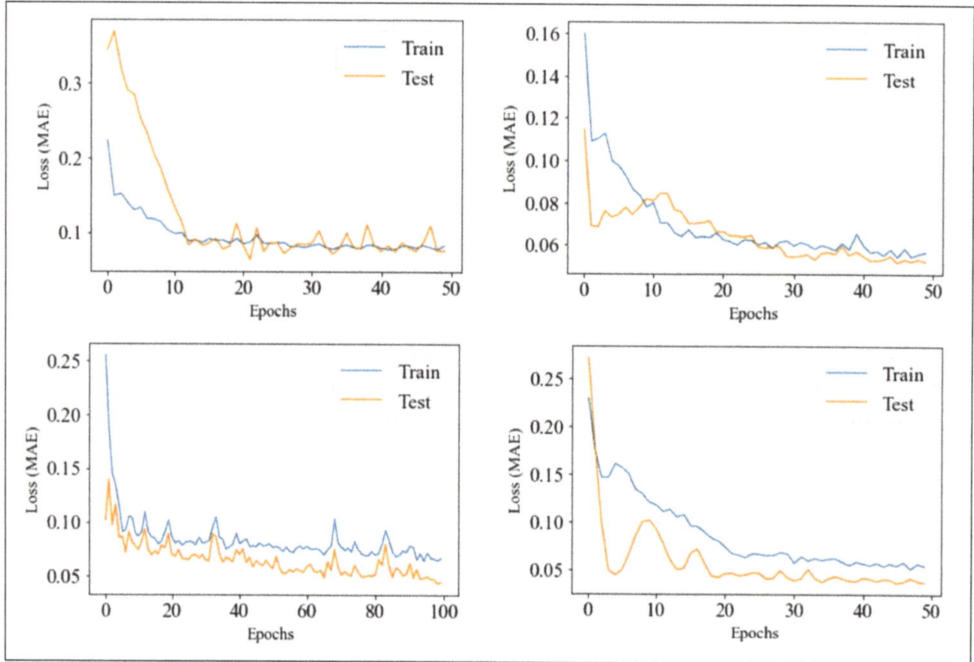

Figure 2. Train and test loss function (*Mean Absolute Error*) versus the number of epochs.

Lookback (k-value): The number of time steps looked back in the prediction is a key value in multi-step ahead forecasting. The weight trend is strongly influenced by the previous values. Hence, considering previous time steps in the forecasting of the weight is necessary to reduce as much as possible the errors committed in the prediction. However, a higher value could bring unwanted results, such as decreasing the performance of the forecaster both in terms of accuracy and computational speed. Following these considerations, we considered for tuning, as a trade-off, the range: 7, 6, 5, 4, and 3.

Dropout rate: The dropout layer is a regularization layer. As its name suggests, it randomly drops a certain number of neurons in a layer. The dropped neurons are not used anymore. The percentage of neurons to drop is set in the dropout rate. A high value may be too severe for the application. To avoid this problem, the dropout rate was chosen from the following range: 0.2, 0.4 and 0.6 [30].

Seasonal terms: For each user, the seasonal term could influence the weight variation. For this reason, in the tuning, we considered whether the addition of the week cosine and week sine terms among the input variables would lead to an increase in the performance of the model or not.

Metrics and optimization algorithm: In the tuning, *"Mean Absolute Error"* (MAE) was used as the GRU loss function, and *"ADAM"* as the optimization algorithm.

2.7. Walk-Forward Validation and Simulation

In time series modeling, the predictions over time become less and less accurate. Walk-forward validation (WFV) is a more realistic approach consisting of continuously re-training the model with actual data as they become available for further predictions. Since the training of GRU neural networks is not too time-consuming, WFV is the most preferred solution to obtain the most accurate results.

Following the same criteria of the WFV, we defined the walk-forward simulation (WFS). The only difference between the two approaches is that in WFS, we used forecasted values as input rather than actual data. The WFS's workflow is shown in Table 1.

Table 1. Concept of WFS.

t	Input $(t-k)$	...	Input $(t-3)$	Input $(t-2)$	Input $(t-1)$	...	w $(t-4)$	w $(t-3)$	w $(t-2)$	w $(t-1)$	w (t)
$t+1$	known	...	known	simulated	simulated	...	known	known	known	known	predict
$t+2$	known	...	known	simulated	simulated	...	known	known	known		predict
$t+3$	known	...	simulated	simulated	simulated	...	known	known			predict
$t+4$	known	...	simulated	simulated	simulated	...	known				predict
...	...	...	...	...	...	...	...	...	...	...	...
$t+n$	simulated	...	simulated	simulated	simulated	...					predict

Columns represent input values at time t. Input $(t-k)$; ... ; Input $(t-1)$ represent covariates, while w $(t-k)$; ... ; w (t) represent the target variable (weight). Rows represent predictions at time $t+1, t+2, \ldots, t+n$. 'known' means that the value is taken from the dataset of actual values, 'simulated' indicates that the value is an input of a simulated diet plan, 'predict' indicates that the value is predicted from the neural network.

A limit of WFV and WFS is the fact that the re-training phase forces the start of forecasting or simulation only from the last acquired time step. However, if there was a need to simulate effects of variations of *EB* or food composition beginning from other starting points, our approach was to avoid the re-training phase. This approach is particularly useful when input data are scarcely sampled in the training set and WFS cannot give correct responses.

2.8. Computer Performance

For the study, a PC with the following characteristics was used: Windows 10 Enterprise, Intel(R) Core(TM) i5-8500 CPU @ 3.00 GHz, 8 GB RAM, Intel(R) UHD Graphics 630.

2.9. Python Libraries

The setup used for this study was composed of the following libraries: tensorflow CPU == 2.8.0 (https://pypi.org/project/tensorflow-cpu/, accessed 5 July 2022), keras == 2.8.0 (https://keras.io/, accessed 5 July 2022), pandas == 1.0.5 (https://pandas.pydata.org/, accessed 5 July 2022), numpy == 1.22.2 (https://numpy.org/, accessed 5 July 2022), matplotlib == 3.5.2 (https://matplotlib.org/, 5 July 2022), seaborn == 0.10.1 (https://seaborn.pydata.org/, accessed 5 July 2022), pymongo == 3.11.4 (https://pymongo.readthedocs.io/en/stable/, accessed 5 July 2022) and scikit-learn == 0.24.2 (https://scikit-learn.org/stable/, accessed 5 July 2022).

3. Results

3.1. Selection of the Optimal Models through Grid Search of GRU Parameters and RMSE Overall Minimization on the Cohort of Users

As a starting point, we selected the four time series and carried out HP tuning (Section 2.6) following reduction of the RMSE of the values predicted using the test-train method with a 7-day test dataset. The optimal hyperparameters (HP) defined the individual model, called PMA, which is reported in Table 2 for each user:

Table 2. Results of the hyperparameter tuning for each user.

User	Number of Neurons	Activation Function	Dropout Rate	Epochs	Batch Size	Lookback	Seasonal Terms	RMSE
0	100	*ReLU*	0.2	50	32	7	No	0.47
1	200	*ReLU*	0.2	200	128	4	No	0.49
2	150	*ReLU*	0.2	50	64	5	No	0.31
3	100	*ReLU*	0.2	50	128	5	No	0.4

We can observe from the table that the PMA differed among users with the exception of the activation function ('*ReLU*'), the dropout rate (0.2), and the seasonal terms that gave no additional improvement to PMA. This is probably because the size of the training set spanned through a time period (i.e., winter and summer) during which well-defined habits did not arise. We also checked the test-train plots for all users (Figure 2). They showed no evident presence of overfitting, guaranteeing the goodness of the model.

3.2. Weight Forecasting: Model Results, WFV and WFS

In this section, we report the forecasting results of the most performant GRU for the weight forecasting and for the WFV and WFS.

The training set for the weight forecasting was selected as 90% of the overall dataset (about 330 days), yielding an RMSE averaged on the four users of 0.59 ± 0.076.

However, predictions of 30 days, albeit with good results, could be subjected to additional uncertainty because they did not account for additional variables that could affect actual weight variations over such a long period of time (abdominal bloating due to excess food ingestion, water retention, constipation). Therefore, we carried out train-test forecasting for each user considering an interval of 7 days. The results are shown in Figure 3.

Test-train RMSE carried out with a test dataset length of one week yielded an averaged value for the four users of 0.41 ± 0.05, showing a 30% decrease. Moreover, RMSE for each user stayed below 0.5. Despite these improved results, it is well known that in time series modeling, the predictions over time become less and less accurate (Section S6). Therefore, WFV was the most preferred solution to obtain the most accurate results by re-training the model with actual data as they became available for further predictions. This technique could be used to perform simulations, namely WFS (Section 2.7).

The WFV and WFS for the PMA were thus performed within a week to evaluate the RMSEs with respect to the true values (Figure 4). A major improvement was obtained with this validation method, yielding an average RMSE of 0.42 ± 0.1 for the WFV and 0.48 ± 0.18 for the WFS. As expected, the results from the WFV were better than those from the WFS ($\text{RMSE}_{\text{WFV}} < \text{RMSE}_{\text{WFS}}$). Nevertheless, the WFS showed optimal results allowing it to be used with specific applications, such as, for example, the simulation of diet plans.

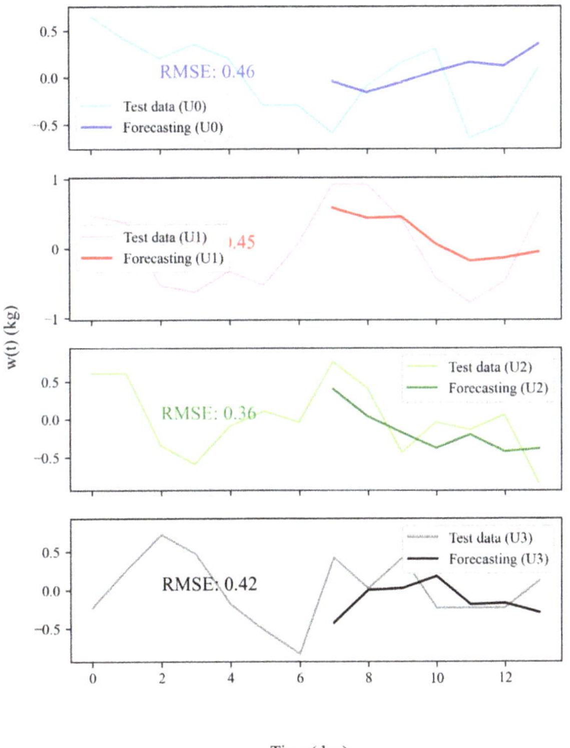

Figure 3. Test-train forecasting for all users (U0, U1, U2 and U3) with the relative root mean squared value.

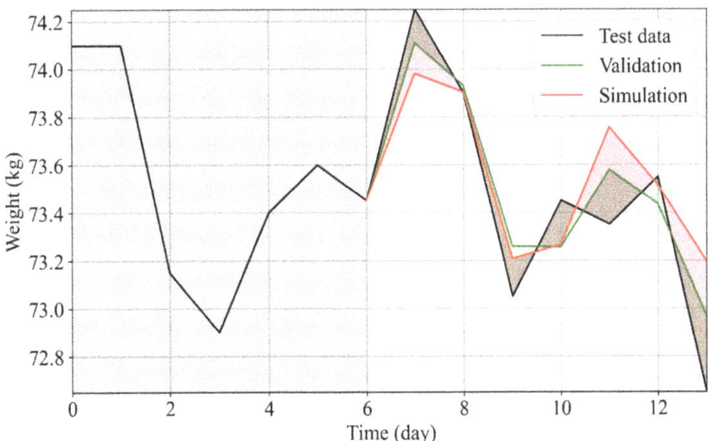

Figure 4. Comparison between actual data and WFV and WFS results for User 2.

3.3. Simulation of the Personalized Effects of Diet Plans on Weight

WFS can be used to simulate personalized diet plans and to predict metabolic responses after the introduction of new food and PA habits (determining variations in *EB* and macronutrient composition). To test the performance of the model in new simulated conditions, dietary plans were obtained by constraining the *EB* value to be constant at a particular level, and the effect of these variations on the weight of each user was simulated.

In detail, a basic simulation was carried out varying EB in the following range: $-1000, -500, 0, 500, 1000$ kcal (Figure 5A), with standard percentage contributions of carbohydrate, protein, and lipid intake (50/20/30%), respectively, included in acceptable macronutrient distribution ranges (AMDR) [31]. The values of macronutrient intakes were calculated by converting their percentages into grams [32]; then, the total caloric intake was evaluated by inverting Equation (1). From the simulations, we can observe that an energy deficit of 500 kcal per day yielded an average weight loss of -0.4 ± 0.2 kg in a week, while an energy surplus of 500 kcal yielded an average weight gain of 0.77 ± 0.63 kg in a week, and that differences existed among users. To summarize simulation results and to cancel out random effects in the daily weight variation due to water retention or constipation, we fitted the simulated trends with a parabolic fit as shown in Figure 5A and estimated the w value representing the weight value at the end of the week. In Figure 5B, individual weight variations in function of the simulated EB values are reported. These differences could be parametrized for each user by retrieving the coefficient of the relation $\Delta w = m \cdot EB + q$ (Table 3). Here, q represents the weight variation at $EB = 0$, which is, therefore, expected to be equal to zero. The q value can furnish an average value of eventual residual biases in data collection, yielding a systematic error in the determination of *EB*. It provided a quality factor of food insertion, which was the highest for User 2. m is a parameter linked to metabolic plasticity, expressed in $\frac{Kg}{Kcal}$, representing the rate of weight variation per unbalanced calorie. A higher value indicates a higher metabolic plasticity and/or a more active metabolism. This parameter can thus be used to develop a metabolic taxonomy of the users. In our use case, users 0 and 2 showed higher metabolic plasticity than users 1 and 3.

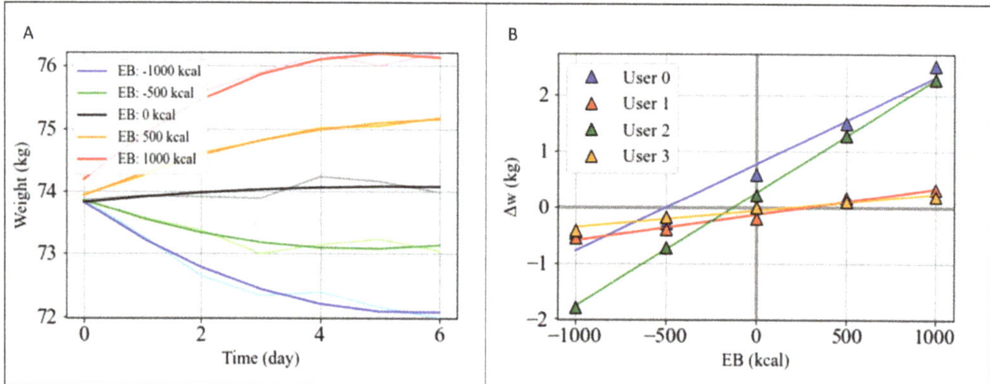

Figure 5. Effects of diet plans on user metabolism. (**A**) WFS performed at different EB values on the data of User 2, keeping constant the percentage of macronutrient intake (50%, 20%, 30%, respectively). Weight data were fitted with a second order polynomial. (**B**) Weight variation Δw calculated from the first and last values of the fit of the second grade versus the EB value and for each user.

Table 3. Metabolic plasticity m and quality factor q for each user.

User	Metabolic Plasticity (m) $[\frac{kg}{kcal}]$	Quality Factor (q) [kg]
0	$1.56 \cdot 10^{-3}$	0.77
1	$0.47 \cdot 10^{-3}$	-0.13
2	$2.03 \cdot 10^{-3}$	0.26
3	$0.30 \cdot 10^{-3}$	-0.06

3.4. Personalized Diet Plan: Use Case

In this use case, rather than performing a toy diet plan, we tested an actual personalized diet plan on User 2 to achieve weight loss in a healthy way supervised by a professional nutritionist considering blood analyses, food and activity habits.

In Figure 6, the actual weight variations (black), the prediction made by WFS using as exogenous data the actual data (red), and the WFS using as exogenous values the data retrieved from the personalized diet plan (green) are shown. The RMSE of the prediction was 0.26 (showing that the technique had good performance), and the weight loss approximately of $\Delta w = 1.5$ kg following the tailored diet plan provided to the user, which was in accordance with the predetermined goal defined by the nutritionist (rapid weight loss). This tool can thus allow us to compare the expected and actual effects of the diet on the weight variations and to test several nutritional plans in terms of energy balance and macronutrient composition.

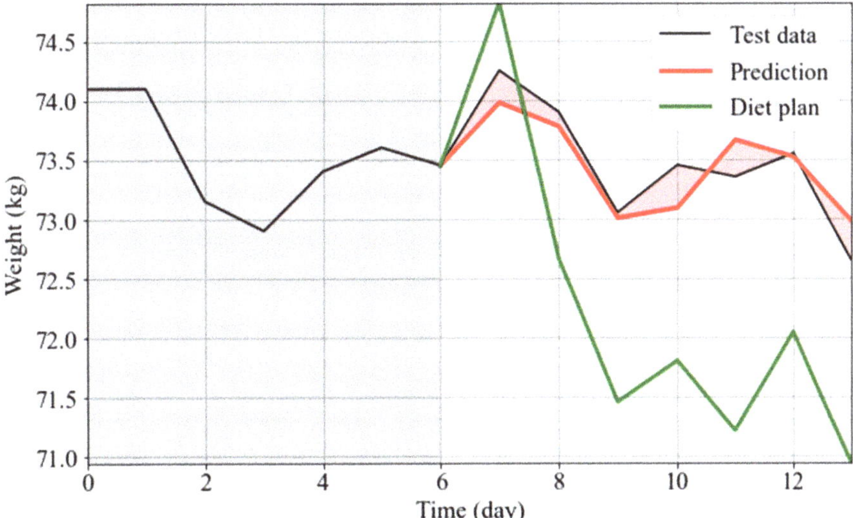

Figure 6. Personalized nutritional intervention plan for User 2. In the first 7 days, the actual weight trend is shown (black line, gray shaded area). Along this trend, WFS for the personalized plan is reported (green line). As a control, WFS when covariates retained the actual values is reported (red line).

4. Discussion

Obesity and its metabolic complications are the most serious public health challenges of the 21st century. The prevalence of obesity has tripled in many countries of the EU [33]. In the current pandemic, the issue of obesity has become more prominent [34], highlighting the need for its prevention. Evidence that relates to obesity is biased towards its causes rather than strategies for prevention, which have not yet been widely replicated or delivered at a scale offering clear options for public health strategies. Finding and implementing solutions require new models able to implement healthy lifestyles and prevent illness by relying on devices that can be used in daily life, reducing the burden on hospitals. Here, we relied on an application able to retrieve, pre-process and analyze spontaneous and voluntary PA, diet, and anthropometric quantities from a set of wearables and home-portable devices provided to the end-user. These data drove the development of a personalized model of the end-user metabolism, the PMA, able to estimate his/her personalized reactions to diet, PA, and environmental and psychological factors. The PMA was integrated into the IoT-reliant infrastructure, allowing it to perform simulations and predictions to gradually improve adoption of healthy habits.

In this manuscript, we have shown how GRU-based deep neural networks are a good solution to predict in an accurate way the weight for the day after (the WFV showed an average RMSE lower than 0.5), and to simulate personalized diet plans to help reach ideal weights in an healthy way, avoiding excessive variations in habitual diet or PA and keeping weight and nutrient balance in the normal range following guidelines. We tested the PMA by using WFS to predict the weekly weight variations of four users subjected to varying energy balance constraints, and we also converted a true nutritional plan developed by a professional nutritionist in a WFS to test the effect on a user, with the aim of evaluating if the metabolic response of the subject could achieve weight loss.

The principal strength of the PMA with respect to established knowledge-driven models resides in the fact that the developed data-driven model can take into account all of the processes involved in metabolism having an influence on weight variations, from genetic predispositions to current microbiome composition. Nutrigenomics (also known as nutritional genomics) is broadly defined as the relationship between nutrients, diet, and gene expression [35] having a deep influence on individual metabolism [36]. 'Microbiome', also called 'gut microbiota' [37], is a complex and dynamic population of microorganisms that exert a marked influence on the host metabolism during homeostasis and disease. Multiple factors contribute to the establishment of the human gut microbiota during infancy, and diet is considered as one of the main drivers in shaping the gut microbiota across one's lifetime. The data-driven nature of the PMA allows it to integrate the complexity of these metabolic processes without the requirement of deterministic or statistical models, which make generalizable claims in trying to describe human metabolism for all human subjects, or for certain subsets of the population. If this is the objective, the distribution needs to be accurately sampled from the population on which the claim is made, and the number of subjects has to be adjusted to improve the significance of the prediction. Here, the claim is different, because we did not realize a single general model, but four distinct models of metabolism, personalized for each individual. We modeled personal metabolism as a black box in which the input was energy balance and macronutrient composition, and the output was weight. In this framework, the statistical unit, rather than the subject, is the daily response of individual weight to the different input stimuli. This allowed us to make forecasts based on a high number of available data (~300 per person). We were able to gain a feel for these peculiar PMA features by comparing its performance with available weight predictors [11]. Nowadays, available weight predictors use general information such as age, sex, height and current weight to forecast weight variations by setting a predefined value of energy balance. As shown in Table 4, these types of data, based on a statistical model describing average features of the analyzed sample population, intrinsically do not allow an actual personalized prediction. The PMA instead allows descriptions of personalized metabolic responses for users, as quantified by the standard deviation of the predictions (0.2 kg), which is almost 10 times that of weight predictors (0.034 kg).

In Figure 7A,B, we can observe how the statistical model (blue points) predicts a weekly weight loss for an $EB = -500$ kcal, which shows slight variations with starting BMI, age or sex. While Users 0, 1 and 3 were well aligned with the general population, we observed that User 2 deviated from the general trend. This was indeed the subject with the highest metabolic plasticity in the systematic simulation performed in Section 3.3.

Table 4. Comparison of weight predictions between statistical and data-driven models (PMA).

User	Age	Sex	Height (cm)	w_i [kg]	Δw (PMA) [kg]	Δw (Statistical Model) [kg]
0	27	M	183	88.7	0.3 ± 0.37	0.31 ± 0.031
1	52	M	186	74.25	0.4 ± 0.25	0.38 ± 0.038
2	44	M	175	73.45	0.72 ± 0.12	0.35 ± 0.035
3	51	F	160	55.25	0.27 ± 0.21	0.4 ± 0.04

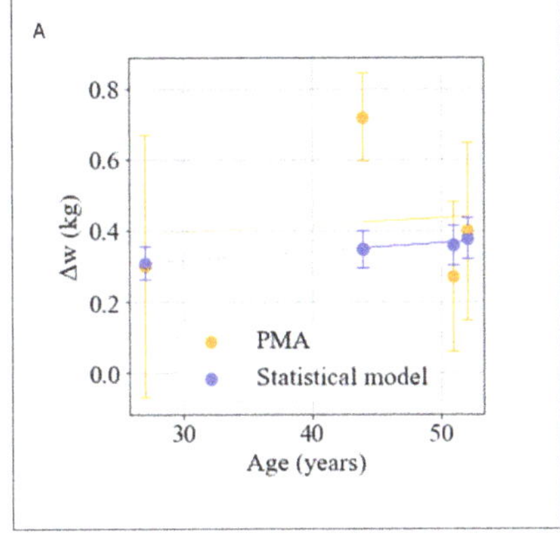
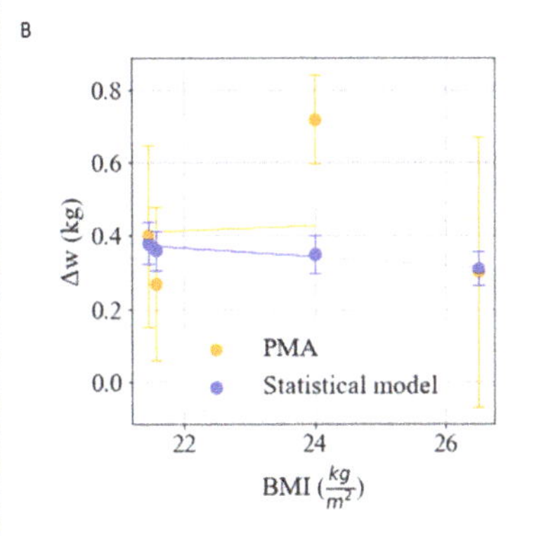

Figure 7. (**A**) Δw calculated with PMA and with statistical models with respect to age of users. (**B**) Δw calculated with PMA and with statistical models with respect to BMI of users. For the statistical model, an error of 15% was considered, while for PMA, it was considered as an error of the RMSE of the WFS.

These anomalous values of metabolic plasticity can be due to several factors, ranging from microbiome diversity to a different nutri-genotype. Additionally, hormonal equilibria and systemic diseases can have a huge influence [38]. This difference with the general trend highlights how a personalized approach, in this particular case, is fundamental in assessing tailored weight loss in response to nutritional treatments. A correction of the metabolic plasticity with microbiome composition and diversity or with nutrigenomic characteristics would be an important advancement in understanding the factors leading to the reshaping of individual metabolism. The clinical relevance of the results presented in the manuscript resides in the possibility to understand if metabolic adaptations due to microbiome variation or general metabolism reprogramming due to treatments or nutritional interventions are occurring, and how to change them through simulations in order to fulfill desired results. Applications can be envisioned for obesity and nutritional disorder treatments, and to generate diet plans in synergy with treatments in cancer and other diseases.

Other than personalization, an additional strength of the PMA resides in the informative content of the inputs: information such as food composition allows better prediction of the metabolic response. Indeed, to reach ideal goals such as weight loss, a correct subdivision of the basic nutrients is fundamental in the generation of a diet plan. It is in principle possible also to include other important variables, going from micronutrient composition of the diet and the use of integrators to sleep quality.

The PMA is also scalable not only in terms of its inputs, but also in terms of outputs, allowing it to contextually predict changes in variables of interest other than weight (e.g., fat and lean mass, resting heart rate).

Therefore, the PMA could become a powerful support tool for nutritionists, dieticians, physicians, etc. Hence, it has the potential to lay the foundations for truly 'personalized nutrition' approaches, using these predictions to identify metabolic impairments and plan actions in advance, and to simulate the metabolic response to several diet plans to achieve the desired results without compromising the body's wellness. The generated diet and activity plans could be delivered to users by front-end components with a virtual assistant helping patients to monitor their behavior and improve their adherence to optimal actions. However, the PMA has some critical issues. First, the prediction of weight with unknown conditions, such as for extreme diet plans (e.g., the ketogenic type), could lead to inaccurate predictions because the PMA may lack training on that data. The PMA could overcome these problems by relying on continuous training day after day. Moreover, noise caused from wrong data insertion could alter the quality of predictions. Another point is that, at the current stage of development, the PMA requires data collection for at least 2 months to achieve good performance. To improve data collection, automatic food detection methods through mobile phones [39], which are continuously evolving, could overcome this limit by reducing manual compilation and decreasing the burden on users.

5. Conclusions

This study shows that the integration of several IoT devices and a diet registry into a single web app able to merge all acquisitions into a single visualization dashboard, with a deep learning analysis of user metabolism through the realization of PMA, provides important information to realize optimal weight forecasting and the personalized generation of diet and activity plans. Relying on this information, appropriate clues can be obtained to empower citizens with scientific knowledge and validated instruments, augmenting their self-awareness with the aim to achieve long-lasting results in the pursuit of a healthy lifestyle. An important advancement could be the integration, as input in the PMA, of novel developed biomarkers of lipid metabolism (such as membrane lipids and membrane fluidity of red blood cells) to study the effects and influence of dietary molecules on their outcomes [40–44]. Moreover, innovative and promising anthropometric markers tracked with wearable devices, such as VO_2max and heart rate variability (HRV), can improve the performance of weight forecasting [45–47]. These integrations could explain and cluster the different responses given by the PMA, furnishing insights into the factors able to shape individual metabolism.

Supplementary Materials: The following supporting information can be downloaded at: https://www.mdpi.com/article/10.3390/nu14173520/s1, Figure S1: Violinplot of the EB for all days in a week; Figure S2: Distribution of weekly w versus weekly EB for all users; Figure S3: Distribution of weekly w versus weekly EB for all users; Figure S4: Activation functions considered in the hyperparameter tuning: hyperbolic tangent and Rectified Linear Unit; Section S1: Seasonality analysis; Section S2: EB Correction; Section S3: Theory of RNN: GRU and LSTM models; Section S4: Regularization techniques; Section S5: Activation functions; Section S6: Performance vs days predicted; Table S1: ANOVA Analysis Results.

Author Contributions: Conceptualization, G.M.; Data curation, A.A., G.B., C.S. and G.M.; Formal analysis, A.A., G.B., C.S. and G.M.; Funding acquisition, Cosimo Ardito and Daniele Malta; Investigation, A.A., G.B., C.S., C.F.A., D.M., M.D.S. and G.M.; Methodology, A.A. and G.M.; Project administration, G.M.; Resources, M.D.S.; Software, A.A., C.F.A., D.M. and M.D.S.; Supervision, G.M.; Validation, A.A. and G.M.; Visualization, A.A.; Writing—original draft, A.A.; Writing—review and editing, G.M. All authors have read and agreed to the published version of the manuscript.

Funding: This project was supported in part by a research grant awarded to GM from Regione Lazio PO FSE 2014-2020 "Intervento per il rafforzamento della ricerca nel Lazio—incentivi per i dottorati di innovazione per le imprese", cofunded by RAN Innovation, and by a research grant awarded to GM from Università Cattolica del Sacro Cuore-Linea D1 2021.

Institutional Review Board Statement: The study was conducted in accordance with the Declaration of Helsinki and approved by the Ethics Committee of Università Cattolica del Sacro Cuore (Protocol Code diab_mf, 16 March 2017).

Informed Consent Statement: Informed consent was obtained from all subjects involved in this study. Written informed consent was obtained from the participants to publish this paper.

Data Availability Statement: Data and codes are available upon reasonable request at https://github.com/Metabolicintelligence/PMA, accessed on 25 August 2022.

Acknowledgments: Thanks to Roberta Martinoli and Silvia Barbaresi for useful discussions about connections between microbiome and metabolism and for the generation of the diet plan.

Conflicts of Interest: The authors declare no conflict of interest.

References

1. Sharma, A.M. Obesity and Cardiovascular Risk. *Growth Horm. IGF Res.* **2003**, *13*, S10–S17. [CrossRef]
2. Klop, B.; Elte, J.W.F.; Cabezas, M.C. Dyslipidemia in Obesity: Mechanisms and Potential Targets. *Nutrients* **2013**, *5*, 1218–1240. [CrossRef] [PubMed]
3. Shahi, B.; Praglowski, B.; Deitel, M. Sleep-Related Disorders in the Obese. *Obes. Surg.* **1992**, *2*, 157–168. [CrossRef] [PubMed]
4. Shields, M.; Tremblay, M.S.; Gorber, S.C.; Janssen, I. Abdominal Obesity and Cardiovascular Disease Risk Factors within Body Mass Index Categories. *Health Rep.* **2012**, *23*, 7–15. [PubMed]
5. Kim, H.Y.; Kim, J.K.; Shin, G.G.; Han, J.A.; Kim, J.W. Association between Abdominal Obesity and Cardiovascular Risk Factors in Adults with Normal Body Mass Index: Based on the Sixth Korea National Health and Nutrition Examination Survey. *J. Obes. Metab. Syndr.* **2019**, *28*, 262–270. [CrossRef]
6. Aggarwal, M.; Bozkurt, B.; Panjrath, G.; Aggarwal, B.; Ostfeld, R.J.; Barnard, N.D.; Gaggin, H.; Freeman, A.M.; Allen, K.; Madan, S.; et al. Lifestyle Modifications for Preventing and Treating Heart Failure. *J. Am. Coll. Cardiol.* **2018**, *72*, 2391–2405. [CrossRef]
7. Chung, M.K.; Eckhardt, L.L.; Chen, L.Y.; Ahmed, H.M.; Gopinathannair, R.; Joglar, J.A.; Noseworthy, P.A.; Pack, Q.R.; Sanders, P.; Trulock, K.M. Lifestyle and Risk Factor Modification for Reduction of Atrial Fibrillation: A Scientific Statement from the American Heart Association. *Circulation* **2020**, *141*, e750–e772. [CrossRef]
8. Maruthur, N.M.; Wang, N.Y.; Appel, L.J. Lifestyle Interventions Reduce Coronary Heart Disease Risk Results from the Premier Trial. *Circulation* **2009**, *119*, 2026–2031. [CrossRef]
9. Wingerter, R.; Steiger, N.; Burrows, A.; Estes, N.A.M. Impact of Lifestyle Modification on Atrial Fibrillation. *Am. J. Cardiol.* **2020**, *125*, 289–297. [CrossRef]
10. Hill, J.O.; Wyatt, H.R.; Peters, J.C. The Importance of Energy Balance. *Eur. Endocrinol.* **2013**, *9*, 111–115. [CrossRef]
11. Thomas, D.M.; Martin, C.K.; Heymsfield, S.; Redman, L.M.; Schoeller, D.A.; Levine, J.A. A Simple Model Predicting Individual Weight Change in Humans. *J. Biol. Dyn.* **2011**, *5*, 579–599. [CrossRef] [PubMed]
12. Brunk, E.; Sahoo, S.; Zielinski, D.C.; Altunkaya, A.; Dräger, A.; Mih, N.; Gatto, F.; Nilsson, A.; Preciat Gonzalez, G.A.; Aurich, M.K.; et al. Recon3D Enables a Three-Dimensional View of Gene Variation in Human Metabolism. *Nat. Biotechnol.* **2018**, *36*, 272–281. [CrossRef]
13. Opdam, S.; Richelle, A.; Kellman, B.; Li, S.; Zielinski, D.C.; Lewis, N.E. A Systematic Evaluation of Methods for Tailoring Genome-Scale Metabolic Models. *Cell Syst.* **2017**, *4*, 318–329.e6. [CrossRef] [PubMed]
14. van Zuuren, E.J.; Fedorowicz, Z.; Kuijpers, T.; Pijl, H. Effects of Low-Carbohydrate- Compared with Low-Fat-Diet Interventions on Metabolic Control in People with Type 2 Diabetes: A Systematic Review Including GRADE Assessments. *Am. J. Clin. Nutr.* **2018**, *108*, 300–331. [CrossRef] [PubMed]
15. Johnston, B.C.; Kanters, S.; Bandayrel, K.; Wu, P.; Naji, F.; Siemieniuk, R.A.; Ball, G.D.C.; Busse, J.W.; Thorlund, K.; Guyatt, G.; et al. Comparison of Weight Loss among Named Diet Programs in Overweight and Obese Adults: A Meta-Analysis. *JAMA* **2014**, *312*, 923–933. [CrossRef] [PubMed]
16. Wycherley, T.P.; Moran, L.J.; Clifton, P.M.; Noakes, M.; Brinkworth, G.D. Effects of Energy-Restricted High-Protein, Low-Fat Compared with Standard-Protein, Low-Fat Diets: A Meta-Analysis of Randomized Controlled Trials. *Am. J. Clin. Nutr.* **2012**, *96*, 1281–1298. [CrossRef]
17. Bianchetti, G.; Abeltino, A.; Serantoni, C.; Ardito, F.; Malta, D.; de Spirito, M.; Maulucci, G. Personalized Self-Monitoring of Energy Balance through Integration in a Web-Application of Dietary, Anthropometric, and Physical Activity Data. *J. Pers. Med.* **2022**, *12*, 568. [CrossRef]
18. Ainsworth, B.E.; Haskell, W.L.; Leon, A.S.; Jacobs, D.R., Jr.; Montoye, H.J.; Sallis, J.F.; Paffenbarger, R.S., Jr. Compendium of Physical Activities: Classification of Energy Costs of Human Physical Activities. *Med. Sci. Sports Exerc.* **1993**, *25*, 71–80. [CrossRef]
19. Hao, Y.; Ma, X.K.; Zhu, Z.; Cao, Z.B. Validity of Wrist-Wearable Activity Devices for Estimating Physical Activity in Adolescents: Comparative Study. *JMIR Mhealth Uhealth* **2021**, *9*, e18320. [CrossRef]
20. Lam, Y.Y.; Ravussin, E. Analysis of Energy Metabolism in Humans: A Review of Methodologies. *Mol. Metab.* **2016**, *5*, 1057–1071. [CrossRef]
21. Calcagno, M.; Kahleova, H.; Alwarith, J.; Burgess, N.N.; Flores, R.A.; Busta, M.L.; Barnard, N.D. The Thermic Effect of Food: A Review. *J. Am. Coll. Nutr.* **2019**, *38*, 547–551. [CrossRef] [PubMed]
22. Freire, R. Scientific Evidence of Diets for Weight Loss: Different Macronutrient Composition, Intermittent Fasting, and Popular Diets. *Nutrition* **2020**, *69*, 110549. [CrossRef] [PubMed]
23. Fahey, M.C.; Klesges, R.C.; Kocak, M.; Talcott, G.W.; Krukowski, R.A. Seasonal Fluctuations in Weight and Self-Weighing Behavior among Adults in a Behavioral Weight Loss Intervention. *Eat. Weight Disord.* **2020**, *25*, 921–928. [CrossRef] [PubMed]

24. Cheson, K.J.A. Methods in Nutrition the Measurement of Food and Energy Intake in Man-an Evaluation of Some Techniquess 3. *Am. J. Clin. Nutr.* **1980**, *33*, 1147–1154. [CrossRef] [PubMed]

25. Ruineihart, D.E.; Hint, G.E.; Williams, R.J. *Learning Internal Representations by error Propagation Two*; MIT Press: Cambridge, MA, USA, 1985.

26. Livieris, I.E.; Pintelas, P. A Novel Multi-Step Forecasting Strategy for Enhancing Deep Learning Models' Performance. *Neural Comput. Appl.* **2022**, 1–18. [CrossRef]

27. Srivastava, N.; Hinton, G.; Krizhevsky, A.; Salakhutdinov, R. Dropout: A Simple Way to Prevent Neural Networks from Overfitting. *J. Mach. Learn. Res.* **2014**, *15*, 1929–1958.

28. Pachitariu, M.; Sahani, M. Regularization and Nonlinearities for Neural Language Models: When Are They Needed? *arXiv* **2013**, arXiv:1301.5650.

29. Egger, R.; Gretzel, U. *Tourism on the Verge Series Editors*; Springer: Berlin/Heidelberg, Germany, 2022.

30. Lai, S.-H.; Lepetit, V.; Nishino, K.; Sato, Y. (Eds.) *Computer Vision—ACCV 2016*; Lecture Notes in Computer Science; Springer International Publishing: Cham, Switzerland, 2017; Volume 10112, ISBN 978-3-319-54183-9.

31. Manore, M.M. Exercise and the Institute of Medicine Recommendations for Nutrition. *Curr. Sports Med. Rep.* **2005**, *4*, 193–198. [CrossRef]

32. Park, H.; Kityo, A.; Kim, Y.; Lee, S.A. Macronutrient Intake in Adults Diagnosed with Metabolic Syndrome: Using the Health Examinee (HEXA) Cohort. *Nutrients* **2021**, *13*, 4457. [CrossRef]

33. Marques, A.; Peralta, M.; Naia, A.; Loureiro, N.; de Matos, M.G. Prevalence of Adult Overweight and Obesity in 20 European Countries, 2014. *Eur. J. Public Health* **2018**, *28*, 295–300. [CrossRef]

34. Gammone, M.A.; D'Orazio, N. COVID-19 and Obesity: Overlapping of Two Pandemics. *Obes. Facts* **2021**, *14*, 579–585. [CrossRef] [PubMed]

35. Chadwick, R. Nutrigenomics, Individualism and Public Health. *Proc. Nutr. Soc.* **2004**, *63*, 161–166. [CrossRef] [PubMed]

36. Ordovas, J.M.; Ferguson, L.R.; Tai, E.S.; Mathers, J.C. Personalised Nutrition and Health. *BMJ* **2018**, *361*, bmj.k2173. [CrossRef]

37. Flint, H.J. The Impact of Nutrition on the Human Microbiome. *Nutr. Rev.* **2012**, *70*, S10–S13. [CrossRef] [PubMed]

38. Christensen, P.; Meinert Larsen, T.; Westerterp-Plantenga, M.; Macdonald, I.; Martinez, J.A.; Handjiev, S.; Poppitt, S.; Hansen, S.; Ritz, C.; Astrup, A.; et al. Men and Women Respond Differently to Rapid Weight Loss: Metabolic Outcomes of a Multi-Centre Intervention Study after a Low-Energy Diet in 2500 Overweight, Individuals with Pre-Diabetes (PREVIEW). *Diabetes Obes. Metab.* **2018**, *20*, 2840–2851. [CrossRef]

39. Sazonov, E.S.; Fontana, J.M. A Sensor System for Automatic Detection of Food Intake through Non-Invasive Monitoring of Chewing. *IEEE Sens. J.* **2012**, *12*, 1340–1348. [CrossRef]

40. Bianchetti, G.; Viti, L.; Scupola, A.; di Leo, M.; Tartaglione, L.; Flex, A.; de Spirito, M.; Pitocco, D.; Maulucci, G. Erythrocyte Membrane Fluidity as a Marker of Diabetic Retinopathy in Type 1 Diabetes Mellitus. *Eur. J. Clin. Investig.* **2021**, *51*, e13455. [CrossRef]

41. Maulucci, G.; Cohen, O.; Daniel, B.; Ferreri, C.; Sasson, S. The Combination of Whole Cell Lipidomics Analysis and Single Cell Confocal Imaging of Fluidity and Micropolarity Provides Insight into Stress-Induced Lipid Turnover in Subcellular Organelles of Pancreatic Beta Cells. *Molecules* **2019**, *24*, 3742. [CrossRef]

42. Maulucci, G.; Cohen, O.; Daniel, B.; Sansone, A.; Petropoulou, P.I.; Filou, S.; Spyridonidis, A.; Pani, G.; de Spirito, M.; Chatgilialoglu, C.; et al. Fatty Acid-Related Modulations of Membrane Fluidity in Cells: Detection and Implications. *Free Radic. Res.* **2016**, *50*, S40–S50. [CrossRef]

43. Cordelli, E.; Maulucci, G.; de Spirito, M.; Rizzi, A.; Pitocco, D.; Soda, P. A Decision Support System for Type 1 Diabetes Mellitus Diagnostics Based on Dual Channel Analysis of Red Blood Cell Membrane Fluidity. *Comput. Methods Programs Biomed.* **2018**, *162*, 263–271. [CrossRef]

44. Bianchetti, G.; Azoulay-Ginsburg, S.; Keshet-Levy, N.Y.; Malka, A.; Zilber, S.; Korshin, E.E.; Sasson, S.; de Spirito, M.; Gruzman, A.; Maulucci, G. Investigation of the Membrane Fluidity Regulation of Fatty Acid Intracellular Distribution by Fluorescence Lifetime Imaging of Novel Polarity Sensitive Fluorescent Derivatives. *Int. J. Mol. Sci.* **2021**, *22*, 3106. [CrossRef] [PubMed]

45. Lee, J.; Zhang, X. Is There Really a Proportional Relationship between VO2max and Body Weight? A Review Article. *PLoS ONE* **2021**, *16*, e0261519. [CrossRef]

46. Serantoni, C.; Zimatore, G.; Bianchetti, G.; Abeltino, A.; de Spirito, M.; Maulucci, G. Unsupervised Clustering of Heartbeat Dynamics Allows for Real Time and Personalized Improvement in Cardiovascular Fitness. *Sensors* **2022**, *22*, 3974. [CrossRef] [PubMed]

47. Zimatore, G.; Gallotta, M.C.; Innocenti, L.; Bonavolontà, V.; Ciasca, G.; de Spirito, M.; Guidetti, L.; Baldari, C. Recurrence Quantification Analysis of Heart Rate Variability during Continuous Incremental Exercise Test in Obese Subjects. *Chaos* **2020**, *30*, 033135. [CrossRef] [PubMed]

Article

Putting the Personalized Metabolic Avatar into Production: A Comparison between Deep-Learning and Statistical Models for Weight Prediction

Alessio Abeltino [1,2], Giada Bianchetti [1,2], Cassandra Serantoni [1,2], Alessia Riente [1,2], Marco De Spirito [1,2] and Giuseppe Maulucci [1,2,*]

[1] Neuroscience Department Biophysics Section, Università Cattolica del Sacro Cuore, 00168 Rome, Italy
[2] Fondazione Policlinico Universitario A. Gemelli IRCSS, 00168 Rome, Italy
* Correspondence: giuseppe.maulucci@unicatt.it

Abstract: Nutrition is a cross-cutting sector in medicine, with a huge impact on health, from cardiovascular disease to cancer. Employment of digital medicine in nutrition relies on digital twins: digital replicas of human physiology representing an emergent solution for prevention and treatment of many diseases. In this context, we have already developed a data-driven model of metabolism, called a "Personalized Metabolic Avatar" (PMA), using gated recurrent unit (GRU) neural networks for weight forecasting. However, putting a digital twin into production to make it available for users is a difficult task that as important as model building. Among the principal issues, changes to data sources, models and hyperparameters introduce room for error and overfitting and can lead to abrupt variations in computational time. In this study, we selected the best strategy for deployment in terms of predictive performance and computational time. Several models, such as the Transformer model, recursive neural networks (GRUs and long short-term memory networks) and the statistical SARIMAX model were tested on ten users. PMAs based on GRUs and LSTM showed optimal and stable predictive performances, with the lowest root mean squared errors (0.38 ± 0.16–0.39 ± 0.18) and acceptable computational times of the retraining phase (12.7 ± 1.42 s–13.5 ± 3.60 s) for a production environment. While the Transformer model did not bring a substantial improvement over RNNs in term of predictive performance, it increased the computational time for both forecasting and retraining by 40%. The SARIMAX model showed the worst performance in term of predictive performance, though it had the best computational time. For all the models considered, the extent of the data source was a negligible factor, and a threshold was established for the number of time points needed for a successful prediction.

Keywords: metabolism; deep learning; gated recurrent unit; long short-term memory; transformer; wearables; forecasting; diet plans; digital nutrition; digital twin; SARIMAX

Citation: Abeltino, A.; Bianchetti, G.; Serantoni, C.; Riente, A.; De Spirito, M.; Maulucci, G. Putting the Personalized Metabolic Avatar into Production: A Comparison between Deep-Learning and Statistical Models for Weight Prediction. *Nutrients* **2023**, *15*, 1199. https://doi.org/10.3390/nu15051199

Academic Editor: Javier Gómez-Ambrosi

Received: 8 February 2023
Revised: 24 February 2023
Accepted: 24 February 2023
Published: 27 February 2023

1. Introduction

Over the past few decades, precise diagnosis and personalized treatment have become increasingly important in healthcare [1]. Nutrition, as an important factor of personalized treatments, has a huge impact on health, from cardiovascular disease to cancer [2,3]. Nutritional habits have been linked to stronger immunity, a lower risk of noncommunicable diseases (such as diabetes and cardiovascular disease) and increased life expectancy [4,5].

Increased knowledge of the effects of nutrition on pathophysiologies of diseases, achieved with new diagnostic and monitoring technologies spanning from -omics [6,7] to wearable devices [8], has offered innovative solutions for personalized treatments. Among the most striking innovations, digital twins (DTs), which are digital replicas of human physiology, represent an emergent solution for prevention and treatment of many diseases [9,10]. DT technology holds the promise of starkly reducing the cost, time and manpower required to test effects of dietary and physical-activity plans, to run clinical trials and to create personalized diets for citizens and patients. DT models are built on data flows sourced from connected biomedical devices on the Internet of Things (IoT) and collected through digital web-based applications integrating dietary, anthropometric and physical activity data, such as the one developed by our research group [11]. Artificial intelligence algorithms have shown good performance in analysis of biometric signals [12,13]. The data streams provided by these data acquisition platforms can be analyzed with data-driven models of human metabolism, such as the personalized metabolic avatar (PMA) [14] developed by our group, to estimate personalized reactions to diets. The PMA consists of a gated recurrent unit (GRU) deep-learning model trained to forecast personalized weight variations according to macronutrient composition and daily energy balance. This model can perform simulations and evaluations of diet plans, allowing definition of tailored goals for achieving ideal weight. However, putting PMAs into production and transforming them in a reliable, fast and continuously updating model for predictive analytics is a difficult task. Among the principal issues, challenges can arise from changes to data sources, models and model parameters, which introduce room for error and overfitting and can lead to abrupt variations in computational time. To overcome these issues, here, we selected the best strategy for deployment in terms of predictive performance and computational time. Among statistical models, we selected, as a representative, the SARIMAX (Seasonal Auto-Regressive Integrated Moving Average with eXogenous factors) model: the most complete for multivariate forecasting. Among deep-learning models, we selected recurrent neural networks (RNNs), such as gated recurrent units (GRUs) [14] and long short-term memory (LSTM) networks, and a new model recently introduced, the Transformer model [15], which has shown great results both in natural language processing and in time-series forecasting [16]. Moreover, we have tested the influence of the data number retrieved, which, in real settings, can vary in range from user to user, on the models. These efforts are necessary to put these models into production to augment citizens' self-awareness, with the aim of achieving long-lasting results in pursuing a healthy lifestyle.

2. Materials and Methods

2.1. Study Population

In this single-arm, uncontrolled-pilot prospective study, a group of 10 voluntary adults (60% females and 40% males, 3 overweight and 7 normal), recruited among our lab staff, self-monitored daily their weight, diet and activities completed for at least 100 days, as explained in a previous work [11]. The participants shared their personal data after signing informed consent.

2.2. Wearables and Devices

To track anthropometric data, the following devices were used:

- The MiBand 6, a smartband of Xiaomi Inc.® (Beijing, China), for estimating calories burned during exercise (walking, running, etc.).
- The Mi Body Composition Scale, an impedance balance of Xiaomi Inc.® (Beijing, China), for tracking weight and RMR.

These devices were already used in 4 studies on PubMed, and 11 clinical trials have been performed using the MiBand1. Validation results in estimating RMR can be retrieved in recent publications [4]. For tracking the food diary for each participant, a website app (ArMOnIA, https://www.apparmonia.com, accessed on 7 February 2023) developed by our group was used for the storing of food data. These data had already been validated in two other studies [11,14].

2.3. Datasets

As already shown in [14], for the development of the deep-learning models implementing PMAs, the following data were used:

- var_1: Weight: $w(t)$ [kg]

- var_2: Energy Balance (EB): $Eb(t)$ [kcal]
- var_3: Carbohydrates: $m_c(t)$ [g]
- var_4: Proteins: $m_p(t)$ [g]
- var_5: Lipids: $m_l(t)$ [g]

Where var_j stands for variable j, with $j = 1, \ldots, 5$.

In Figure 1a,b, the representative time series of the five selected quantities are reported.

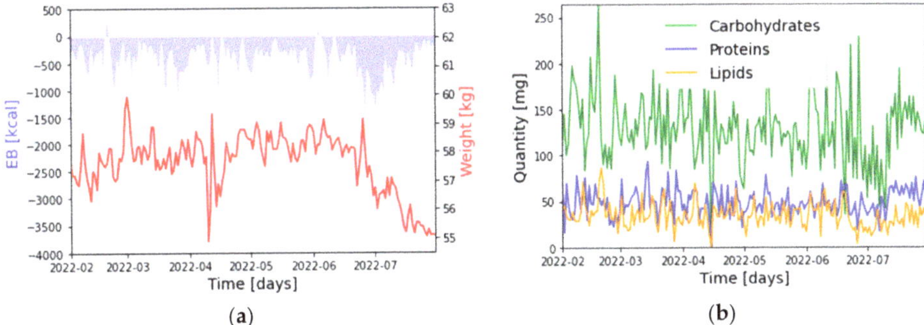

Figure 1. Weight, EB (**a**) and food composition (**b**) time series (for user 2).

We reframed the time-series forecasting problem as a supervised learning problem, using lagged observations (including the seven days before the prediction, e.g., $t-1$, $t-2$, $t-7$) as input variables to forecast the current time step (t), as already explained in [12]. The inputs of our model were $var_1(t-7), \ldots, var_5(t-k), \ldots, var_j(t-i), \ldots, var_5(t-1)$, with $i = 1, \ldots, 7$ indicating the lagged observation and $j = 1, \ldots, 5$ indicating the input variable. Therefore, the total number of inputs for the PMA was $7 \cdot 5 = 35$. In this notation, the output of the PMA is $var_1(t)$, i.e., the weight at time t.

The dataset fed to the SARIMAX model is described in the next section.

2.4. Description of Models

As explained in the introduction, DDMs are divided into two types: statistical and deep-learning models. To select the best option for the development of the PMA, we chose to compare 4 different models:

SARIMAX:

The SARIMAX model (Seasonal Auto-Regressive Integrated Moving Average with eXogenous factors) is a linear regression model: an updated version of the ARIMA model. It is a seasonal equivalent model, like the SARIMA (Seasonal Auto-Regressive Integrated Moving Average) model, but it can also deal with exogenous factors, which are accounted for with an additional term, helping to reduce error values and improve overall model accuracy. This model is usually applied in time-series forecasting [17].

The general form of a $SARIMA(p,d,q)(P,D,Q,s)$ model is

$$\Theta(L)^P \theta(L^s)^P \Delta^d \Delta_s^D w_t = \Phi(L)^q \varphi(L^s)^Q \Delta_s^D \epsilon_t \tag{1}$$

where each term is defined as follows:

1. $\Theta(L)^P$ is the nonseasonal autoregressive lag polynomial;
2. $\theta(L^s)^P$ is the seasonal autoregressive lag polynomial;
3. $\Delta^d \Delta_s^D w_t$ is the time series, differenced d times and seasonally differenced D times;
4. $\Phi(L)^q$ is the nonseasonal moving average lag polynomial;
5. $\varphi(L^s)^Q$ is the seasonal moving average lag polynomial.

When dealing with n exogenous values, each defined at each time step, t, denoted as x_t^i for $i \leq n$, the general form of the model becomes

$$\Theta(L)^p \theta(L^s)^P \Delta^d \Delta_s^D w_t = \Phi(L)^q \varphi(L^s)^Q \Delta_s^D \epsilon_t + \sum_{i=1}^{n} \beta_i x_t^i, \tag{2}$$

where β_i is an additional parameter accounting for the relative weight of each exogenous variable.

In Supplementary Materials (Section S1), additional details about the model are reported.

We implemented this model on Python using the StatsModels library (https://www.statsmodels.org/stable/index.html, accessed on 7 February 2023), with the SARIMAX (https://www.statsmodels.org/0.9.0/generated/statsmodels.tsa.statespace.sarimax.SARIMAX.html, accessed on 7 February 2023) class.

For the SARIMAX model, var_2, var_3, var_4 and var_5 (i.e., *EB* and macronutrients) are considered as exogenous variables, with the weight as output. Considering our dataset structure, the exogenous variables at time t correspond to the inputs for the forecasting of the weight at time $t + 1$. However, in the SARIMAX equation, the exogenous term is considered at the same time, t, with respect to the output. To overcome this issue, we shifted the exogenous values of $\Delta T = 1$ day with respect to weight. In this way, the exogenous term changed as follows: $\sum_{i=1}^{n} \beta_i x_{t-1}^i$.

LSTM:

Long short-term memory (LSTM) networks [18], a variant of the simplest recurrent neural networks (RNNs), can learn long-term dependencies and are the most widely used for working with sequential data such as time-series data [19–21].

The LSTM cell (Figure 2) uses an input gate, a forget gate and an output gate (a simple multilayer perceptron). Depending on data's priority, these gates allow or deny data flow/passage. Moreover, they enhance the ability of the neural network to understand what needs to be saved, forgotten, remembered, paid attention to and output. The cell state and hidden state are used to gather data to be processed in the next state.

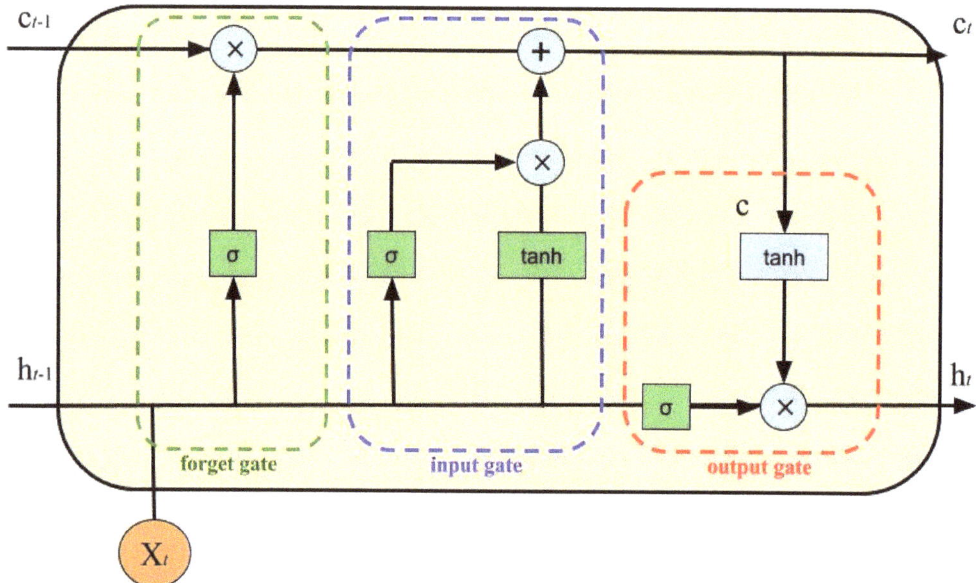

Figure 2. An LSTM cell, where σ is the sigmoid function.

The gates have the following equations:

1. Input Gate:

$$i = \sigma(W_i h_{t-1} + W_i h_t),\tag{3}$$

2. Forget Gate:

$$f = \sigma\left(W_f h_{t-1} + W_f h_t\right),\tag{4}$$

3. Output Gate:

$$o = \sigma(W_o h_{t-1} + W_o h_t),\tag{5}$$

4. Intermediate Cell State:

$$g = tanh\left(W_g h_{t-1} + W_g h_t\right),\tag{6}$$

5. Cell State (Next Memory Input):

$$c_t = (g*i) + (f*c_{t-1}),\tag{7}$$

6. New State:

$$h_t = o * tanh(c_t),\tag{8}$$

with X_t as the input vector, h_t as the output vector, W and U as parameter matrices and f as the parameter vector.

We implemented the LSTM network using the TensorFlow Keras library (https://www.tensorflow.org/api_docs/python/tf/keras, accessed on 7 February 2023), which implements an LSTM cell as an available class on Python (https://www.tensorflow.org/api_docs/python/tf/keras/layers/LSTM, accessed on 7 February 2023), which we added into a model as a monolayer neural network.

GRU:

The gated recurrent unit, just like the LSTM network, is a variant of the simplest RNN but with a less complicated structure. It has an update gate, z, and a reset gate, r. These two variables are vectors that determine what information passes or does not pass to output. With the reset gate, new input is combined with the previous memory while the update gate determines how much of the last memory to keep.

The GRU has the following equations:

1. Update Gate:

$$z = (W_z h_{t-1} + U_z x_t),\tag{9}$$

2. Reset Gate:

$$r = (W_r h_{t-1} + U_r x_t),\tag{10}$$

3. Cell State:

$$c = tanh(W_c(h_{t-1}*r) + U_c x_t),\tag{11}$$

4. New State:

$$h_t = (z*c)((1-z)*h_{t-1}),\tag{12}$$

A GRU cell is shown in Figure 3.

A more accurate description can be found in the Supplementary Materials (Section S3) of a previous work [14].

As for the LSTM network, we implemented the GRU in TensorFlow using a GRU cell (https://www.tensorflow.org/api_docs/python/tf/keras/layers/GRU, accessed on 7 February 2023) implemented into a monolayer neural network.

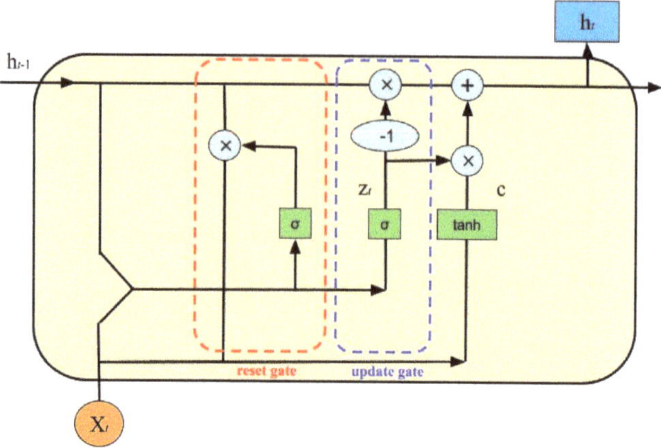

Figure 3. A GRU cell, where σ is the sigmoid function.

Transformer:

LSTM and GRUs have been strongly established as state-of-the-art approaches in sequence modeling and transduction problems such as language modeling and machine translation [22–26] because of their ability to memorize long-term dependency. Since they are inherently sequential, there is no parallelization within training examples, which makes batching across training examples more difficult as sequence lengths increase. Therefore, to allow modeling of dependencies for any distance in the input or output sequences, attention mechanisms have been integrated in compelling sequence modeling and transduction models in various tasks [24,27]. Commonly [28], such attention mechanisms are used in conjunction with a recurrent network. In 2017, a team at Google Brain® developed a new model [15], called "Transformer", with an architecture that avoids recurrence and instead relies entirely on an attention mechanism to draw global dependencies between inputs and outputs. This architecture uses stacked self-attention and pointwise, fully connected layers for both the encoder and the decoder, shown in the left and right halves of Figure 4, respectively. In Supplementary Material (Section S2), a more accurate description of the model is reported.

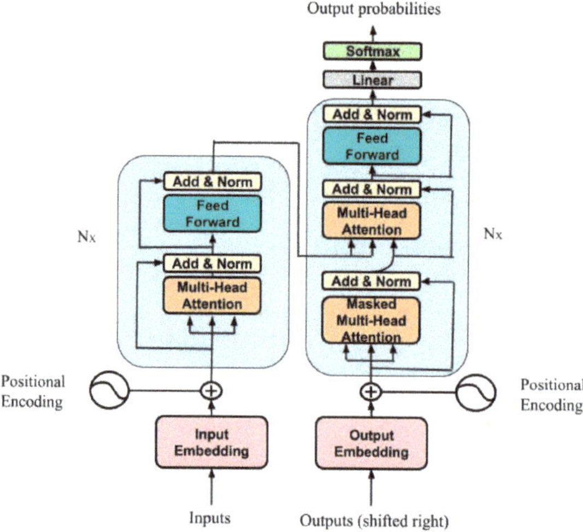

Figure 4. Transformer architecture. In supplementary material (Section S2), an accurate description of the architecture is reported.

The implementation of the model in Python followed the Transformer starting code shared by the Google Brain team (https://keras.io/examples/timeseries/timeseries_transformer_classification/, accessed on 7 February 2023).

2.5. Model Selection and Comparison

2.5.1. Implementation and Selection of Models

For each selected model, parameter scanning was performed, and the best model was selected. Below, procedures are indicated according to models.

SARIMAX:

Augmented Dickey–Fuller (ADF) tests, applied to a weight time series, yielded *p*-values larger than $\alpha = 0.05$ for 90% of the overall participants. Therefore, we transformed the weight time series into a stationary one that performed first-order differentiation. The ADF test, repeated on preprocessed series, confirmed stationarities for all of the transformed time series. Following this adjustment, the terms d and D were each set to 1.

We started with fitting a SARIMAX model for all the datasets available, considering the ranges in Table 1.

Table 1. Parameter range for the selection of the best SARIMAX model.

p	q	d	P	Q	D	S
(1–9)	(1,2)	(1,2)	(1–5)	(1,2)	(1,2)	(7)

In the literature, the most common way to find the best parameters for SARIMAX models is based on a simple grid search following the Akaike information criterion (AIC) and the Bayesian information criterion (BIC), respectively. These criteria help to select the model that explains the greatest amount of variation using the fewest possible independent variables, using maximum likelihood estimation (MLE) [29], and they both penalize a model for having increasing numbers of variables, to prevent overfitting.

Therefore, we ranked the models according to the lowest AIC values. The first 5 models were then trained on the datasets, and the root mean squared error (RMSE) scores were calculated. The model with the lowest RMSE was then selected.

The LSTM and GRU Models:

Hyperparameter tuning with the aim of minimizing loss function was carried out to select the best deep-learning model [14]. Typically, in time-series forecasting, tuning is carried out to reduce the RMSE of test-training forecasting.

Considering that the LSTM and GRU models had the same configuration and the same hyperparameters, we proceeded with both to parameter scanning in the range shown in Table 2.

Table 2. Hyperparameter range for the selection of the best GRU and LSTM models.

Units	Epoch	Batch Size	Dropout	Activation Function	Optimizer
(50, 100, 150)	(50, 100, 150)	(8, 16, 32)	(0.2)	('ReLU')	('adam')

We selected the best model via considering the lowest RMSE obtained from a prediction on the same training-test sets.

Transformer:

The implementation of this model into Keras was like that of the other two neural networks, with some exceptions for the hyperparameters. We considered a grid search that would take into account the range of the hyperparameters shown in Table 3.

Table 3. Hyperparameter range for the selection of the best Transformer model.

Head Size	Num Heads	Epoch	Batch Size	Dropout	Activation Function	Optimizer
(64, 128, 256)	(2, 4, 8)	(50, 100, 150)	(8, 16, 32)	(0.2, 0.25)	('ReLU')	('adam')

Differently from LSTM and GRU, there are two more parameters: head size, which is the dimensionality of the query, key and value tensors after the linear transformation, and num heads, which is the number of attention heads.

In this case, we also chose the best model via minimizing the RMSEs on the training-test sets.

2.5.2. Performance of Models with Datasets of Varying Length

Following model selection and parameter optimization, we compared the models, considering, as a quality index, the RMSE, which indicates errors in weight prediction with a test-set length of 7 days, considering a training set of more than 100 days (mean ± SD = 161.3 ± 22.4) for each participant.

In addition, since scarcity of data is a common problem in deployment of PMAs in production, we tested the models in more realistic settings. We thus divided the dataset of each participant into 9 independent groups of 15 days. Then, we evaluated the RMSE on a test set with a length of 1 day for each group (with a training set of 14 elements). The final RMSE was the average of these 9 RMSEs. An ANOVA followed by a Tukey test was applied for pairwise comparison of RMSEs.

2.5.3. Computational Time

In addition to prediction performance, the computational times were calculated for the retraining and prediction phases for the four models.

A Kruskal–Wallis test followed by a Dunn test was applied for pairwise comparison of computational times.

2.6. Computational Requirements and Python Libraries

Computational requirements were minimal in order to allow deployment on virtual machines available on the web. The code for the development of the models was run in Google Colab with the default settings (free plan). The code requires the following libraries: *tensorflow* = 2.9.2 (https://pypi.org/project/tensorflow/, accessed 7 February 2023), *pandas* = 1.3.5 (https://pandas.pydata.org/, accessed 7 February 2023), *numpy* = 1.21.6 (https://numpy.org/, accessed 7 February 2023), *matplotlib* = 3.2.2 (https://matplotlib.org/, 7 February 2023), *seaborn* = 0.11.2 (https://seaborn.pydata.org/, accessed 7 February 2023), *statsmodels* = 0.12.2 (https://www.statsmodels.org/stable/index.html, accessed on 7 February 2023), *scipy* = 1.7.3 (https://pypi.org/project/scipy/, accessed on 7 February 2023), *bioinfokit* = 2.1.0 (https://pypi.org/project/bioinfokit/0.3/, accessed on 7 February 2023), *scikit-learn* = 1.0.2 (https://scikit-learn.org/stable/, accessed 7 February 2023) and *scikit-posthocs* = 0.7.0 (https://scikit-posthocs.readthedocs.io/en/latest/, accessed on 7 February 2023).

3. Results

3.1. Selection of the Optimal Model

We started with optimizing parameters for each selected model and each participant, as explained in par. 2.6.1.

For the GRU, LSTM and Transformer models, we considered an Adam optimizer and, as a loss function, the mean absolute error (MAE), defined with the formula

$$MAE = \frac{\sum_{i=1}^{n} |y_i - x_i|}{n},$$

(13)

where y_i is the actual value and x_i is the prediction.

In Table S1, the selected parameters for each user are reported for each type of model. As shown in [14], we trained a model for each user to adapt it to the personalized characteristics of metabolism.

3.2. Comparison between Models

As explained in par. 2.6.2, to compare model performance, we used the RMSE of the prediction of the test set for each participant. Datasets were structured to make the training and test set homogeneous, ensuring that the models learned from the same data and tested their knowledge under identical conditions. In Figure 5, we report the forecasting with each model for a single participant.

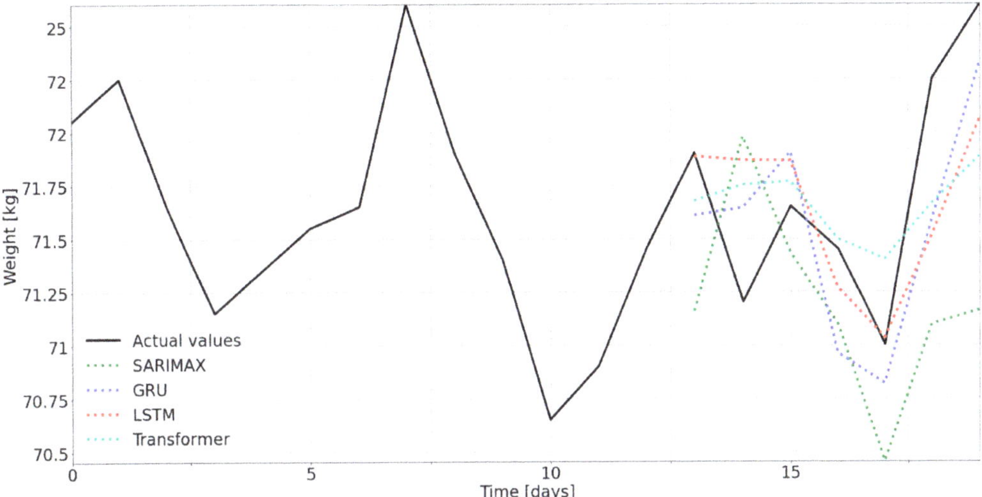

Figure 5. Predictions of test set of user 0, for all models.

From a visual inspection, the GRU and LSTM models follow the variations of weight more accurately while the SARIMAX model shows the worst result. In Figure 6, we show the RMSEs, grouped based on models, for each participant. Indeed, there is an evident difference between the SARIMAX model and the others, confirming that neural networks outperform statistical models in time-series forecasting. On the other hand, the deep-learning models show comparable RMSEs to each other.

Hence, the Transformer model did not demonstrate improvement with respect to the GRU or LSTM models, having, on the contrary, slightly worse results.

To quantify these observations, we carried out an ANOVA among the RMSEs of the models, showing a p-value lower than $\alpha = 0.05$ $\left(4.31 \cdot 10^{-4}\right)$ and confirming that there was at least one model different from the others. We then performed a Tukey test for pairwise comparison, and the results, reported in Table 4, confirmed that the SARIMAX model is different from the others (adjusted p-value lower than α), while there is no statistical difference among the other three models, yielding a p-value bigger than α.

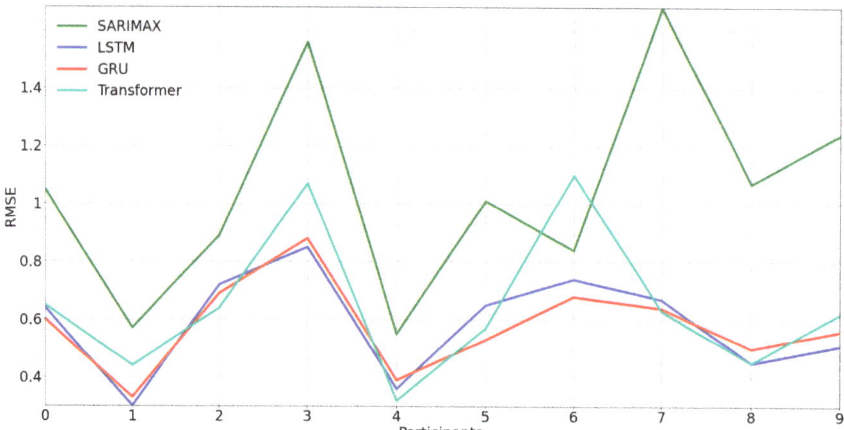

Figure 6. RMSEs for each model versus users.

Table 4. Tukey test results: Diff represents the mean difference between the pair of groups; Lower and Upper the lower and upper difference between the pair of groups, respectively; q-value is a value that provides a means to control the positive false discovery rate (pFDR); and *p*-value is the probability of obtaining test results at least as extreme as the result observed, under the assumption that the null hypothesis is correct.

Group 1	Group 2	Diff	Lower	Upper	q-Value	*p*-Value
SARIMAX	LSTM	0.458859	0.152600	0.765117	5.706676	0.001490
SARIMAX	GRU	0.467519	0.161261	0.773778	5.814384	0.001196
SARIMAX	Transformer	0.397567	0.091309	0.703826	4.944415	0.006675
LSTM	GRU	0.008660	−0.297598	0.314919	0.107708	0.900000
LSTM	Transformer	0.061291	−0.244967	0.367550	0.762261	0.900000
GRU	Transformer	0.069952	−0.236307	0.376210	0.869969	0.900000

3.3. Analysis of the Performance with a Limited Dataset Length

As explained in Section 2.6, PMAs often operate on datasets with limited length. For example, diet diaries are often compiled for a limited amount of time. Therefore, we carried out a test to show the performances of these models, considering a limited dataset of 15 days. The model was trained to predict the weight for the day afterward.

To acquire a reliable index of the performance of each model, we tested the models on nine subsets of data in the original dataset for each participant. In this way, we could refer to a mean RMSE for each model and for each participant.

In Figure 7, the RMSE distribution of each model is reported (each point represents a user). From a visual inspection, we can conclude that, again, the SARIMAX model displays the worst results, while the others have similar performances. To confirm this observation, we carried out an ANOVA (*p*-value = 0.019) followed by a Tukey test. The pairwise comparison showed that only the SARIMAX model presented accuracy that was statistically different from that of the deep-learning models.

3.4. Performance versus Data Length

The results reported in the previous section show how the model provided accurate solutions for few data. In this section, we analyze changes in performance with decreasing data length. We considered the following subsets: 100% of the dataset and 100, 80, 60, 40 and 30 days. In Figure 8, we report the RMSE versus the data length for each participant and for each model, with error bars representing the standard deviations (SDs).

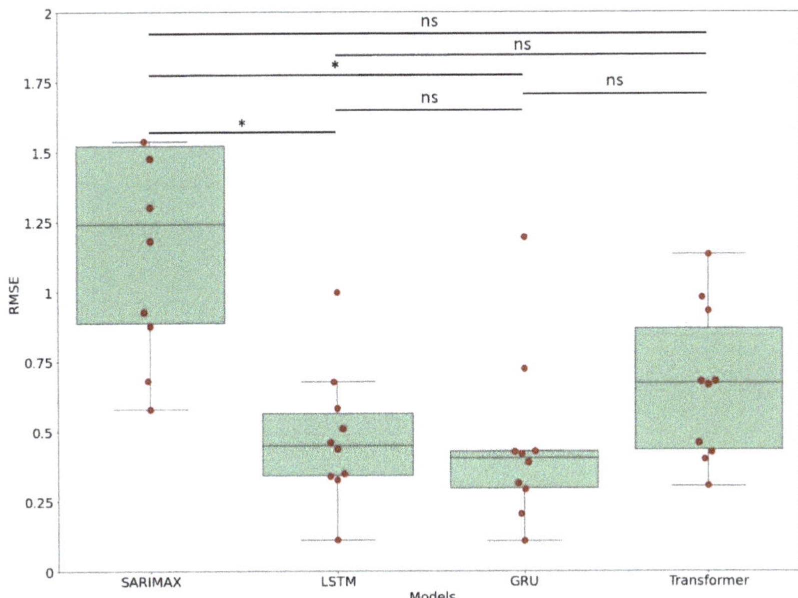

Figure 7. Distribution of RMSEs for each model and for each participant for a forecasting of 1 day, with a training set of 14 days for nine subsets, where * stands for a *p*-value < 0.05. ns: not significant.

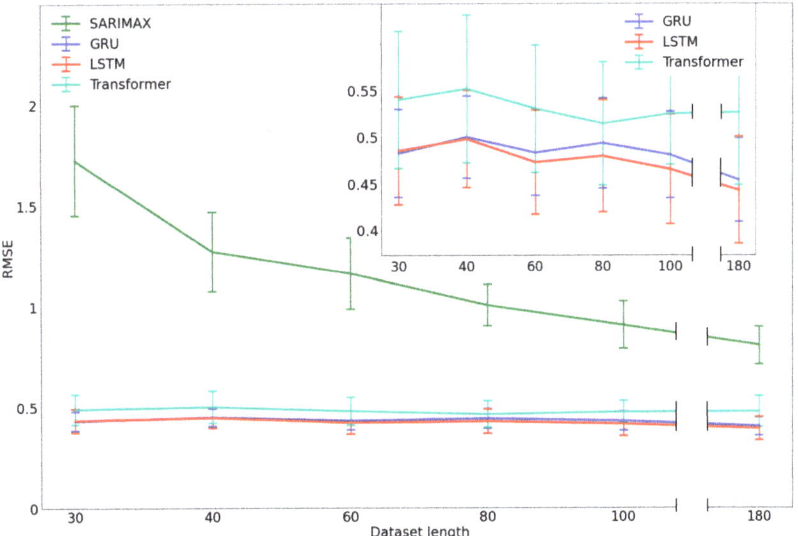

Figure 8. Mean RMSEs and deviation standards for each model for different dataset lengths, with a 7-day forecasting. In the inset at right, an enlargement of the deep-learning RMSEs is reported.

While the SARIMAX model showed an important decrease in performance as data length decreased, the others were characterized with stable performances, also with data collected only for thirty days.

3.5. Computational Time

In the evaluation of the performance of a model in a production environment, we must consider another important parameter: the computational cost, expressed in time. This computational time is the sum of the (re)training time and the forecasting time, since in a production environment, the model must be retrained every time and data are gathered in real time. In Table 5, we report, for each model, the computational time, the (re)training time and the forecasting time. The times are averaged based on the number of participants.

Table 5. Time costs for the training and forecasting (of 7 days) mean for each model, averaged based on 10 participants.

Model	Computational Time	Standard Deviation	Retraining Time	Forecasting Time
SARIMAX	1.83 s	1.06 s	1.56 ± 1.05 s	0.29 ± 0.03 s
LSTM	13.5 s	3.60 s	12.6 ± 3.30 s	0.92 ± 0.33 s
GRU	12.7 s	1.42 s	12.0 ± 1.22 s	0.86 ± 0.26 s
Transformer	48.6 s	10.7 s	47.9 ± 10.7 s	0.85 ± 0.13 s

It is possible to observe how the GRU and LSTM models each require about 1/5 of the time requested to retrain and forecast with the Transformer model but $10\times$ more time than that of the SARIMAX model. Therefore, a major burden of the Transformer model resides in the retraining time, since it requires more complex operations than the others.

To quantify these observations, a Kruskal–Wallis test was carried out [30] among the models, since a Shapiro–Wilk test [30] had confirmed that distributions would not be normal. The test yielded a p-value < 0.05, showing the presence of a statistical difference between the models. Therefore, a posthoc test (Dunn test [31]) was carried out to investigate the pairwise comparison. The test showed no statistical difference between the GRU and LSTM models, confirming that they have similar performances (Figure 9), which are better than that of the Transformer model.

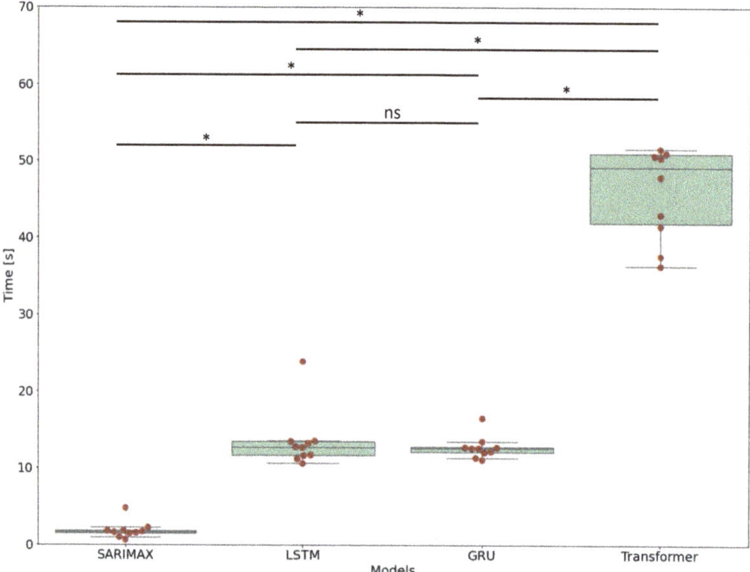

Figure 9. Computational time distribution for each model, with pairwise comparison, where * stands for a p-value < 0.05. ns: not significant.

4. Discussion

Obesity and cardiovascular disease, as the most serious public health challenges of the 21st century, are strongly conditioned through dietary habits. Digital health can help people to monitor themselves and prevent these diseases. The advent of wearable devices and the evolution of smartphone technology have allowed the development of an infrastructure able to retrieve data that could be used for the development of what is defined as a "Digital Twin": a digital representation of human physiology. With this technology, it is possible to import digital health into the lifestyles of citizens, promoting a healthy lifestyle, since people would be in conditions to better know their own physiologies and responses to nutrition and physical activity. Here, we relied on the ArMOnIApp application, which is able to fetch, preprocess and analyze spontaneous and voluntary physical activities (PAs), dietary measures and anthropometric measures from a set of commercial wearables and other smart devices provided to the end user [11]. These data led to the development of a model, the PMA, that is able to give personalized responses for each end user, such as personalized reactions to the introduction of a particular food in their diet [14]. Here, we compared predictive and computational performances of several models, with the aim of providing useful parameters to put the PMA into production. Moreover, we tested the influence of the data number retrieved, which, in real settings, can vary in range from user to user, on the four models. In a production environment, the practice of automating deployment, integration and monitoring of machine-learning (ML) models is called MLOps [32], and this automation is crucial to increase the speed at which organizations can release models into production. MLOps also involves ensuring continuous quality and dynamic adaptability of projects throughout the entire model lifecycle [32].

To make an efficient and accurate PMA, data must be retrieved in real time. Therefore, web applications must be structured to continuously fetch new data as they are made available with devices, and to control data quality using algorithms. To include these functionalities, we relied on our web application, ArMOnIApp [11]. Moreover, ML models require automation of model retraining, and in this framework, the time cost for this procedure has an important role to optimize end performance. To this aim, we evaluated the time necessary for retraining of and prediction for the most used and reliable forecasting models. The results (summarized in Table 6) show how the GRU and LSTM models require about 1/5 of the computational time of the Transformer model, despite this time being more than 10 times that of the SARIMAX model.

Table 6. In this table, the overall results are summed up for each model, reporting each mean value and its associated SD.

Model	RMSE	RMSE$_{reduced}$	Computational Time	Retraining Time
SARIMAX	0.85 ± 0.37	1.95 ± 2.30	1.83 ± 1.06 s	1.56 ± 1.05 s
LSTM	0.39 ± 0.18	0.48 ± 0.24	13.5 ± 3.60 s	12.6 ± 3.30 s
GRU	0.38 ± 0.16	0.45 ± 0.30	12.7 ± 1.42 s	12.0 ± 1.22 s
Transformer	0.45 ± 0.25	0.66 ± 0.28	48.6 ± 10.7 s	47.9 ± 10.7 s

On top of these optimizations, there is a need to monitor quality of predictions. To this aim, we outlined a workflow to evaluate the performances of different models with varying data lengths. We found out that the SARIMAX model, though being the fastest, had the worst RMSE, with a great variability among users. This RMSE, being four times higher than that of the GRU or LSTM model, penalizes the SARIMAX model in the deployment of the PMA. In terms of the RMSE, the Transformer model had a better performance than the SARIMAX model as well, but was comparable with RNNs. However, the time cost was the highest (four times higher than for the GRU/LSTM model), and this criticality has a strong impact on production development.

According to the performances and computational times, we can conclude that the PMAs built on the GRU or LSTM model show optimal predictive performances with acceptable computational time, making them the best candidates for a production environment.

Another issue is the need to compare the effectiveness of training several ML models specialized in different groups versus training one unique model for all the data. To address this issue, planning to create a unique model accounting for the metabolisms of a cohort of participants will require an increased number of participants.

Before these models are put into production, several ethical concerns must be moreover addressed. In regard to privacy concerns, collection and storage of personal health data by wearable devices can potentially compromise users' privacy if this information is shared with third parties without their consent. In this study, we retrieved health data from the Zepp API, where users have explicitly consented to data sharing. The privacy policy can be retrieved on the Zepp website (https://www.zepp.com/privacy-policy, accessed on 24 February 2023). There are, in any case, security risks. Wearable devices are often connected to the internet, making them vulnerable to hacking and data breaches. This can result in sensitive personal data being stolen or compromised, potentially leading to identity theft or other forms of fraud. In addition, discrimination is an issue to be addressed, since use of wearable devices and data collected can potentially lead to discrimination against individuals with pre-existing health conditions or disabilities. This can result in denial of insurance coverage or job opportunities. Finally, there are social implications: use of wearable devices to track personal data can promote unhealthy obsessions with self-monitoring. These issues have been constantly monitored in pilot and clinical studies, but protocols must be developed and optimized before the use of these systems on a large scale is allowed. Some of these protocols already exist or are under research [33,34].

5. Conclusions

Putting the PMA into production can produce diets and activity regimens that are specifically tailored to users' needs. Thanks to the PMA, pertinent hints can be found to provide citizens and nutritionists with scientific knowledge and reliable tools, enhancing their self-awareness and assisting them in their quests for healthy lifestyles. An important development might be inclusion of newly developed lipid metabolism indicators (such as membrane lipids and fluidity of red-blood-cell membranes) as input in the PMA to research the impacts and influences of dietary components on their results [35–39]. Additionally, cutting-edge and promising anthropometric markers, such as VO2max and heart rate frequency, monitored using wearable technology can enhance the accuracy of weight predictions [40]. These integrations could group and cluster various PMA responses, providing insights into these variables that could affect an individual's metabolism. Another important advancement may come from the advent of quantum computing and the achievement of quantum supremacy [41], which will revolutionize ML models, including the PMA, via increasing their performances and reducing their computational times.

Supplementary Materials: The following supporting information can be downloaded at: https://www.mdpi.com/article/10.3390/nu15051199/s1, Table S1: The results of the tuning of each model are reported in the table, in particular for the SARIMAX model the parameters are the following: (p,dq)x(P,D,Q,S); for the LSTM and GRU: (units, epochs, batch size, dropout); for the Transformer: (head size, num heads, epochs, batch size, dropout); Section S1: SARIMAX architecture; Section S2: Transformer architecture; Section S3: Models' parameters.

Author Contributions: Conceptualization, G.M. and A.A.; methodology, G.M., A.A., G.B., C.S., A.R. and M.D.S.; software, A.A. and G.M.; validation, G.M., A.A., G.B., C.S. and A.R.; formal analysis, G.M., A.A., G.B., C.S. and A.R.; investigation, G.M., A.A., G.B., C.S., A.R. and M.D.S.; resources, G.M. and A.A.; data curation, A.A. and G.M.; writing—original draft preparation, G.M. and A.A.; writing—review and editing, A.A., G.B. and G.M.; visualization, A.A. and G.B.; supervision, G.M.; project administration, G.M.; funding acquisition, G.M. All authors have read and agreed to the published version of the manuscript.

Funding: This project was supported in part by two research grants awarded to GM from Regione Lazio PO FSE 2014-2020 "Intervento per il rafforzamento della ricerca nel Lazio—incentivi per i dottorati di innovazione per le imprese", one cofunded by RAN Innovation (2021) and another co-funded by BLUSISTEMI S.R.L. (2022), and by a research grant awarded to GM by Università Cattolica del Sacro Cuore-Linea D1 2021.

Institutional Review Board Statement: This study was conducted in accordance with the Declaration of Helsinki and approved by the Ethics Committee of Università Cattolica del Sacro Cuore (Protocol Code diab_mf, 16 March 2017).

Informed Consent Statement: Informed consent was obtained from all subjects involved in this study. Written informed consent to publish this paper has been obtained from the participants.

Data Availability Statement: Data and codes are available upon reasonable request.

Conflicts of Interest: The authors declare no conflict of interest.

References

1. Mold, J. Goal-Directed Health Care: Redefining Health and Health Care in the Era of Value-Based Care. *Cureus* **2017**, *9*, e1043. [CrossRef]
2. Bianchi, V.E. Impact of Nutrition on Cardiovascular Function. *Curr. Probl. Cardiol.* **2020**, *45*, 100391. [CrossRef]
3. Wiseman, M.J. Nutrition and Cancer: Prevention and Survival. *Br. J. Nutr.* **2019**, *122*, 481–487. [CrossRef]
4. Tourlouki, E.; Matalas, A.-L.; Panagiotakos, D.B. Dietary Habits and Cardiovascular Disease Risk in Middle-Aged and Elderly Populations: A Review of Evidence. *Clin. Interv. Aging* **2009**, *4*, 319–330. [CrossRef]
5. Pignatti, C.; D'adamo, S.; Stefanelli, C.; Flamigni, F.; Cetrullo, S. Nutrients and Pathways That Regulate Health Span and Life Span. *Geriatrics* **2020**, *5*, 95. [CrossRef]
6. Kussmann, M.; Raymond, F.; Affolter, M. OMICS-Driven Biomarker Discovery in Nutrition and Health. *J. Biotechnol.* **2006**, *124*, 758–787. [CrossRef] [PubMed]
7. Chaudhary, N.; Kumar, V.; Sangwan, P.; Pant, N.C.; Saxena, A.; Joshi, S.; Yadav, A.N. Personalized Nutrition and -Omics. In *Comprehensive Foodomics*; Elsevier: Amsterdam, The Netherlands, 2020; pp. 495–507, ISBN 9780128163955.
8. Iqbal, S.M.A.; Mahgoub, I.; Du, E.; Leavitt, M.A.; Asghar, W. Advances in Healthcare Wearable Devices. *NPJ Flex. Electron.* **2021**, *5*, 9. [CrossRef]
9. Ahmadi-Assalemi, G.; Al-Khateeb, H.; Maple, C.; Epiphaniou, G.; Alhaboby, Z.A.; Alkaabi, S.; Alhaboby, D. Digital Twins for Precision Healthcare. In *Advanced Sciences and Technologies for Security Applications*; Springer: Berlin/Heidelberg, Germany, 2020; pp. 133–158.
10. Mulder, S.T.; Omidvari, A.H.; Rueten-Budde, A.J.; Huang, P.H.; Kim, K.H.; Bais, B.; Rousian, M.; Hai, R.; Akgun, C.; van Lennep, J.R.; et al. Dynamic Digital Twin: Diagnosis, Treatment, Prediction, and Prevention of Disease During the Life Course. *J. Med. Internet Res.* **2022**, *24*, e35675. [CrossRef]
11. Bianchetti, G.; Abeltino, A.; Serantoni, C.; Ardito, F.; Malta, D.; de Spirito, M.; Maulucci, G. Personalized Self-Monitoring of Energy Balance through Integration in a Web-Application of Dietary, Anthropometric, and Physical Activity Data. *J. Pers. Med.* **2022**, *12*, 568. [CrossRef] [PubMed]
12. Nguyen, T.V.T.; Huynh, N.T.; Vu, N.C.; Kieu, V.N.D.; Huang, S.C. Optimizing Compliant Gripper Mechanism Design by Employing an Effective Bi-Algorithm: Fuzzy Logic and ANFIS. *Microsyst. Technol.* **2021**, *27*, 3389–3412. [CrossRef]
13. Wang, C.N.; Yang, F.C.; Nguyen, V.T.T.; Vo, N.T.M. CFD Analysis and Optimum Design for a Centrifugal Pump Using an Effectively Artificial Intelligent Algorithm. *Micromachines* **2022**, *13*, 1208. [CrossRef] [PubMed]
14. Abeltino, A.; Bianchetti, G.; Serantoni, C.; Ardito, C.F.; Malta, D.; de Spirito, M.; Maulucci, G. Personalized Metabolic Avatar: A Data Driven Model of Metabolism for Weight Variation Forecasting and Diet Plan Evaluation. *Nutrients* **2022**, *14*, 3520. [CrossRef] [PubMed]
15. Vaswani, A.; Brain, G.; Shazeer, N.; Parmar, N.; Uszkoreit, J.; Jones, L.; Gomez, A.N.; Kaiser, Ł.; Polosukhin, I. Attention Is All You Need. In Proceedings of the 31st Conference on Neural Information Processing Systems (NIPS 2017), Long Beach, CA, USA, 4–9 December 2017.
16. Wu, N.; Green, B.; Ben, X.; O'Banion, S. Deep Transformer Models for Time Series Forecasting: The Influenza Prevalence Case. *arXiv* **2020**, arXiv:2001.08317.
17. Alharbi, F.R.; Csala, D. A Seasonal Autoregressive Integrated Moving Average with Exogenous Factors (SARIMAX) Forecasting Model-Based Time Series Approach. *Inventions* **2022**, *7*, 94. [CrossRef]
18. Hochreiter, S.; Urgen Schmidhuber, J. Long Short-Term Memory. *Neural Comput.* **1997**, *9*, 1735–1780. [CrossRef]
19. Lipton, Z.C.; Berkowitz, J.; Elkan, C. A Critical Review of Recurrent Neural Networks for Sequence Learning. *arXiv* **2015**, arXiv:1506.00019.
20. Liu, Y.; Hou, D.; Bao, J.; Qi, Y. Multi-Step Ahead Time Series Forecasting for Different Data Patterns Based on LSTM Recurrent Neural Network. In Proceedings of the 2017 14th Web Information Systems and Applications Conference, WISA 2017, Liuzhou, China, 11–12 November 2017; pp. 305–310.

21. Schmidhuber, J.; Wierstra, D.; Gomez, F.J. Evolino: Hybrid neuroevolution/optimal linear search for sequence prediction. In Proceedings of the 19th International Joint Conferenceon Artificial Intelligence (IJCAI), Scotland, UK, 30 July–5 August 2005.
22. Elman, J.L. Finding Structure in Time. *Cogn. Sci.* **1990**, *14*, 179–211. [CrossRef]
23. Mikolov, T.; Karafiát, M.; Burget, L.; Cernocky, J.H.; Khudanpur, S. Recurrent Neural Network Based Language Model. In Proceedings of the 11th Annual Conference of the International Speech Communication Association, Makuhari, Japan, 26–30 September 2010.
24. Bahdanau, D.; Cho, K.; Bengio, Y. Neural Machine Translation by Jointly Learning to Align and Translate. *arXiv* **2014**, arXiv:1409.0473.
25. Sutskever, I.; Vinyals, O.; Le, Q.V. Sequence to Sequence Learning with Neural Networks. *arXiv* **2014**, arXiv:1409.3215.
26. Cho, K.; van Merrienboer, B.; Gulcehre, C.; Bahdanau, D.; Bougares, F.; Schwenk, H.; Bengio, Y. Learning Phrase Representations Using RNN Encoder-Decoder for Statistical Machine Translation. *arXiv* **2014**, arXiv:1406.1078.
27. Kim, Y.; Denton, C.; Hoang, L.; Rush, A.M. Structured Attention Networks. *arXiv* **2017**, arXiv:1702.00887.
28. Parikh, A.P.; Täckström, O.; Das, D.; Uszkoreit, J. A Decomposable Attention Model for Natural Language Inference. *arXiv* **2016**, arXiv:1606.01933.
29. Myung, I.J. Tutorial on Maximum Likelihood Estimation. *J. Math. Psychol.* **2003**, *47*, 90–100. [CrossRef]
30. Ostertagová, E.; Ostertag, O.; Kováč, J. Methodology and Application of the Kruskal-Wallis Test. *Appl. Mech. Mater.* **2014**, *611*, 115–120. [CrossRef]
31. Dinno, A. Nonparametric Pairwise Multiple Comparisons in Independent Groups Using Dunn's Test. *Stata J.* **2015**, *15*, 292–300. [CrossRef]
32. Kreuzberger, D.; Kühl, N.; Hirschl, S. Machine Learning Operations (MLOps): Overview, Definition, and Architecture. *arXiv* **2022**, arXiv:2205.02302.
33. Ahmad, W.; Rasool, A.; Javed, A.R.; Baker, T.; Jalil, Z. Cyber Security in IoT-Based Cloud Computing: A Comprehensive Survey. *Electronics* **2022**, *11*, 16. [CrossRef]
34. Langone, M.; Setola, R.; Lopez, J. Cybersecurity of Wearable Devices: An Experimental Analysis and a Vulnerability Assessment Method. In Proceedings of the 2017 IEEE 41st Annual Computer Software and Applications Conference (COMPSAC), Turin, Italy, 4–8 July 2017; Volume 2, pp. 304–309.
35. Bianchetti, G.; Viti, L.; Scupola, A.; di Leo, M.; Tartaglione, L.; Flex, A.; de Spirito, M.; Pitocco, D.; Maulucci, G. Erythrocyte Membrane Fluidity as a Marker of Diabetic Retinopathy in Type 1 Diabetes Mellitus. *Eur. J. Clin. Investig.* **2021**, *51*, e13455. [CrossRef]
36. Maulucci, G.; Cohen, O.; Daniel, B.; Ferreri, C.; Sasson, S. The Combination of Whole Cell Lipidomics Analysis and Single Cell Confocal Imaging of Fluidity and Micropolarity Provides Insight into Stress-Induced Lipid Turnover in Subcellular Organelles of Pancreatic Beta Cells. *Molecules* **2019**, *24*, 3742. [CrossRef] [PubMed]
37. Maulucci, G.; Cohen, O.; Daniel, B.; Sansone, A.; Petropoulou, P.I.; Filou, S.; Spyridonidis, A.; Pani, G.; de Spirito, M.; Chatgilialoglu, C.; et al. Fatty Acid-Related Modulations of Membrane Fluidity in Cells: Detection and Implications. *Free Radic. Res.* **2016**, *50*, S40–S50. [CrossRef] [PubMed]
38. Cordelli, E.; Maulucci, G.; de Spirito, M.; Rizzi, A.; Pitocco, D.; Soda, P. A Decision Support System for Type 1 Diabetes Mellitus Diagnostics Based on Dual Channel Analysis of Red Blood Cell Membrane Fluidity. *Comput. Methods Programs Biomed.* **2018**, *162*, 263–271. [CrossRef] [PubMed]
39. Bianchetti, G.; Azoulay-Ginsburg, S.; Keshet-Levy, N.Y.; Malka, A.; Zilber, S.; Korshin, E.E.; Sasson, S.; de Spirito, M.; Gruzman, A.; Maulucci, G. Investigation of the Membrane Fluidity Regulation of Fatty Acid Intracellular Distribution by Fluorescence Lifetime Imaging of Novel Polarity Sensitive Fluorescent Derivatives. *Int. J. Mol. Sci.* **2021**, *22*, 3106. [CrossRef]
40. Serantoni, C.; Zimatore, G.; Bianchetti, G.; Abeltino, A.; de Spirito, M.; Maulucci, G. Unsupervised Clustering of Heartbeat Dynamics Allows for Real Time and Personalized Improvement in Cardiovascular Fitness. *Sensors* **2022**, *22*, 3974. [CrossRef] [PubMed]
41. Terhal, B.M. Quantum Supremacy, Here We Come. *Nat. Phys.* **2018**, *14*, 530–531. [CrossRef]

MDPI
St. Alban-Anlage 66
4052 Basel
Switzerland
Tel. +41 61 683 77 34
Fax +41 61 302 89 18
www.mdpi.com

Nutrients Editorial Office
E-mail: nutrients@mdpi.com
www.mdpi.com/journal/nutrients